T0310672

Social and Economic Effects of Community Wireless Networks and Infrastructures

Abdelnasser Abdelaal
Ibri College of Applied Sciences, Oman

Managing Director:	Lindsay Johnston
Editorial Director:	Joel Gamon
Book Production Manager:	Jennifer Yoder
Publishing Systems Analyst:	Adrienne Freeland
Assistant Acquisitions Editor:	Kayla Wolfe
Typesetter:	Henry Ulrich
Cover Design:	Jason Mull

Published in the United States of America by
Information Science Reference (an imprint of IGI Global)
701 E. Chocolate Avenue
Hershey PA 17033
Tel: 717-533-8845
Fax: 717-533-8661
E-mail: cust@igi-global.com
Web site: http://www.igi-global.com

Library of Congress Cataloging-in-Publication Data

Social and economic effects of community wireless networks and infrastructures / Abdelnasser Abdelaal, editor.
     pages cm
  Includes bibliographical references and index.
  Summary: "This book highlights the successes of community wireless networks but also boldly addresses the potential risk factors and broader socioeconomic concerns, exploring previous successes and failures, various designs, and potential challenges with CWNs"--Provided by publisher.
  ISBN 978-1-4666-2997-4 (hardcover) -- ISBN 978-1-4666-2998-1 (ebook) -- ISBN 978-1-4666-2999-8 (print & perpetual access) 1. Information technology--Economic aspects. 2. Telecommunication. 3. Wireless communication systems. I. Abdelaal, Abdelnasser.
  HC79.I55S598 2013
  384.3'3--dc23
                                  2012043392

British Cataloguing in Publication Data
A Cataloguing in Publication record for this book is available from the British Library.

All work contributed to this book is new, previously-unpublished material. The views expressed in this book are those of the authors, but not necessarily of the publisher.

Table of Contents

Detailed Table of Contents

Chapter 1
Introduction to Social and Economic Effects of Community Wireless Networks 1
Abdelnasser M. Abdelaal, Ibri College of Applied Sciences, Oman

This chapter provides an overview of the topic of community wireless networks and infrastructures. In particular, it compares community wireless networks to municipal and commercial wireless networks. It also provides a brief account of the social and economic benefits of CWNs.

Chapter 2
Challenges Facing Municipal Wireless: Case Studies from San Francisco and Silicon Valley 12
Heather E. Hudson, University of Alaska – Anchorage, USA

This chapter discusses the challenges facing municipal wireless networks in the United States. It articulates a number of case studies from Silicon Valley. The authors explores the demand, context, and limitation of technology. They conclude that selecting a suitable business model is a key success factor.

Chapter 3
Legal and Political Barriers to Municipal Networks in the United States 27
Eric Null, Juris Doctor, USA

The United States has been one of the most active countries in the deployment of municipal broadband networks. In America, many remote areas have no Internet access or are served by a single provider that might not meet local needs. Increasingly, access to the Internet is vital for social and economic development as well as prosperity. Without access, rural areas lose economic competitiveness and have lower quality of life standard of living. An attractive solution for such localities is to provide Internet access themselves, provided they believe that their area will realize significant benefits from it. However, the States' complex legal frameworks are a significant barrier to local network success. Each state makes its own laws governing its municipalities, and states have almost unfettered ability to constrain, either by ban or by lesser restraint, the emergence of local networks. These states are further influenced by the lobbying of incumbent service providers. Moreover, judicial remedies can be used strategically by incumbents to hinder, delay, or prevent local networks from succeeding. Despite all this, local networks can be and have been successful. This chapter discusses various legal and political barriers to municipal networks and explores case studies with the goal of learning from past successes and failures.

Chapter 4

Mahabir Pun, E-Networking Research and Development, Nepal

Information and Communication Technology (ICT) has become a vital instrument for delivering a number of services such as education, healthcare, and public services. Community wireless networks are community-centric telecommunication infrastructures developed to provide affordable communication for those who live in remote areas. This chapter discusses the role of Nepal Wireless in achieving socio-economic development of rural communities by facilitating affordable Internet access. In particular, the authors discuss the philosophy and objectives of the project, used network technology, financial resources, and management structure. In addition, the chapter discusses its key services including e-learning, telemedicine, e-commerce, training, and research support. The authors also analyze the challenges Nepal Wireless faced and articulate on the approaches it took to address those challenges. These challenges include lack of technical skills, selecting appropriate technology, ensuring funding resources, difficult geographical terrains, unstable political situation, and expensive devices. They conclude the chapter with some suggestions for policy makers, community developers, and academicians.

Chapter 5

Ashok Jhunjhunwala, IIT Madras's Rural Technology and Business Incubator (RTBI), India
Janani Rangarajan, IIT Madras's Rural Technology and Business Incubator (RTBI), India
N. Neeraja, IIT Madras's Rural Technology and Business Incubator (RTBI), India

This chapter discusses some attempts over the last decade in using Information and Communication Technology (ICT) to empower rural communities and those who are socially and economically left behind in India. It begins with discussing the drivers of telecommunication growth in India since the mid-nineties. Then, it addresses the role of the village Internet kiosks in bringing the Internet to remote villages and articulates on the challenges facing the kiosk model. It then touches upon the rapid growth of mobile telephony in rural India. Following this, it discusses a number of attempts that use mobile telephony to empower rural communities. The authors also use multiple case studies to explore the role of ICT in supporting agriculture, delivering healthcare, achieving financial inclusion and improving the overall livelihood of rural communities in India. The key lessons learned include that the "one-size fits all" model does not work for all communities. In addition, involving both local and federal governments is crucial for the success of community-focused initiatives. Moreover, engaging communities and educating them about the benefits of delivered services would help in sustaining such community-focused initiatives.

Chapter 6

Barbara Walker, Cisco Systems, Inc., USA
Evelyn Posey, University of Texas at El Paso, USA

The Digital El Paso (DEP) community wireless network was deployed as a public-private business model to achieve digital inclusion, sustain economic development, and enhance government and public services. The design, implementation, and funding of DEP were achieved through the collaboration of local businesses, core community members, non-profit organizations, academic institutions, and government entities. In particular, Cisco Systems, Inc. provided design and planning support to complement Intel Corporation's seed funding for the site survey. El Paso County, the City of El Paso, El Paso Independent School District, the Housing Authority of the City of El Paso, and El Paso Electric provided equipment and services. The purpose of DEP is to provide wireless Internet access to achieve social inclusion and economic development. DEP's main challenges include lack of funds, limited user acceptance, and insufficient user training. The policy implication is that leveraging public/private partnerships enhances collaboration and increases the chances of success of community wireless networks. A family-centric approach to drive the adoption of these emerging networks and increase bandwidth utilization, particularly in rural and underserved communities is also recommended.

Chapter 7

Giovanni Camponovo, University of Applied Sciences of Southern Switzerland, Switzerland

Anna Picco-Schwendener, Università della Svizzera Italiana, Switzerland

Lorenzo Cantoni, Università della Svizzera Italiana, Switzerland

Wireless communities are an interesting alternative to 3G networks to provide mobile Internet access. However, the key success factor for their sustainability is whether they are able to attract and retain a critical mass of contributing members. It is thus important to understand what motivates and dissuades people to join and participate. This chapter analyzes motivations, concerns, usage, and satisfaction of members of Fon. Fon is the largest wireless community in the world. This study employs a mixed research method, combining qualitative exploratory interviews with a quantitative survey. Members are mainly motivated by a mix of utilitarian (getting free connectivity) and idealistic motivations (reciprocity and altruism), whereas intrinsic and social motivations are less relevant.

Chapter 8

Jocelyn Williams, Unitec Institute of Technology, New Zealand

This chapter discusses community outcomes of free home Internet access. It draws on case study research on Computers in Homes (CIH), a scheme established in New Zealand in 2000 for the purpose of bridging the digital divide, particularly for low-income families who have school-aged children. The government-funded CIH scheme aims to strengthen relationships between families and schools, improve educational outcomes for children, and provide greater opportunities for their parents. CIH achieves this by working with many primary (elementary) schools, each of which selects 25 families who will benefit from the program. Each family receives a refurbished computer, software, and six months free Internet, as well as twenty hours of free IT training and technical support so that all adults are equipped to make effective use of the Internet. The scheme has evolved to deliver much more than technology. It has become a contributor to social capital in the communities where it has been established. This chapter uses a case study research approach to demonstrate and theorize this process of community building using a construct of social cohesion, which appears to be strengthened by the CIH intervention. Where stronger social networks, volunteerism, and civic engagement were documented in the research, leader figures also mobilized to act on shared goals. These findings highlight the value of existing social resources within communities for achieving community goals while also maximizing community Internet longevity.

Chapter 9

Dorothy Okello, Makerere University, Uganda

Julius Butime, Makerere University, Uganda

This chapter shares the experiences of the Community Wireless Resource Centre (CWRC) as it embarked on the journey to address affordable connectivity for four telecentres in rural and underserved Uganda via telecentre-based community wireless networks. Telecentres have long played a key role in availing access to Information and Communication Technologies (ICTs) and in supporting the provision of universal access. With falling prices and new technologies increasing individual access to ICTs, the telecentre-based community wireless networks need to continually innovate in order to remain relevant to both the telecentres and the partners that together comprise the community wireless networks.

Chapter 10

Remote and underserved communities do not attract telecommunication companies because of their low income, remote location, and limited capacity. This chapter discusses the challenges small communities face when developing their own Wi-Fi network, even when an investment is made. In particular, this chapter examines the technical, social, and economic challenges faced by the community of Chapleau (Ontario, Canada) while building its Wi-Fi network. The project adopted a public-private partnership in which Bell Canada and Nortel Networks funded its pilot phase. However, the project failed because of unclear and divergent goals, lack of sustainable applications, and insufficient technical skills on the part of the community. Using a change management framework, the chapter identifies key lessons learned and success factors required for public-private partnerships.

Chapter 11

Broadband Internet access is important for rural and remote areas to access e-commerce, e-government, e-learning, e-healthcare, Internet telephony, and other online resources. This chapter discusses the main opportunities and challenges of developing telecommunication infrastructures for rural and remote areas. In addition, affordable high-speed Internet access is important for communication (voice, data, Internet, etc.), community empowerment, job search and career development, and weather and climate monitoring. Expanding Internet access to rural areas, in particular, faces a number of challenges, such as lack of sustainable and affordable power supply, limited funding opportunities, and selecting a suitable technology. The authors discuss these issues using anecdotic evidence from a number of projects and case studies developed in the last 30 years by International Telecommunication Union (ITU). They conclude the chapter with recommendations of successful practices and policy guidelines.

Chapter 12

The Czech Republic (CR) has been ranked the 1st among the countries of the European Union (EU) countries in the growth rate of broadband access. The Internet penetration rate has increased by 48 percent between 2005 and 2011. This high growth rate is driven by the entry of new operators and the proliferation of Community Wireless Networks (CWNs). The CR holds the first place in EU in the number of newly entered operators. There are 1150 companies providing Internet access in 601 Czech towns and 5645 villages. In addition, a number of community wireless networks have emerged as an alternative of these commercial Internet Service Providers (ISPs). Their main purpose is to increase the affordability and penetration of broadband Internet in the country. This chapter discusses the contribution of CWNs to the proliferation and affordability of broadband access in the CR, focusing on the reasons for their success and popularity. Their key success factors include obtaining a non-profit status, engaging academics, and cooperating with government entities. They formed the CZFree.net forum for experts and volunteers to exchange information and best practices with respect to new technologies, design considerations, and technical and social issues. It also articulates on technology options and best practices for building low-cost CWNs. Furthermore, the chapter discusses the role of the Netural czFree eXchange association in aggregating their technical, financial, and personal resources of individual CWNs. Thanks to this association and the CZFree.net forum, CWNs in the CR have become influential competitors in the local telecommunication industry.

Chapter 13

Grassroots groups in a number of European countries are building Community Wireless Networks (CWN) on small budgets. In underserved regions, CWNs are even surfacing as the principal Internet Service Providers (ISPs). These networks have identified and implemented innovative strategies for providing connectivity—encompassing aspects ranging from software development to infrastructure design and skills training. In other words, these grassroots Wi-Fi networks mobilize human, technical, and financial resources to create sustainable alternatives to telephone and cable companies. This chapter provides an understanding of both the strengths and weaknesses of these initiatives. The authors use data from action research and interviews with leaders and participants of six successful community Wi-Fi networks in Europe. The findings show that these ad hoc initiatives are forcing local incumbent ISPs to lower prices and alter terms of service agreements. In addition, these projects broaden the public sphere, create opportunities for civic engagement, and transfer knowledge among community members. The chapter suggests that community wireless networks should be fostered by governments and the European Union in order for them to function as true alternatives to conventional ISPs, particularly in the last mile. They conclude the chapter with key learned lessons and policy implications.

Chapter 14

This chapter examines the limitations and the socio-political effects of the Brazilian National Broadband Plan (PNBL: is its Portuguese acronym). The discussion considers the main transformations witnessed in the telecommunications landscape in Brazil during the second half of the twentieth century. On the one hand, the end of state monopoly of telecommunications services and the provision of such services by the private sector called for greater investments in infrastructure. On the other hand, the Brazilian regulatory agencies have failed to lower prices, promote competition, and spread broadband access to remote and underserved areas. The PNBL was launched in order to deal with these difficulties. The plan, however, has at least three important problems: (1) the low-speed connection offered to users, (2) the unattractive prices, and (3) the lack of reflection on issues such as net neutrality. The text argues that only by taking such issues into consideration will the plan ensure innovation, economic growth, diversity, and freedom of access to information.

Foreword

This book is a welcome addition to the literature in the area of community networks. As its title indicates, the book addresses important social and economic factors related to the development and expansion of community wireless networks. Studying these factors is important for the development of new technologies, formulation of new policies, and education of the public.

The first chapter gives an introduction to Community Wireless Networks and provides an overview of the literature. The second chapter discusses challenges facing Municipal Wireless Networks in the U.S.A. The remaining chapters cover a wide range of related topics including social and economic impact, best practices, policies, implementation models, financial issues, success factors, and lessons learned from key projects all over the world. The chapters discuss projects from different countries including Nepal, Uganda, India, Canada, Brazil, European Union, the Czech Republic, and New Zealand.

I highly recommend this book for those seeking an introduction to the many issues related to Community Wireless Networks and their social and economic factors.

Hesham El-Rewini
University of North Dakota, USA

Hesham El-Rewini *is the Dean of the College of Engineering and Mines at the University of North Dakota. He is the coauthor of five books. His first book with Ted Lewis in 1992 was among the very early books written in the area of Parallel Computing and was widely adopted by universities all over the world. His two latest books in computer architecture have been translated to Chinese. Dr. El-Rewini has supervised the research of many MS and PhD students. His research interests include the areas of parallel processing, energy-aware computer architecture, and scheduling techniques. His research has resulted in numerous publications in journals and conference proceedings. His research projects have been funded by grants from industry and federal agencies. Dr. El-Rewini was the Principal Investigator of a number of international projects funded by the US Agency for International Development (USAID) and Higher Education for Development (HED) to establish partnerships with and offer academic training programs to universities in Mexico and the Middle East. Dr. El-Rewini is the former Chairman of the Department of Computer Science and Engineering at Southern Methodist University (SMU) from 2001 to 2008. He is a registered engineer in the state of Texas.*

Preface

The Internet has become an effective super highway for people to access a broad spectrum of social, economic, educational, political, medical, and personal resources. However, around 72 percent of the world's population does not have access to affordable high-speed Internet, particularly in rural, last-mile, and underserved areas. These people are lagging behind, in the digital sense, because they cannot join the emerging information society. They are digitally isolated mainly because they lack the commercial incentives necessary to attract telecommunication companies. Many cities have built their own wireless networks to support municipal services and achieve digital inclusion. However, some of these networks failed, and the existing ones face severe legal and political obstacles. It is prohibited in many places in the world for public entities to own or run such a business.

Community Wireless Networks (CWNs) have emerged as a third solution, in addition to the commercial and public solutions, for providing ubiquitous Internet access. CWNs are collective telecommunication infrastructures built by community contributions to provide shared, affordable, or free Internet access to community members. Community contributions include sharing access points, donating money and hardware, developing software for the system, and/or providing manpower and technical expertise. CWNs provide a range of benefits that include achieving digital inclusion, building a sense of community, accruing social capital, and generating human capital.

The diffusion of CWNs, particularly in rural and remote areas, faces a number of challenges that include assessing their socioeconomic outcomes, obtaining legal status, assuring power supply, preventing misuse, engaging community members, obtaining government support, and adopting a successful business model.

This book discusses the challenges facing community and municipal wireless networks and articulates on some of the failed projects. Most importantly, it presents a number of case studies from all over the world and discusses their socioeconomic benefits, financial and business models, legal issues, best practices, and learned lessons.

The first chapter provides a conceptual overview of the topic of social and economic effects of community wireless networks and infrastructure. It distinguishes Municipal Wireless Networks (MWNs) from community wireless networks. It also articulates on the applications and socioeconomic effects of both types of networks.

The second chapter discusses key challenges facing MWNs in the U.S.A., using case studies from San Francisco and Silicon Valley. It also discusses the process of selecting proposals from companies to build MWNs in the investigated cases. The key learned lessons include understanding demand, context, and limitations of technology. The author also discusses the importance of selecting a suitable business model for the sustainability of MWNs.

Chapter three discusses legal and political barriers to municipal networks in the United States. In particular, it articulates on the enacted federal laws, regulation of the Federal Communication Commission, barriers of state laws, and tactics of service providers to hinder municipal wireless networks. It also touches on some case studies and identifies lessons learned.

The fourth chapter discusses the role of Nepal Wireless in achieving digital inclusion. It also articulates on other benefits, such as delivering e-healthcare, e-learning, training, and e-commerce services. In addition, the chapter discusses key challenges facing such rural community-focused projects. These challenges include lack of technical skills, technology limitations, lack of funding resources, difficulty of geographical terrains, unsupportive political regime, and high cost of devices.

The fifth chapter discusses some attempts over the last decade in using Information and Communication Technology (ICT) to empower rural communities in India. It also presents multiple case studies to show how ICT is used to support agriculture, deliver healthcare, achieve financial inclusion, and improve the overall livelihood of rural Indians. The authors suggest adopting context-aware models, involving local and federal governments, and engaging community members.

Chapter six discusses the public-private business model of the Digital El Paso community wireless network. It articulates its partners, objectives, sources of funds, design and implementation model, and challenges. The authors suggest a public-private business model to sustain such projects and a family-centric approach to drive the adoption of these emerging networks and increase bandwidth utilization, particularly in rural and underserved communities.

Chapter seven analyzes the motivations and concerns of participants of the Fon wireless network. It uses a mix of qualitative and quantitative research methods to study the motivations, contributions, usage, and satisfaction of users.

Chapter eight discusses social outcomes of bridging the digital divide of low-income families that have school students in New Zealand through the Computers in Homes program. The program provided participants a refurbished computer, necessary software, training sessions, and six-months of free Internet. The collected data shows that the program has improved social cohesion, built stronger social networks, empowered volunteers, and enhanced civic engagement.

Chapter nine discusses the partners, services, and benefits of community tele-centers in Uganda. Authors conclude that tele-centers have decreased Internet fees, introduced new technologies, and increased individual access to ICTs.

Chapter ten identifies the technical, social, and economic challenges faced by the community of Chapleau (in Ontario, Canada) while building its Wi-Fi network. It highlights its failing factors including unclear goals, lack of sustainable applications, and lack of technical skills on the part of the community. It also suggests key lessons learned and success factors required for public-private partnerships.

Chapter eleven discusses the main opportunities and challenges facing the development of telecommunication infrastructures for rural and remote areas. The discussed opportunities include healthcare, education, e-business, communication, community empowerment, e-government, job search and career development, and weather and climate monitoring. It also discusses the challenges facing these projects, such as lack of sustainable and affordable power supply, limited funding opportunities, and choosing a suitable technology. The author uses data from the ITU-D case library.

Chapter twelve explains how CWNs has contributed to the growth of affordable access to broadband in The Czech Republic. It also discusses technology options, importance of non-profit status, success factors, and best practices for building low-cost CWNs.

Chapter thirteen provides an understanding of both the strengths and weaknesses of grassroots wireless networks in Europe. It explains how these projects broaden the public sphere, create opportunities for civic engagement, and transfer knowledge among community members. The authors conclude that these networks enforced ISPs to decrease Internet fees and alter terms of service agreements. The chapter suggests that community wireless networks should be fostered by governments and the European Union in order for them to function as true alternatives to conventional ISPs.

The last chapter examines the limitations and socio-political effects of the Brazilian National Broadband Plan. It articulates specific factors of this plan, such as the low-speed of Internet, high fees, and lack of net neutrality.

To this end, this book compiles different views, lesson learned, and best practices from all over the world to serve community developers, policy makers, practitioners, and academics. We expect that this book will be a key reference to promote CWNs as a viable alternative to commercial and public telecommunication infrastructures.

Abdelnasser M. Abdelaal
Ibri College of Applied Sciences, Oman

Acknowledgment

I would like to thank Prof. Robert Craig, the Director of the IT program in the Ministry of Higher Education in Sultanate of Oman, for his passionate support. Most importantly, I would like to thank the authors, reviewers, and the advisory board for their invaluable support and services in the review committee. My special thanks are due to Mr. Sacsha Mienrath, the Director of the Open Technology Institute, U.S.A., for crafting the idea of the book with me. I am immensely grateful to Dr. Ahmed Alreyami, Dean of Ibri College of Applied Sciences, Dr. Jehad Bani-Younis, Assistant Dean, Dr. Zubeir Izaruku, HoD, and Dr. Hamad Al-Alawi, previous Dean of Sur College of Applied Sciences, for fostering this project. I am also grateful to my dear friends Radwan Jaber, Rustaque College of Applied Sciences, and Faisal Aga, Sur College of Applied Sciences, for their valuable comments and fruitful suggestions. My appreciation and gratitude are due for Ministry of Higher Education in Sultanate of Oman. I would like to thank Joel Gamon, Austin DeMarco, Kristin Klinger, Erika Carter, and Jan Travers for their constructive development support and swift reply. Finally, my gratitude is due to my wife, Marwa, my son, Mahmoud, and my daughter, Lienah, for their patience and unlimited support.

Abdelnasser M. Abdelaal
Ibri College of Applied Sciences, Oman

Chapter 1
Introduction to Social and Economic Effects of Community Wireless Networks

Abdelnasser M. Abdelaal
Ibri College of Applied Sciences, Oman

abstract
ABSTRACT

This chapter provides an overview of the topic of community wireless networks and infrastructures. In particular, it compares community wireless networks to municipal and commercial wireless networks. It also provides a brief account of the social and economic benefits of CWNs.

By itself Internet cannot feed the poor, defend the oppressed, or protect those subject to natural disasters, but by keeping us informed, it can allow those of us who have the opportunity to give whatever help we can.

The Dalai Lama

INTRODUCTION

Pervasive computing and ubiquitous communications are increasingly becoming essential to conduct our daily life affairs. The Internet, in particular, has grown to be a superhighway for accessing tremendous social economic, social, entertaining,

DOI: 10.4018/978-1-4666-2997-4.ch001

Copyright © 2013, IGI Global. Copying or distributing in print or electronic forms without written permission of IGI Global is prohibited.

and personal services and opportunities. However, 85 percent of the world populations do not have access to affordable and reliable Internet services. According to the Internet World Stats (2009), the percentage of people who do not have affordable high speed Internet is 84 in Africa; 78 in Asia, 37 in Europe; and 22 in North America. These people are lagging behind, in the digital sense, because they are not part of the information society. Such communities usually lack viable commercial incentives to attract telecommunication companies. This could be due to their remote location, harsh terrain, high costs of deploying and maintaining infrastructures, low income and willingness to pay, harsh geographical terrain, and insufficient population density and/or limited capacity (Middleton, Longford, Clement, & Potter, 2006; Abdelaal & Ali, 2007; Siochrú & Girard, 2005; Kawasumi, 2004). As a result, the market mechanisms have failed to achieve digital inclusion of the society at large. Such social settings, we believe, require innovative and customized solutions for the digital inequality problem.

Wireless technologies can provide the much needed high-speed Internet access to any community in any location either through terrestrial telecommunication infrastructures or satellite backbones. They are particularly beneficial to a wide range of populations (Cisco, 2007). For example, those who by their nature are quasi nomadic (e.g. healthcare practitioners, real estate brokers, municipal employees, students, the mobile business persons, etc.) would find these emerging infrastructures to be of great benefit. Another group that would benefit from these emerging technologies is those who live in old neighborhoods, rural, rocky and mountainous, and remote areas.

Wireless standards (e.g. Wi-Fi and WiMax) have gained the capabilities to provide a wide range of customized connectivity solutions that suit different social settings. They enable individuals to use laptops, Wi-Fi phones, Personal Digital Assistants (PDAs) security cameras and other portable communications devices. There are many initiatives to make wireless devices more affordable all over the world. The One Laptop Per Child (OLPC) program is a great initiative lead by the MIT Media Lab for the purpose of providing low-cost and low-power laptop to children of developing countries. There are 30 projects all over world working on this program. This laptop consumes very low power (<=7W). Therefore, it could be charged by hand or powered by car or truck battery. It also uses wireless mesh technology, has a camera and audio facilities. In addition, it uses Open Source Software (OSS) such as Mozilla Web browser and Linux.

In addition to providing mobile and flexible real-time communications, these emerging communication technologies achieve significant time, money, and effort savings to their users. Most importantly, wireless technologies enable individuals to share their Internet connections and computer resources with others. As a result, wireless communications have the potential to provide ubiquitous and affordable Internet access and assist all communities to become and remain full participants in the emerging Internet-based "Information Age."

COMMUNITY WIRELESS NETWORKS AND INFRASTRUCTURES

The digital revolution has reached a tipping point after the deregulation of the 2.4 GHz spectrum in many countries. Opening this spectrum for the public provides individuals, private companies, non-profit organizations, and public entities the opportunity to build their own wireless networks using Wi-Fi standards. The following section discusses the main types of wireless networks that have evolved depending on the 2.4 GHz spectrum.

Private Wireless Networks

The most dominant type of wireless networks is the private Wireless Local Area Networks (WLANs) owned by individuals and not shared with their neighbors. These networks do not have a strong societal or community factors, as they are not shared with others.

Public-WiFi Hotspots

The second famous form is the public Wi-Fi hotspots found at restaurants, airports, libraries, train stations, coffee shops, bookstores, and other public places. Usually, these networks are built by private companies for commercial purposes such as revenue generation, advertisement and/or customer attraction (Abdelaal & Ali, 2007). However, this connectivity solution may not be suitable for old neighborhoods, rural communities, or developing societies. Such societies usually lack sufficient economies of scale to attract telecommunication companies. In addition, one of the key drawbacks of this type of networks is lack of roaming. Because of lack of coordination and absence of a central management unit users cannot roam using the same authentication among networks in the area. The only important benefit of this type of networks is increasing the penetration of Internet access, which could be offered for free.

Municipal Wireless Networks

Many cities have started building wireless network projects to provide ubiquitous Internet access (Municipal Wireless Networks Worldwide, 2012). For instance, New York City, has started a program to provide city-wide wireless access all over the city. This form of wireless networks is called "Municipal Wireless Networks" (MWNs) (Vos, 2005; Ford, 2005). In this model, a municipality (e.g. city, county) funds and owns project using public resources and facilities (Abdelaal & Ali, 2007). This model has been implemented in several U.S. cities such as Corpus Christi, Texas; Moorhead, Minnesota; and Saint Cloud, Florida. However, MWNs face severe opposition from telecommunication companies and others. They argue that it is neither efficient nor legal, particularly in the U.S., South Africa, India, European Union, for municipalities to own and operate such business (Ford, 2005). In addition, governments and municipalities of rural communities and developing societies usually either do not have enough public funds for such projects or these projects may not be in their priority list. It is important to note that commercial and private Wi-Fi networks are outside the scope of this book. The development, maintenance, ownership, and sustainability of these networks are not as critical as that of CWNs. In addition, their socioeconomic effects are limited compared to that of CWNs.

Community Wireless Networks

Numerous societies have built autonomous Community Wireless Networks (CWNs) with their own resources, taking advantage of the free 2.4 GHz spectrum and Open Source Software (OSS) (Abdelaal & Ali, 2007; Damsgaard, Parikh, & Rao, 2006; Bina & Giaglis, 2006; Vos, 2005). These grassroots projects have been recognized as a form of collective actions or wireless commons (Abdelaal, Ali, & Khazanchi, 2009b; Damsgaard, et al., 2006; Sicker, Grunwald, Anderson, Doerr, Munsinger, & Sheth, 2006). The term "collective actions" refers to the voluntary cooperation and contribution of a group of people to the construction of a common project or initiative (Olson, 1971, p. 7). Some CWNs provide Internet access for free or at affordable rate. For instance, Wireless Leiden provides free service while Nepal Wireless charges users. This means the first saves the communities money. Therefore, we should distinguish between free and paid wireless networks, when addressing their impact on their societies. It is also important, we believe, to differentiate between CWNs from similar networks to allow for comparisons between units of analysis and accumulate knowledge.

CWNs have emerged as a third solution, in addition to the public and commercial solutions, that can provide ubiquitous high-speed Internet access. When two neighbors, family members, or roommates rent a DSL link and share a wireless router with each other, they just create a simple CWN. A more complex form of CWNs is created when a big community (or neighborhood) shares a few access points developing a mesh of wireless network (Abdelaal & Ali, 2009b). Such networks could grow or concatenate with other wireless clouds and Wi-Fi hotspots to cover the entire city (Akyildiz, Wang, & Wang, 2005).

CWNs are socio-technical crucibles where community resources are shared, mobilized, and reproduced to build a common telecommunication infrastructure.

For the diversity of their beneficiaries and the broad range of their outcomes, CWNs have been recognized as a form of collective actions or wireless commons (Abdelaal & Ali, 2007; Bina & Giaglis, 2006). The term "collective actions" refers to the voluntary cooperation and contribution of a group of people to construct a common project or initiative (Olson, 1971). Instinctively, these collective or common networks serve as a third solution, besides the public and private, to the digital inequality problem. CWNs have been established in New York, Austin, Seattle, San Francisco, USA; Paris, France; Hamburg, Germany; Turku, Finland; and many other cities all over the globe. These informal clusters of wireless networks are created and maintained by the contributions of local communities and other advocates. This particular form of networks is the focus of our book. CWNs have captured the attention of communities, OSS developers, businesses, technology vendors, civil rights groups, nonprofit organizations, developing agencies, academics, and other formal and informal entities.

CWNs are particularly important for rural and developing societies. These societies usually suffer form limited public facilities, scarcity of technical personnel, difficult geographical terrain, bad climate conditions, low income, limited business opportunities, and insufficient population density (Kawasumi, 2004).

For the purpose of this book, we define CWNs as "common wireless networks built and maintained collectively by community members to provide free or affordable Internet for the purpose of achieving digital inclusion." Digital inclusion, in this context, compromises both contributors to and beneficiaries from CWNs. This is because we view CWNs as venues that provide participants the services and opportunities to be active actors in the information society. In other words, CWNs are informal broadband telecommunications systems built by the community for the community. Community contributions to the development of these networks may include sharing access points, donating money, donating old hardware, hosting access points, developing software for the system, and/or providing manpower and technical support to build and maintain the network. Therefore, the shared resources of these contributors (e.g. time, efforts, money donations, skills, beliefs, and computing resources) are the main resources of CWNs (Abdelaal & Ali, 2009; Quinn, 2006).

There is a variety of funding methods adopted by different CWNs (Abdelaal & Ali, 2007). Some of these networks (e.g. such as Zgwireless and OmahaWireless) were built through volunteerism and donations. Others (e.g. Red Libre de Ometepe) provide the service for free to community members but charge local businesses. Other networks are built on the notion of cost sharing. Examples of this cost-sharing model are PTAWUG, Seattle Wireless, and Court Housing Coop. Another important observation is that the majority of paid networks (e.g. Jhai Networks, Dharamsala, Wireless Ghana, and Nepal Wireless) are in developing societies. In addition, some networks (e.g. Nepal Wireless and Batam Wireless) provide the service at a small flat rate.

One of the famous CWNs is Austin Wireless[1] that is built through collaboration between community members. The key attributes of participants of Austin Wireless, according to Wilco (2004), are openness, tolerance, and willingness to share knowledge with others. According to Wilco, the success of CWNs depends on engaging those who may benefit from the project. The key partners of Austin Wireless include Dell Recycling Program and Image Microsystems. In addition, they collaborate with open source software groups such as Linux, Apache, NoCatAuth, and Postnuke. Moreover, local schools provide student interns to help with network installation in exchange for experience. In addition, job seekers volunteer to obtain expertise and build their resumes. Furthermore, public libraries host the facilities to obtain free Wi-Fi access and attract more visitors. These are some examples of stakeholders of CWNs and their specific contributions to, and benefits from, these collective projects.

The Tribal Digital Village[2] network started in 2001 with a $5 million grant from (Hewlett-Packard) HP Inc. through its e-inclusion initiative. Building this network through sponsorship was important, as there was no feasible commercial solution to deploy infrastructure in the reservations of Native Americans in the San Diego California area. The network is owned and operated by tribal people. In particular, the infrastructure has been designed, built, and managed by trained tribe members, particularly high school students. The network has empowered the tribal community, created training opportunities, enabled to access to online resources, supported local businesses, flourished their culture, and facilitated access to e-government services.

The academic community may be another population that could utilize these wireless infrastructures for campus based access to Web resources and expand beyond the campus to serve the surrounding community. A typical example of this model is the Omaha Wireless project. This project is built and maintained by voluntary students from the Peter Kiewit Institute (PKI) at the University of Nebraska at Omaha's College of Information Science and Technology[3]. This project started as an extension of a research project, aimed at achieving "digital inclusion" in the area, in addition to serving the educational process. It provides free Wi-Fi access at several public places such as the university campus and its neighborhoods, Elmwood Park, The Washington Library, and The Rosenblatt Stadium.

These are three different types of CWNs, with different implementation models, target populations, stakeholders, and social settings.

CWNs can also build the sense of community and bring community together to give for the community cause. For instance, the Seattle Community Technology Alliance (SCTA) is a collaborative group whose objective is to develop community technologies and services. The alliance includes Washington University, Seattle city, Seattle Public Library, Cisco Systems, One Economy, Microsoft, Seattle Public Schools, Seattle Public Library, and many others. It provides Internet access to hundreds of community members, has established training centers, facilitates internship programs, and supports e-commerce programs.

Driving Forces for Community Wireless Networks

Again, CWNs represent a very unique and elegant form of telecommunications infrastructures built collectively to serve as community connectivity solution. The rise of this innovation is driven primarily by the following factors (Damsgaard, et al., 2006; Sirbu, 2006; Sandvig, 2004; Meinrath, 2005):

1. The contributions of volunteers, OSS developers, technology vendors, students, content providers and other actors. One of the main themes of this dissertation is addressing the role of these categories of actors in developing CWNs. In particular, the concept of

social capital is used to address the social inputs and benefits and the concept of business models is used to address the economic inputs and outputs.

2. The recent advances of wireless standards (e.g. Wi-Fi and WiMax) in terms of improved signal processing, sophisticated routing algorithms, advanced antenna technologies and the convergence between wireless and wired data networks. These advances have improved the network bandwidth, signal coverage, service affordability, and Quality of Service (QoS) parameters such as packet delay, jitter, and packet drop. In fact, these technical developments are shaping the telecommunications landscape and the overall social sphere.

3. Deregulation of the 2.4 GHz frequency that allows any entity (e.g. public, private, non-profit) to establish its own wireless infrastructure without permission from public authorities;

4. The availability of OSS and Open Source Hardware (OSH) that is used to build and run the network.

5. The explosive growth of home networks and public Wi-Fi hotspots.

Therefore, CWNs have the potential to complement commercial and public solutions and provide the much needed connectivity to a wide range of populations and to bring them to the information age. As with similar emerging phenomena, the wisdom of CWNs carried out thus far has unclear and immature definitions, lack of conventional instruments, and absence of careful empirical analysis. For instance, researchers and practitioners usually confuse CWNs with municipal wireless networks and public Wi-Fi hotspots. In addition, there is no conventional research stream where CWNs is situated. Moreover, the socioeconomic issues of CWNs are well studied. Such challenges encourage us to develop this book, focusing on their socioeconomic effects.

THE SOCIAL AND ECONOMIC BENEFITS OF CWNS

CWNs represent an emerging telecommunication revolution that reconciles the interactions between economic factors, social structures and technology (Abdelaal & Ali, 2012). This is because they serve as engines for sustainable development, particularly for rural and developing societies (Best & Maclay, 2002; Sinan, Escobari, & Nishina, 2001). The key benefit of CWNs is providing free or affordable Internet access to the society at large. In addition, they boost volunteerism and civic engagement, broker knowledge and expertise, foster innovation, and improve the business environment (Siochrú & Girard, 2005; Bina & Giaglis, 2006; Sandvig, 2004). CWNs provide a wide range of outcomes such as providing affordable Internet access, generating social capital, accruing human capital, and improving the business opportunities in the area. The following section discusses the key benefits of CWNs.

Achieving Digital Inclusion

The key objective of CWNs is achieving digital inclusion. Digital inclusion, in this book, is a broad term referring to the digital services, practices, and opportunities that enable different segments of the community (e.g., Internet users, volunteers, donors) to participate in the information society (Abdelaal & Ali, 2009a). In other words, the ritual of digital inclusion, in this context, includes not only Internet access but also other digital benefits the community obtains. We measure the role of CWNs in achieving digital inclusion by their size and their capacity. The size could be measured by the number of supported users, volunteers, and contributors. Their capacity is measured by the number of access points. Abdelaal and Ali use the following factors to measure the role of CWNs in achieving digital inclusion: the network size, the capacity of the network (Abdelaal & Ali, 2009b). Wong (2007) views CWNs as open networks built to improve the availability and affordability of Internet access through community ownership of telecommunication infrastructure. The author notes the differences (e.g., objectives and scope) between different types of networks. For instance, according to Wong, the aim of municipal networks may be providing ubiquitous Internet in densely populated cities, while rural networks may serve as a last-mile broadband backhaul in underserved areas.

Creating Human Capital

The literature provides anecdotic evidence on the role of CWNs in creating human capital. For instance, Sweetland (1996) notes human capital is generated through formal or informal education, on-the-job training and/or apprenticeships. Another set of scholars state that human capital is formed by interaction of individuals and the exchange of ideas (Laroche, Merette, & Ruggeri, 1999). Providing evidence from 25 networks, Abdelaal and Ali (2012) conclude that CWNs create human capital through supporting formal and informal education, facilitating healthcare delivery, and improve access to information. When volunteers collaborate to develop OSS, they learn new ideas, transfer competences and share skills. In other words, developing OSS is considered a form of informal education, according to Abdelaal and Ali (2012). Abdelaal and Ali (2012) suggest evaluating human capital created in the domain of CWNs by its opportunity cost.

Similarly, Kawasumi (2004) explores the applications of wireless networks in rural areas of Japan. He identifies their key applications that include e-education, e-healthcare, e-administration, and environmental and monitoring services. According to Kawasumi, the benefits of using these networks in healthcare include on-demand and real-time access to system resources and information, efficiency and accuracy in operations, and patient satisfaction. Varshney (2007) discusses the role of these networks in supporting applica-

tions of pervasive healthcare. These applications include health monitoring, intelligent emergency management system, data access, and ubiquitous mobile telemedicine. Healthcare practitioners can use mobile devices to track, update, and store patient information for the purpose of improving the overall quality of healthcare (Camponovo, Heitmann, Stanoevska-Slabeva, & Pigneur, 2003). According to Camponovo et al., these networks provide on-site staff training, research, and end-user education.

Forming Social Capital

A number of scholars point out the role of CWNs in the creation of social capital and attempt to treat their social outcomes as such. For instance, Kavanaugh and Patterson (2001) note that community computer networks positively impact social capital because they increase access to the information society. In previous work, we treated the contributions of participants to CWNs as the collective actions' dimension of social capital (Abdelaal, Ali, & Khazanchi, 2009c). Another research thread investigates the dual relationship between CWNs and social capital. In particular, Simpson (2005) argues that social capital enhances the sustainability of community informatics projects. According to Simpson, social capital is important for realizing the capacity of the community, building the shared understanding of the benefits of these projects, and exploring the capability of such networks in building new forms of social structures. It also enables community ownership of the infrastructure, according to Simpson. Additionally, it fosters social inclusion, increases community interaction, and boosts community cohesiveness. Like others, Simpson does not provide empirical evidence on his claims. It is important to note that there is no agreed upon wisdom about the relationship between the Internet and social capital (Quan-Haase & Wellman, 2002).

According to authors, the Internet may decrease social capital, transform social capital, supplement social capital, increase social capital, or accrue a new form of social capital that cannot be measured using mainstream standards. Despite the plausibility of these studies, they have conceptual and instrumental limitations. Similarly, Powell (2009) argues that CWNs build social capital.

Damsgaard *et al.* (2006) describe CWNs as "wireless commons" where groups of people use their private wireless network to create common resources for the community. Similarly, Powell (2009) describes CWNs as "networks of aid" and considers them sites for civic participation. Powell also points out that CWNs attract community activists who advocate and facilitate the development of these networks. Crow, Miller, and Powell (2007) all agree that CWNs engage users and enable them to share their skills and knowledge with other CWNs around the world. In other words, the authors indicate that CWNs share resources with other CWNs. A multiple case study by Sandvig (2004) attributes CWNs as cooperative actions created by developers to obtain technical expertise. Its findings show that these networks have little impact on democracy, social capital, or economic development. Sandvig, however, notes that these networks generate social ties and promote knowledge in the community.

Producing Physical Capital

Physical capital includes machinery, factories, plants, raw materials, inventories, and means of transportation and communication (Laroche, Merette, & Ruggeri, 1999). CWNs generate physical capital when they serve as venues for contributors to build computer networks, develop applications and services, develop OSS and repair computers. Sandvig, Young, and Meinrath (2004) note the most significant aspect of CWNs is providing users the opportunity to create new services and applications. Similarly, Powell (2009) points out CWNs provide community members the opportunity to develop innovative applications. Participants of CWNs develop OSS and share it with other CWNs (Crow, Miller, & Powell, 2007).

Improving the Quality of Life

Some scholars argue that community-based Internet initiatives have the capability to improve the quality of life of the rural poor by creating new opportunities for education, healthcare, and economic development (Sinan, Escobari, & Nishina, 2001). Many CWNs provide their services free of charge. A study by Abdelaal and Ali (2009b) have collected data about the role of 26 CWNs in achieving digital inclusion. This study shows that 77 percent of the investigated networks provide free Internet access to their communities. Powell (2009) notes CWNs improve economic relationships. In addition, local ownership of telecommunication infrastructure has some advantages. For instance, it will save money for the community as they will keep the money within the community and will not pay to major ISPs. They will also gain more support and less local opposition from local communities. For instance, the Native American tribes in San Diego California have agreed on building a wireless network in their land because they would own and the run the network by themselves. Otherwise, they would not allow ISPs to expand their networks in their lands.

We identify a number of aspects that set CWNs apart from other wireless networks:

1. One difference between CWNs and MWNs is that the first serve particular communities within a specific proximity. The later, however, usually serve the entire city or town.
2. Another unique aspect of CWNs is that they are collective and common systems built by community resources to be used for their benefits.
3. The service of CWNs is usually provided for free or at affordable rates.
4. Most of CWNs are open and loosely affiliated organizations. Therefore, their sustainability and governance are at stake, unlike MWNs and commercial Wi-Fi networks.
5. CWNs are distinctive because these open networks strengthen the social ties among

participants; build the sense of community, form social and economic capital for communities, and enable community ownership of modern telecommunication infrastructures.

6. Unlike MWNs, CWNs do not have contractual obligations to cover specific areas.

The main objective of this book is to discuss the socioeconomics of CWNs, particularly their role in achieving digital inclusion and sustaining community development. Due to their informal structure and non-profit status, it is important for these collective projects to adopt a successful and proven business model (Abdelaal & Ali, 2007). Exploring the business models of CWNs is another key objective of this book. Furthermore, it is essential to identify and measure the benefits (e.g. free Internet access, accruing human capital, and generating social capital) and contributions (e.g. time, money, computer resources, and skills) that community members can obtain from, or provide to, these collective projects. This is important in order for us to provide policy makers and community developers with best practices and learned lessons necessary to empower similar communities to replicate and sustain such collective projects.

REFERENCES

Abdelaal, A., & Ali, H. (2007). A typology for community wireless network business models. In *Proceedings of the Thirteenth Americas Conference on Information Systems*. Keystone, CO: IEEE.

Abdelaal, A., & Ali, H. (2009a). Analyzing community contributions to the development of community wireless networks. In *Proceedings of the European Conference for Information Systems*. IEEE.

Abdelaal, A., & Ali, H. (2009b). *Community wireless networks: Collective projects for digital inclusion*. Paper presented at the International Symposium on Technology and Society (ISTAS 2009). Tempe, AZ.

Abdelaal, A., & Ali, H. (2012). Human capital in the domain of community wireless networks. In *Proceedings of the 45th Hawaii International Conference on System Sciences*, (pp. 3338-3346). IEEE Press.

Abdelaal, A., Ali, H., & Khazanchi, D. (2009c). *The role of social capital in the creation of community wireless networks*. Paper presented at the 42nd Hawai'i International Conference on Systems Sciences. Waikoloa, HI.

Adeyeye, M., & Gardner-Stephen, P. (2011). The village telco project: A reliable and practical wireless mesh telephony infrastructure. *EURASIP Journal on Wireless Communications and Networking*, 78. doi:10.1186/1687-1499-2011-78

Akyildiz, I. F., Wang, X., & Wang, W. (2005). Wireless mesh networks: A survey. *Computer Networks*, 47(4), 445–487. doi:10.1016/j.comnet.2004.12.001

Best, M. L., & Maclay, C. M. (2002). *Community internet access in rural areas: Solving the economic sustainability puzzle*. Retrieved from http://unpan1.un.org/intradoc/groups/public/documents/apcity/unpan008658.pdf

Bina, M., & Giaglis, G. M. (2006). Unwired collective action: Motivations of wireless community participants. In *Proceedings of the 5th International Conference on Mobile Business (ICMB)*. ICMB.

Camponovo, G., Heitmann, M., Stanoevska-Slabeva, K., & Pigneur, Y. (2003). *Exploring the WISP industry Swiss case study*. Paper presented at the 16th Bled Electronic Commerce Conference eTransformation. Bled, Slovenia.

Cisco. (2007). *Municipalities adopt successful business models for outdoor wireless networks*. Retrieved from http://www.cisco.com/en/US/netsol/ns621/networking_solutions_white_paper0900aecd80564fa3.shtml

Crow, B., Miller, T., & Powell, A. (2006). *A case study of ISF 'free' hotspot owners and users*. Paper presented to the Canadian Communications Association. Ottawa, Canada.

Damsgaard, J., Parikh, M. A., & Rao, B. (2006). Wireless commons: Perils in the common good. *Communications of the ACM*, *49*(2). doi:10.1145/1113034.1113037

Ford, G. (2005, February). Does municipal supply of communications crowd-out private communications investment? An empirical study. *Applied Economic Studies*.

Internet World Usage Statistics. (2009). *Website*. Retrieved February 28, 2009, from http://www.internetworldstats.com/stats.htm

Kavanaugh, A., & Patterson, S. (2001). The impact of community computer networks on social capital and community involvement. *The American Behavioral Scientist*, *45*(3), 496–509. doi:10.1177/00027640121957312

Kawasumi, Y. (2004). Deployment of WiFi for rural communities in Japan and ITU's initiative for pilot projects. In *Proceedings of the 6ᵗʰ International Workshop on Enterprise Networking and Computing in Healthcare Industry, HEALTHCOM*, (pp. 200-207). HEALTHCOM.

Laroche, M., Merette, M., & Ruggeri, G. C. (1999). On the concept and dimensions of human capital in knowledge- based economy context. *Canadian Public Policy*, *25*(1), 87–100. doi:10.2307/3551403

Meinrath, S. (2005). Community wireless networking and open spectrum usage: A research agenda to support progressive policy reform of the public airwaves. *The Journal of Community Informatics*, *1*(2), 204–209.

Middleton, C., Longford, G., Clement, A., & Potter, A. B. (2006). ICT infrastructure as public infrastructure: Exploring the benefits of public wireless networks. In *Proceedings of the 34th Research Conference on Communication, Information and Internet Policy*. IEEE.

Municipal Wireless Networks. (2012). *Wikipedia*. Retrieved from http://en.wikipedia.org/wiki/Municipal_wireless_network

Olson, M. (1971). *The logic of collective actions, public goods and the theory of groups*. Cambridge, MA: Harvard University Press.

Powell, A. (2009). *Last mile or local innovation: Canadian perspectives on community wireless networking as civic participation*. Retrieved from http://www3.fis.utoronto.ca/iprp/cracin/publications/workingpapersseries.htm

Quan-Haase, A., & Wellman, B. (2002, November 12). How does the internet affect social capital. *IT and Social Capital*.

Quinn, P. (2006). Community wireless and the digital divide. *Center for Neighborhood Technology*. Retrieved from http://www.cnt.org/repository/WCN-AllReports.pdf

Sandvig, C. (2004). An initial assessment of cooperative action in wi-fi networking. *Telecommunications Policy*, *28*, 579–602. doi:10.1016/j.telpol.2004.05.006

Sandvig, C., Young, D., & Meinrath, S. (2004). *Hidden interfaces to "ownerless" networks*. Paper presented to the 32nd Conference on Communication, Information, and Internet Policy. Washington, DC.

Sicker, D. C., Grunwald, D., Anderson, E., Doerr, C., Munsinger, B., & Sheth, A. (2006). Examining the wireless commons. In *Proceedings of the 34th Research Conference on Communication, Information*. IEEE.

Simpson, L. (2005). Community informatics and sustainability: Why social capital matters. *The Journal of Community Informatics, 1*(2), 102–119.

Sinan, A., Escobari, M., & Nishina, R. (2001). *Assessing network applications for economic development.* Cambridge, MA: Harvard University.

Siochrú, S. O., & Girard, B. (2005). community-based networks and innovative technologies: New models to serve and empower the poor. *The United Nations Development Program.* Retrieved from http://propoor-ict.net

Sirbu, M. (2006). Evolving wireless access technologies for municipal broadband. *Government Information Quarterly, 23*, 480–502. doi:10.1016/j.giq.2006.09.003

Sweetland, S. R. (1996). Human capital theory: Foundations of a field of inquiry. *Review of Educational Research, 66*(3), 341–359.

Varshney, U. (2007). Pervasive healthcare and wireless health monitoring. *Mobile Networks and Applications, 12*(2-3), 113–127. doi:10.1007/s11036-007-0017-1

Vos, E. (2005). *Reports on municipal wireless and broadband projects.* Retrieved from http://www.muniwireless.com

Wilco, R. (2004). *Austin wireless city project.* Retrieved from http://www.lessnetworks.com/static/AustinWirelessCityProject.ppt

Wong, M. (2007). Wireless broadband from backhaul to community service: Cooperative provision and related models of local signal access. In *Proceedings of the 35th Research Conference on Communication, Information and Internet Policy.* Arlington, VA: IEEE.

ENDNOTES

[1]   www.austinwirelesscity.org
[2]   http://www.sctdv.net/
[3]   http://omahawireless.unomaha.edu/

Chapter 2
Challenges Facing Municipal Wireless:
Case Studies from San Francisco and Silicon Valley

Heather E. Hudson
University of Alaska – Anchorage, USA

ABSTRACT

This chapter discusses the challenges facing municipal wireless networks in the United States. It articulates a number of case studies from Silicon Valley. The authors explores the demand, context, and limitation of technology. They conclude that selecting a suitable business model is a key success factor.

Wireless broadband for San Francisco: "It was the most impossible deal you could imagine. Every mayor in every small town would be famous for at least a day if they stood up on a soapbox and said, 'We are going to have this free service.'" (Settles quoted in Skidmore, 2008)

INTRODUCTION

Providing affordable broadband access to residents in the United States has become a key public policy issue. Residential access to broadband in the U.S. is behind that of many other industrialized countries. Specifically, the U.S. currently ranks 15th among industrialized countries in broadband access per

DOI: 10.4018/978-1-4666-2997-4.ch002

100 inhabitants, according to the Organization for Economic Co-operation and Development (OECD) (OECD, 2011). Adoption of broadband in the U.S. depends highly on socio-economic status. There are only 93 percent of households with annual incomes above $150,000 that have broadband access. In addition, about 45 percent of households with incomes below $25,000 have access to broadband (NTIA, 2011).

Some U.S. cities have responded to limited availability or affordable access to broadband by undertaking initiatives to provide free or low cost broadband via wireless technology (Abdelaal & Ali, 2007; Vos, 2005). To implement a citywide wireless network, also referred to as municipal wireless, cities have several different options when it comes to ownership and business models (Abdelaal & Ali, 2007; Mandviwalla, Jain, Fesenmaier, Smith, Weinberg, & Meyers, 2006). The most common model used by some small cities in and outside the U.S is the city to own and operate the wireless network by itself. Typically, the city contracts with private vendors to supply and install the equipment then operates the network. In a second model, the city selects a private entity (individual company or consortium) to build, own, and operate the network under specific terms required by the city. These terms are likely to include coverage parameters (ranging from public areas such as squares, libraries, parks, and community centers) to complete coverage of the entire city including business and residential areas. They are also likely to include pricing requirements such as providing free access to specific areas or target users, sometimes coupled with another fee-based option with higher bandwidth and/or better security.

In December 2005, the city of San Francisco issued a Request For Proposal (RFP) for a community wireless broadband network of its own. This initiative received significant national and even international attention, largely because of San Francisco's visibility in the high tech world, and the involvement of Google in the consortium

selected to build and operate the network. However, the project died without any network being built.

This chapter analyzes municipal broadband in the U.S. through multiple case studies from San Francisco and Silicon Valley. In particular, it examines the reasons for its demise and compares San Francisco's approach with other models for municipal wireless adopted by nearby Silicon Valley communities. It concludes with lessons learned and unresolved issues.

SAN FRANCISCO TECHCONNECT

The Concept of Affordable and Ubiquitous Broadband

The city of San Francisco has a population of about 805,000 and area of 49 square miles, with an average population density of about 16,500 people per square mile. It is highly ethnically diverse, with a population that is 33.3 percent Asian, 15.1 percent Hispanic, and 6.1 percent African American, according to the 2010 census. Median household income is about $70,000, but about 12 percent of the population lives below the poverty line (US Census, 2010). Some 44.3 percent speak a language other than English at home. Providing ubiquitous Internet access to these ethnic groups will enable them to benefit from a rich pool of online resources, create their own content, and communicate with their peers and relatives.

In mid 2005, the City and County of San Francisco (both City and County have the same geographic boundaries and administration) established TechConnect initiative. TechConnect is a "strategy to promote digital inclusion by ensuring affordable internet access, affordable hardware, community-sensitive training and support, and relevant content to all San Franciscans, especially low-income and disadvantaged residents" (San Francisco, 2008). In September 2005, TechConnect released a Request For Information and Content (RFI/C) which stated: "Universal, affordable wireless broadband Internet access is essential to connect all residents of San Francisco to the social, educational, informational, and economic opportunities they deserve" (San Francisco, 2005a).

Between the announcement of the initial strategy and the release of the RFI/C, "affordable Internet access" had become "affordable *wireless broadband* Internet access." Yet there was little rationale for the emphasis on wireless technology as a citywide solution. San Francisco has broadband available over DSL and cable provided primarily by AT&T and Comcast. In addition, fixed wireless and fiber cables are available in some areas and buildings.

No data were provided on where broadband was unavailable, nor on broadband subscribership by zip code, neighborhood, income, ethnicity, or any other variables. If broadband usage was lowest among low-income and other disadvantaged residents, as appeared likely, no studies were available to show whether the primary reason was pricing of broadband services, or whether other factors were also important such as lack of computers, lack of computer and Internet skills, perception that content was irrelevant or harmful, etc. TechConnect set up a Task Force on Digital Inclusion with representatives from many community and ethnic organizations, but this was not done until April 2006, *after* the RFP was written and the winning proposal was selected.

However, following the RFI/C process, the City issued a Request For Proposal (RFP) in December 2005 with the goal of providing "*universal, affordable wireless broadband* access for all San Franciscans, especially low-income and disadvantaged residents" (italics added). The RFP stated that the network was to be built, operated, and maintained at no cost to the city. In addition, a basic level of service should be free, and that the entire city (including the county) should be covered. Other specifications included:

- Premium services can be fee-based, but should be priced lower than existing service alternatives;
- Outdoor coverage shall be provided for a minimum of 95 percent of the city's area;
- Indoor coverage shall be provided for ground and second floors of a minimum of 90 percent of all residential and commercial buildings in the city; and
- Indoor perimeter room coverage above the second floor shall be provided for a minimum of 90 percent of all residential and commercial buildings (San Francisco, 2005b).

Given San Francisco's topography with its numerous hills, and its high urban density including many areas with multistory residential buildings and office buildings, these were very demanding specifications. Also, although "existing service alternatives" were not defined, a version of DSL at the time was offered by AT&T for $13 per month and cable modem access for $20 per month.

The Process of Selecting Proposals

The city received six proposals in total, five complete and one vague and incomplete. The five complete proposals were from a consortium headed by EarthLink and Google (the eventual winner); MetroFi, a privately held firm founded by former Covad executives; nextWLAN, a privately funded wireless LAN company founded in 2003 and headquartered in the Bay Area; Razortooth, a grassroots Internet company headquartered in the Mission District of San Francisco, and doing business as RedTAP; and a consortium headed by a nonprofit called Seakay, partnering with Cisco and IBM.

The submitted proposals have the following distinguishing characteristics:

- **RedTAP:** This proposal was submitted by Razortooth. It proposed a cooperative model, building community access centers, also offering training and technology for residents. Their proposal noted: "Without a real strategy to provide technologically underserved residents with Wi-Fi enabled laptops or desktops, the disadvantaged will be further left behind" (RedTAP, 2006).

- **Cisco-IBM-Seakay:** Their proposal discussed finances only. In particular, this consortium proposed that the project would be financed through nonprofit fundraising, with cash and in-kind donations. Another unique aspect of this consortium is including a nonprofit organization with Seakay. Incidentally, theirs was the longest and most technically detailed proposal (Seakay, et al., 2006).

- **NextWLAN:** It proposed that the higher speed premium service would be financed by deploying about 100,000 micronode repeaters that subscribers would rent and attach to an existing estimated 40,000 DSL lines (NextWLAN, 2006).

All three proposals were evaluated by five reviewers, four of whom were city employees with various IT responsibilities. The written proposals were scored on a scale of 80 points, 20 being for firm qualifications and 60 for degree of compliance with the city's specifications. The top two were EarthLink/Google and MetroFi separated by only 4 points with 260 and 256 out of a possible 400 aggregated from the five reviewers, SeaKay came in a distant third with 148 total points. Three of the reviewers ranked EarthLink higher on firm qualifications, while four ranked MetroFi higher on compliance with city specifications.

The evaluation process also included oral interviews with the competing companies. Interviews covered a wide range of issues including technical solutions, costs to the city, user interfaces, digital inclusion, experience in other municipal Wi-Fi projects, etc., that turned out to be critical to the outcome. EarthLink/Google won the combined review, but the aggregate scores differed by only 16 points out of a possible 500 point total (San Francisco, 2006). After the winner was announced, the press coverage highlighted the involvement of Google.

However, the announcement of the winner of the RFP process turned out to be far from the final step in the process. More than six months passed before the final contract was negotiated. It was finally signed in January 2007. However, more hurdles remained. The city's Public Utilities Commission (PUC) and the Board of Supervisors (San Francisco's city council) also had to approve the project. The contract stated that the city had to finalize approvals within six months of signing; later, an extension until October 2007 was negotiated (San Francisco, 2007).

The Supervisors held several public hearings during spring of 2007. Their objections ranged from the aesthetics of wireless antennas on city power poles to user privacy (even free users had to sign up with Google), to whether the city itself should build and operate a municipal broadband network. The PUC did eventually approve the project, but by summer of 2007, approval by the Board of Supervisors was stalled.

To escape this morass, the Mayor's office announced in July 2007 that it would put the project on the November election ballot (which included local and state initiatives, as well as local elections). However, in August 2007, EarthLink announced that it was withdrawing from the project. As noted below, EarthLink appeared to have reservations about the commercial viability of municipal wireless. In addition, in August 2007, managers began restructuring the company in response to financial losses (Reardon, 2007). Although moot at this point, the proposition to approve the wireless project remained on the ballot. Meanwhile, the Mayor announced that a new initiative would be announced within weeks.

In 2009, San Francisco's TechConnect continued to provide training to low income groups, and its Network of Community Networks (NCN) provided residents of more than 3750 low income housing units with free access to the Internet. (San Francisco TechConnect, 2011) The City and County of San Francisco also commissioned a feasibility study for municipal fiber optics. This study was released in 2009, but no further action was taken (Columbia Communications, 2009).

OTHER INITIATIVES IN THE BAY AREA

Wireless Silicon Valley

The San Mateo County Telecommunications Authority (SAMCAT) released a RFP for municipal wireless in April 2006. Its objective was to provide service to some 35 communities in four counties in the Bay Area. The project became known as "Wireless Silicon Valley." The RFP differed in several aspects from that of San Francisco's TechConnect. For instance, the area and population covered was much larger, the combined population of San Mateo County and Santa Clara County is about 2.4 million, and land area is about 1740 square miles. In addition, population density is about 1400 per square mile, and about 30 percent of residents live in multi-dwelling units. Multi-dwelling units means that the social gains of the project would be high because a single computer or Internet connection would serve more than one person. Only outdoor wireless coverage of the region was required. The provided service could be either for free or at low cost. Other services were "desired" but not required, including enhanced (e.g. security, and speed) outdoor service, indoor guaranteed service, government service, and public safety services. All of these other services could be fee-based (Wireless Silicon Valley, 2006).

Seven proposals were received. Neither written proposals nor evaluations by reviewers were made available online. The winner of the Wireless Silicon Valley competition was Silicon Valley MetroConnect. It is a consortium consisting of nonprofit Seakay, Cisco and IBM, and Azulstar. The total cost was estimated to be $200 million, or about $125,000 per square mile. The first phase was to be two pilot projects (Moura & Fearey, 2006). However, by the end of 2007, the $500,000 needed for the two pilots had not been raised, and the only contributor was Cisco. A *San Jose Mercury* columnist commented: "Half a million dollars is peanuts around here—Intel takes in that much cash every 8 minutes. So Wireless Silicon Valley's inability to scrape up the money says a lot about our civic and corporate leaders as well as our place as the purported center of the technology universe" (Goel, 2007).

In February 2008, Wireless Silicon Valley announced that it had a new partner, Covad Communications, which would construct a test network in downtown San Carlos. San Carlos is a small city of about 28,000 residents on the San Francisco Peninsula. It stated: "Covad's entry reinvigorates the project, which was offline for most of 2007 while the team searched for a workable business model that didn't require taxpayer revenue. Covad has a long history of providing business-class communication services using wireline and wireless technologies, and provides a promising complement to Cisco and the other members of the MetroConnect team." Covad, based in San Jose, was one of the original CLECs (Competitive Local Exchange Carriers) that struggled to survive reselling communication services. Wireless broadband startup MetroFi was founded by former Covad executives (Wireless Silicon Valley, 2008). However, the San Carlos network was never built.

Interestingly, EarthLink chose not to respond to the RFP. Its rationale, submitted in a letter to SAMCAT, identified some of the key issues in developing sustainable business models for municipal wireless. EarthLink stated: "…we have not been able to reconcile the RFP's strong desire for a basic free layer of access throughout the coverage area." It pointed out the Quality of Service (QoS) problems that may result from low node deployment densities on some free systems: "Some of the operators ... have attempted to obscure these problems by switching to free or advertisement-supported business models, hoping that because end users are no longer required to pay for the service, they would be willing to overlook the poor performance and poor coverage of the networks" (Reinwand, 2006).

However, this was the same EarthLink that won the San Francisco competition which required free citywide service and penetration within buildings for premium service, and that had signed a ten-year agreement with Philadelphia (see below). Was EarthLink simply stating that the less rigorous RFP would enable bidders with cheaper designs to win, regardless of sustainability? Or was it implying that Google's deep pockets were the only reason it chose to partner with EarthLink in the bid in San Francisco? In its presentation for San Francisco, Google noted that it had over $6 billion in revenue in 2005, and that, as of December 2005, it had $8 billion in cash and equivalents and no debt (EarthLink/Google Proposal, 2006).

MetroFi's Networks in Silicon Valley

Other Silicon Valley communities, including Cupertino, Santa Clara, and Sunnyvale, contracted with MetroFi to provide wireless broadband. However, the business models varied in such small cities, which differ from big cities like San Francisco in terms of geographical characteristics and population density.

The city of Sunnyvale (population of approximately 138,000) approved a five-year, non-exclusive franchise agreement with MetroFi in late 2004. They agreed to provide free service with revenue based on advertisement. The city rented out access to street light poles at a discount, losing an estimated $11,300 per year (Wilson & Kraatz, 2008).

Santa Clara, located between Sunnyvale and San Jose, has close to 110,000 residents, and covers 19.3 square miles. In addition, it is the home of Santa Clara University, a convention center, and numerous high tech companies including Intel (Santa Clara, 2008). Santa Clara authorized MetroFi to install a wireless citywide network. Neither RFPs (if any) nor contracts between the cities and MetroFi were made publicly available. MetroFi started with a coverage of half the city, and planned to cover the whole city including about 40,000 households by the end of 2006. MetroFi stated: "Those with a Wi-Fi enabled device ... can now easily access the Internet or eliminate their existing service provider's charges by accessing the high-speed MetroFi network" (MetroFi, 2006).

It appears from early press releases that MetroFi originally intended to use a subscription model, charging $19.95 per month. However, it then changed to an advertising-based business approach: "The MetroFi network also brings a new opportunity for local businesses to reach the community through a truly local Internet advertising medium. Customers that are accessing the network will be shown a banner advertisement in the frame of the browser. Local businesses can take advantage of the local and regional nature of the network by providing links to their website, coupons or announcements to those that are guaranteed to be near their establishment" (MetroFi, 2006).

Cupertino, home of Apple Computer, is an upscale community of about 52,000, of whom about 50 percent are Caucasian and 44 percent are Asian; the median income is just over $100,000 per household (Cupertino, 2008). The contract between MetroFi and the city was a non-exclusive installation and service agreement. The business model here was subscription-based; MetroFi acted as a no-frills provider and open access wholesaler. The basic service for $19.95 per month included a wireless modem, but no email or other services, so that the user could continue with an existing ISP. MetroFi also wholesaled access to ISPs such as EarthLink (which offered enhanced service for $24.95 per month). Coverage was about 75 percent of Cupertino (nealy 15,000 households). MetroFi stated the service in Santa Clara and Cupertino offered "DSL-like speeds" (about one megabit per second). MetroFi claimed it was able to provide municipalities with "all-in" pricing of $50,000 per square mile, inclusive of site surveys, network design, equipment, and installation (MetroFi, 2007).

However, MetroFi abruptly ended service to its customers in Cupertino, Sunnyvale, Santa Clara, and other cities on June 10, 2008, as evidenced by the following notice to MetroFi customers on its website:

Dear MetroFi User: It is with regret that we notify you of our intentions to discontinue offering MetroFi FREE (sic) and MetroFi Premium services in your area. We are in the process of negotiating with your city to keep the network in place, under city ownership, and during this time your services will not be affected. As soon as we know the outcome of these negotiations, we will provide you with further information (MetroFi, 2008.)

The company stated that it intended to sell the networks to the cities or to a third party, or to shut them down altogether (Wilson & Kraatz, 2008).

When the commercial providers pulled out, cities could not secure funds to buy the networks. They did not even have the desire to run them if facilities were donated or bought by a benefactor. Many cities were apparently in no position to take on these expenses or responsibilities. For example, MetroFi offered to sell the Cupertino network to the city for $135 per node, or approximately $135,000. However, a city spokesman stated "We don't have the financial dexterity to get into the municipal Wi-Fi business. The city does not have the resources to do it." Sunnyvale's representative took a similar position: "I think we're somewhat dismayed at the loss of this service … we're not planning on doing anything. It's a private business that the city has nothing to do with" (Skidmore,

2008). Where there was no interest in buying the network, MetroFi faced the cost of removing equipment on poles throughout the cities. The city of Portland, Oregon, where MetroFi was also to build a wireless network, seized 600 antennas and related equipment after MetroFi determined that its advertising-based business model would not be viable and ceased construction.

Google's Wi-Fi in Mountain View

While involved in the San Francisco bid, Google donated a free Wi-Fi network to its home town of Mountain View. This city is located between San Francisco and San Jose and has a population of about 71,000. Mountain View signed a five-year agreement with Google, which included installation and operation of the free network. The agreement also required Google to pay the city an annual fee for using its streetlight poles and equipping the city's library services vehicle with mobile wireless equipment (Mountain View, 2008). Costing nearly a million dollars, the network began service in August 2006, covering 11.5 square miles with 380 access points and a link to Google headquarters, known as the Googleplex. Users could access 1 megabit service for free by registering with Google (Malik, 2006). As of mid-2011, the Mountain View network was apparently still in operation (Google, 2011).

Was this project about philanthropy or a market test bed? Google's Wi-Fi website states "Google Wi-Fi is a free wireless Internet service we're offering to the city of Mountain View as part of our ongoing efforts to reach out to our hometown" (Google, 2008). However, in 2006, a Google representative said the goal was to turn Mountain View into a large-scale test bed for various Wi-Fi enabled devices that are coming to market, and that Google had no intentions of responding to additional municipal wireless RFPs. (Sacca, quoted in Malik, 2006). A similar mix of interests may also have influenced Google's intent to invest in San Francisco, always publicly described as a philanthropic initiative, but clearly a major urban market research opportunity for wireless advertising and services. Yet, while the San Francisco project was floundering in political quicksand, Google had already created a testbed in its own backyard.

LESSONS LEARNED AND UNRESOLVED ISSUES

Given these case studies, we can identify the following lessons and issues of municipal wireless networks:

Understanding Demand

San Francisco and other communities seeking to achieve the goal of affordable broadband for low-income and disadvantaged populations need more information on why so few subscribe. San Francisco appeared to assume that affordability was the major barrier, as its RFI stated: "Fees for access to the Network must be priced lower than existing alternatives...." Yet it did not have data that could be used to estimate demand and identify barriers. For example, were there areas of the city that did not have broadband available by DSL or cable or some other means? Where service was available, what percentage of households in each neighborhood subscribed? Were the barriers to access strictly financial, or were there other barriers, such as lack of computers, lack of skills or confidence to use computers and the Internet, lack of appropriate content or applications? Since these wireless proposals and projects were undertaken, more research has been done on broadband access and barriers to adoption in the U.S., but it is not disaggregated to the city level (See, for example, Horrigan, 2009; NTIA, 2010; NTIA, 2011).

We also know from other community access experience that many people who are not connected to the Internet will need computers, and supported software and hardware, training and awareness of how the Internet could be useful to them and their families. Community outreach will also likely be needed to engage local residents in the project. As Strover et al. note: "The formidable access issues surrounding mounting meaningful access programs, reaching populations unused to and involved with computers, and creating economic advantages for individuals and businesses require strong communities and a leadership cognizant of community dynamics" (Strover, Chapman, & Waters, 2004).

If availability or affordability of computers is a key barrier, cities could consider initiating a program for subsidizing computer purchase for low-income and disadvantaged residents. This could be a lease-to-own scheme for a small monthly down payment. Cities could also adopt a program for refurbishing and recycling donated computers for needy individuals. For example, San Francisco's TechConnect provided donated computers and offered training, although the wireless project was not implemented (San Francisco TechConnect, 2008).

If price of broadband service is a significant barrier (and if most unconnected households already have computers), the goal of affordable access for the low-income and disadvantaged could be achieved through a discount or voucher for those who meet low-income criteria. In San Francisco and many other U.S. cities, affordability rather than availability does seem to be the primary issue, as DSL and cable broadband have become widely deployed, even in low-income areas. Yet the rollout of "DSL-lite" for $15 to $20 per month suggests that monthly charges may not be out of reach, although upfront installation costs and modem rental or purchase may be barriers.

Understanding the Context

It is important for developers to understand the bureaucratic, legal, and cultural context of their city or community. The contrast of business and bureaucratic cultures was very evident in San

Francisco. It took six months to negotiate the contract after the EarthLink/Google proposal was selected. However, as noted above, city regulations required that the Board of Supervisors (equivalent of a city council) also had to approve the project. Several Supervisor hearings were held where neighborhood and minority community groups praised the project and asked for implementation as soon as possible, while some speakers questioned the technical design and aesthetics of wireless transmitters on utility poles, and others were concerned about privacy of data collected by Google. Yet the subtext among the supervisors who opposed the project appeared to be "not invented here"—that the project was the Mayor's initiative and not their idea. While the Google connection brought national media attention, some Supervisors favored instead a city-owned and operated network. By summer 2007, the project was stalled, even before EarthLink's withdrawal.

Evidence of these different cultures can be seen in other cities as well. When announcing the decision to sell the Philadelphia network, EarthLink's CEO stated: "It quickly became evident that we would have a really difficult time changing the perception by some of the cities that we owed them a free network rather than the city stepping up to make the business model viable for both them and for our shareholders" (Huff, quoted in Lakshmipathy, 2007).

Technology Limitations

In specifying Wi-Fi for municipal broadband, some cities tended to confuse the technological means with the goals of achieving universally available and affordable broadband for their residents. The San Francisco RFP stated: "Universal, affordable *wireless* broadband Internet access is essential to connect all residents of San Francisco to the social, educational, informational, and economic opportunities they deserve." This pronouncement confuses the means with the ends. If the goal is universal and affordable access to broadband, should the wireless technology be only one so-

lution? Wireless technology is very appropriate for outdoors and public spaces. It is likely to be less suitable for individual households, multiunit dwellings, high rises, office buildings, etc. Besides, many of these could already be served by commercial cable systems or DSL.

Key considerations in evaluating the suitability of Wi-Fi are assumptions about availability of service—should the wireless network penetrate deep into buildings or high floors? In addition, what is the quality of service requirements? How strong and reliable a signal do users require? Settles notes: "Wi-Fi's drawback as a means of digital inclusion is that its signal doesn't travel well through walls or most other obstacles; connecting the outdoor signal to a home's indoor network requires boosters or other costly add-ons" (Settles, quoted in Skidmore, 2008). A city official in Cupertino saw Wi-Fi as a problematic technology in Cupertino's densely treed neighborhoods. "We've had issues with antennas right outside the buildings, but the signal can't always penetrate stucco and trees" (Kittson quoted in Wilson & Kraatz, 2008).

EarthLink believed that quality of service requirements for Wireless Silicon Valley dictated a dense mesh wireless design of 36 Wi-Fi nodes per square mile. This mesh density is required to push the signal further into the consumer's home. It noted that this design would require a higher investment than estimates proposed by competitors, and concluded that advertising revenue alone would not cover its capital costs or provide a sufficient revenue stream (Reinwand, 2006). In Philadelphia, EarthLink was forced to install double the number of routers it had planned for because of inconsistent reception. Project costs rose to $20 million compared to the $12 to $15 million that was originally budgeted (Skidmore, 2008).

Business Models and Sustainability

Choosing a suitable and working business model is a critical success factor of municipal wireless networks. Several business models have been proposed for municipal broadband (Abdelaal & Ali,

2007; Mandviwalla, et al., 2006). More than 2000 U.S. communities have public power systems; many have their own optical fiber for managing their networks. Some of these municipalities have decided to provide broadband access to the public over their networks. Their typical anchor tenant is the local government; some expand to serve local businesses. If they choose to serve residential customers, they may tie into a local telephone company for long distance services, or expand their networks to homes using fiber, hybrid fiber-coax, or Broadband over Power Line (BPL). They may become an ISP or a conduit for multiple ISPs.

For wireless, municipalities typically contracted with the private sector to build and operate the network, which could be owned by the municipality or the contractor (the latter investment model is known as Build, Own, and Operate or "BOO"). The key suggested revenue models are:

- Free service, advertising-supported;
- Subscription service, often with various tiers or options;
- Hybrid with free service in some areas or some users, and fee-based services elsewhere.

Given the insistence of many communities to include free service, advertisement-based model seems to be necessary, particularly for highly populated areas and business districts. This model assumes that subscription fees would not be sufficient to cross-subsidize free service. The requirement of many municipalities to keep subscription fees low, or even to specify that they must be lower than available alternatives, limited operators' pricing flexibility. In addition, some cities require revenue sharing, in return for providing free or discounted access to their infrastructure and public space such as electric poles, public parks, and libraries.

The verdict is still out on which, if any, business models are sustainable, although the demise of EarthLink and MetroFi projects described above casts doubt on the commercially operated free service model. In its letter outlining why it declined to bid on the SAMCAT RFP, EarthLink also notes that free or advertising-supported networks typically ignore other items that "comprise a comprehensive broadband solution" such as CPE (customer premises equipment) and technical support. It also foresees the needs to update and upgrade the network: "We do not believe that user needs five years from now will be the same as they are today" (Reinwand, 2006).

In Philadelphia, where EarthLink had built a Wi-Fi network, it offered to transfer the network to Philadelphia for free and to donate new Wi-Fi equipment in 2008. However, city officials stated that the cost of network operating would be too expensive. Then, EarthLink shut down the Philadelphia project after it failed to find a buyer for the $17 million unfinished network. Network Acquisition Corporation (NAC) later purchased the network abandoned by EarthLink. They planned to resume offering consumers free Wi-Fi service and to charge businesses for service on a new wired and wireless network (Skidmore, 2008). In 2009, the city of Philadelphia announced it was exercising its option with NAC to purchase the network, for $2 million, and would invest an additional $17 million between 2011 and 2015 to build out its existing core fiber network and the wireless mesh network acquired from NAC. The city plans to create a "multi-purpose public safety and municipal wireless network that will improve government operations as well as providing free Internet to citizens in targeted public spaces" (Albanesius, 2009).

Some city-owned municipal systems have survived, and are apparently sustainable. In general, these rely on revenue streams predominantly from large subscribers, or they are designed primarily or exclusively for city government use such as in Minneapolis and Oklahoma Cities. In 2006, the City of Minneapolis signed a 10-year contract with USI Wireless to own, build, and manage its wireless network. The network provides free Internet access through 117 "Wireless Minneapolis" hotspots throughout the city. In addition, the network is used for city services. For example, the City used the network for emergency response and rescue efforts when a major freeway bridge collapsed on Aug. 1, 2007 (Wireless Minneapolis, 2011).

Another sustainable network is that of Oklahoma. This network is dedicated only for public safety and municipal applications. The city invested $5 million from public safety capital sales tax and city capital improvement funds to construct and operate the network. City departments and agencies such as police, firefighters, building inspectors and utilities were to be major users of the network. The IT director of Oklahoma City stated: "We designed our system to meet our goals at a specific price point, which it continues to do with great aplomb. While it may not meet the expectations and vision of someone with a different set of goals, it has and continues to be a very successful model to build on for us and our needs" (Mark Meier, quoted in Vos, 2008).

INDUSTRY STRATEGIES

Cities should consider the strategies and tactics of telecommunication industries. The following is a discussion of these strategies:

A Third Pipe?

Given the uncertainties of municipal politics and the lack of proven business models, why would the private sector want to get into municipal wireless? There appear to be two major drivers for the operators and content providers (as opposed to the vendors, who have a clear interest in selling equipment for this new market). New entrants may see municipal wireless as a means to compete with incumbents by building their own relatively inexpensive networks, to provide the illusive "intermodal competition" championed by the FCC (Federal Communications Commission). It appears that MetroFi and EarthLink fall

into this camp. MetroFi was founded by former executives of Covad, which struggled as a CLEC to build a customer base by leasing capacity from phone companies. Covad's consumer business eventually failed, and it survives as a modest enterprise-services business.

In Congressional testimony, EarthLink saw municipal broadband as an antidote to facilities-based duopoly. EarthLink believes that municipal Wi-Fi will serve as a viable third broadband alternative to connect homes at prices that will spur competition and choice in progressive cities (Reinwand, 2006). It made the analogy to the U.S. mobile industry before and after its expansion from a duopoly, noting that the 20 million mobile subscribers in 1994 had grown to 200 million subscribers in 2005. EarthLink also pointed out that the average price of mobile service had dropped 85 percent between 1993 and 2004 (Puttala, 2006).

As described above, in Philadelphia, EarthLink stated that it planned to invest $10 to $15 million, with the intent to offer 1 mbps broadband Internet service for as low as $10 per month for low income homes and $20 for others. As in San Francisco, EarthLink appeared to see this investment as an infrastructure play, a means of getting into broadband without having to lease capacity from other providers. However, in 2007, Verizon was offering 768 kbps for $14.95 per month in Philadelphia.

Content and service providers were looking for new markets and testbeds. For content providers and applications providers such as Google, municipal wireless could provide opportunities to develop strategies and content that could be transferred to 3G networks. Eric Schmidt, then CEO of Google, stated: "We can make more money on mobile than we do on the desktop, eventually" (Schmidt, 2008). Of course, this scenario would also affect the business model of municipal wireless. As mobile broadband expands, 4G mobile networks could be the "killer response" that wipes out revenue for municipal wireless once mobile users have access to broadband on their cell phones, tablets, notebook computers, and other devices. If so, wireless broadband will be available, but not free or likely cheap, as city officials had hoped.

Incumbent Perspectives

The incumbent telecom industry provides various perspectives about the growth of municipal broadband, including wireless. Typically, they claim that municipal broadband is not the answer to increasing broadband access. For example, some industry analysts are skeptical about whether municipal wireless is a viable means of bridging the digital divide. Pyramid Research questioned "whether the dream of 'cheap Internet for everyone everywhere' will ever materialize as expected." Their view was echoed by incumbents: "We expect that municipal Wi-Fi networks will not match other offerings' unique content, security features, and reliability" (Pyramid Research, 2005).

Another incumbent argument is that public investment in communication networks crowds out private investment. The telecommunications industry has lobbied vigorously at the state level to prohibit or severely limit municipal broadband. Baller noted that in 2004: "Not just small rural communities, but even large cities, such as Philadelphia, San Francisco and Minneapolis, had become intensely interested in developing citywide wireless projects. The incumbents saw this as a much more significant threat than the relatively small number of municipalities that were operating or pursuing wireline options" (Baller, 2004). As of August 2006, seventeen states had passed legislation to prohibit or hinder municipal entry into communications (Baller, 2006).

A Florida study by Ford found no evidence to support the "crowding out" hypothesis, but strong support for a stimulation hypothesis—municipal-run networks typically provide wholesale access to key components of telecommunications infrastructure. Ford's empirical model, using data on the number of CLECs in particular markets in Florida, indicates that municipal communications

actually increase private firm entry. The inference from this research is that legislation restricting or precluding municipal provision of communication services reduces the overall level of competition (Ford, 2005).

Yet some incumbents also apparently concluded in mid-decade that they needed to get in the game, if only to stave off the likes of new competitors such as EarthLink partnered with major investors such as Google or possibly Yahoo or Microsoft. One consulting firm predicted that municipal networks (not only wireless) could grab up to 35 percent of the market share for video, fixed voice and high-speed Internet services, and up to 20 percent of the mobile services market. It noted that "The competitive impacts will be especially threatening to incumbents to the extent that municipal networks can be cost-justified by increasing efficiencies, cost-savings and other 'internal' or 'social' benefits captured by local governments, schools and other public institutions." Thus, "... broadband incumbents may have to deploy their own competitive Wi-Fi network offerings, and offer mobility as a differentiation tool" (Pyramid Research, 2005).

CONCLUSION

The experiences of San Francisco and other Bay Area communities seeking citywide wireless broadband offer several lessons. They also suggest that there remain many unanswered questions about the demand for ubiquitous broadband, the barriers to broadband usage, and the financial and technical sustainability of municipal wireless.

San Francisco's emphasis on wireless technology alone was misplaced. Its goal should have been *universal (i.e. available and affordable) access to broadband for all*. Adopting wireless networks is definitely a major part of the solution to achieve this goal, but it should not be the only solution. Cities should be seeking wireless coverage for outdoor public spaces and other community and public access locations such as public squares, parks, community centers and other community meeting places.

If price is a barrier for low-income areas and disadvantaged populations that do have other potential options such as DSL or cable, other solutions such as targeted subsidies or vouchers could be considered at local or state levels.

The expected long-term impact and outcomes of municipal wireless remains unclear. Will it simply be seen as a stepping-stone to broadband mobile (3G and 4G) networks, that are already offering Web access and video downloads? If so, the goal of achieving broadband wireless coverage of U.S. cities will have largely been achieved, but the goal of providing free access will remain out of reach.

However, is free access a necessary long-term goal? Perhaps municipal wireless networks will be viewed as the "freenets" of this decade. They will stimulate demand, but eventually they will die or be absorbed by commercial ISPs. Freenets were ISPs provided by nonprofit organizations and community groups that provided free dial-up services such as email and information services in the early days of the Internet. Freenets introduced their users to online services. Eventually, freenets disappeared, as their users migrated to commercial ISPs for higher bandwidth, better quality of service, and more security. Such may be the fate of most municipal wireless networks.

REFERENCES

Abdelaal, A., & Ali, H. (2007). Typology for community wireless network business models. In *Proceedings of the Thirteenth Americas Conference on Information Systems*. Keystone, CO: IEEE.

Albanesius, C. (2009). *Philadelphia repurchases city wi-fi network for $2M*. Retrieved December 21, 2009, from http://www.pcmag.com/article2/0,2817,2357395,00.asp#fbid=Y0Rf0aYjKVe

American Public Power Association. (2004). State barriers to community broadband services. *APPA Fact Sheet*. Retrieved from http://www.baller.com/pdfs/Barriers_End_2004.pdf

Baller, J. (2006a, September-October). Quoted in state broadband battles. *Public Power Magazine*. Retrieved from http://www.appanet.org

Baller, J. (2006b). *Proposed barriers to state entry*. Retrieved from http://www.baller.com/pdfs/Baller_Proposed_State_Barriers.pdf

Clara, S. (2011). *Website*. Retrieved from http://www.ci.santa-clara.ca.us

Columbia Telecommunications Corporation. (2009). *Enhanced communications in San Francisco: Phase II fiber optics feasibility report*. Kensington, MD: Columbia Telecommunications Corporation.

Cupertino. (2008). *Wesbsite*. Retrieved from http://www.cupertino.org

EarthLink/Google. (2006). *Presentation*. Paper presented at San Francisco TechConnect. San Francisco, CA.

Flamm, K., & Anindya, C. (2005). *An analysis of the determinants of broadband access*. Paper presented at the Telecommunications Policy Research Conference. Washington, DC.

Ford, G. S. (2005, February). Does municipal supply of communications crowd-out private communications investment? *Applied Economic Studies*.

Francisco, S. (2005a). Request for information and comment (RFI/C). *San Francisco techconnect community wireless broadband initiative*. Retrieved from http://www.sfgov.org/site/uploadedfiles/dtis/tech_connect/BroadbandFinalRFIC.doc

Francisco, S. (2005b). *RFP 2005-19: Request for proposals: TechConnect community wireless broadband network.* Retrieved from http://www.sfgov.org/site/uploadedfiles/dtis/tech_connect/TechConnectRFP_2005-19_12-22-05Rev1-17-06.pdf

Francisco, S. (2011). *San Francisco TechConnect.* Retrieved from http://www.sfgov3.org/index.aspx?page=1432

Goel, V. (2007, November 4). Vindu's view from the valley. *San Jose Mercury News.*

Google. (2012). *Free wi-fi access for mountain view.* Retrieved from http://Wi-Fi.google.com/

Hiner, J. (2007). *US cities are jumping off the municipal wireless bandwagon.* Retrieved from http://blogs.techrepublic.com.com/hiner/?p=545

Holson, L., & Helft, M. (2008, August 14). T-Mobile to offer first phone with Google software. *New York Times.*

Horrigan, J. (2009). *Broadband adoption and use in America.* Working Paper. Washington, DC: Federal Communications Commission.

Hudson, H. E. (2005). *Comments on request for information and comment (RFI/C): San Francisco TechConnect community wireless broadband initiative.* San Francisco, CA: Government of San Francisco.

Hudson, H. E. (2007, January 30). Why San Francisco should approve the EarthLink/Google wireless contract now. *San Francisco Examiner.*

Hudson, H. E. (2010). Municipal wireless broadband: Lessons from San Francisco and Silicon Valley. *Telematics and Informatics, 27*(1), 1–9. doi:10.1016/j.tele.2009.01.002

Joint Venture. (2012). *Census data listed in the SAMCAT RFP.* Retrieved from http://www.joint-venture.org/programs-initiatives/smartvalley/projects/wirelesssv/documents

Lakshmipathy, N., Meinrath, S., & Breitbart, J. (2007, December 11). The Philadelphia story: Learning from a municipal wireless pioneer. *New America Foundation.*

Malik, O. (2006). *Google launches wi-fi network in Mountain View.* Retrieved from http://gigaom.com/2006/08/15/google-launches-wi-fi-network-in-mountain-view/

Mandviwalla, M., Jain, A., Fesenmaier, J., Smith, J., Weinberg, P., & Meyers, G. (2008). Municipal broadband wireless networks: Realizing the vision of anytime, anywhere connectivity. *Communications of the ACM, 51*(2), 72–80. doi:10.1145/1314215.1314228

MetroFi. (2008). *Notice to customers.* Retrieved from http://www.metrofi.com/press/20060130b.html

Minneapolis, W. (2011). *Website.* Retrieved from http://www.ci.minneapolis.mn.us/wirelessminneapolis/

Moura, B., & Fearey, S. (2006). *Joint venture: Silicon Valley network and SAMCAT announce vendor for wireless Silicon Valley.* Retrieved from http://www.jointventure.org/inthenews/pressreleases/090506wirelessvendor.html

National Telecommunications and Information Administration. (2010). *Digital nation: 21st century America's progress toward universal broadband internet access.* Washington, DC: NTIA.

National Telecommunications and Information Administration. (2011). *Exploring the digital nation: Computer and internet use at home.* Washington, DC: US Department of Commerce.

NextWLAN. (2006). *Proposal in response to San Francisco tech connect RFP.* San Francisco, CA: City of San Francisco.

Organization for Economic Cooperation and Development. (2011). *OECD broadband statistics.* Retrieved from http://www.oecd.org/sti/ict/broadband

Putala, C. (2006, June 14). *Testimony before the committee of the judiciary, US Senate, hearing on reconsidering our communications laws: Ensuring competition and innovation.* Washington, DC: US Senated.

Pyramid Research. (2005). Municipality wi-fi: Despite EarthLink, Google, viability remains unclear. *Pyramid Research Analyst Insight.* Retrieved from http://www.pyramidresearch.com/documents/AI-Wi-Fi.pdf

Reardon, M. (2007). *Earthlink to lay off 900.* Retrieved from http://news.cnet.com/8301-10784_3-9767410-7.html

RedTap. (2006). *Executive summary of proposal in response to San Francisco TechConnect RFP.* San Francisco, CA: RedTap.

Reinwand, C. (2006, June 26). *Letter to Brian Moura, Chairman, SAMCAT.* San Mateo, CA: San Mateo County Telecommunications Authority.

Seakay, Cisco, & IBM. (2006). *Proposal in response to San Francisco TechConnect RFP.* San Francisco, CA: Government of San Francisco.

Shapiro, M. (2006). Municipal broadband: The economics, politics and implications. *Pike and Fisher.* Retrieved from http://www.pf.com/marketResearchPDInd.asp?repId=397

Skidmore, S. (2008, July 19). MetroFi ending wi-fi service in Calif., Ore., Ill. *San Jose Mercury News.* Retrieved from http://www.mercurynews.com/ci_9637287?IADID

Strover, S., Chapman, G., & Waters, J. (2003, September). Beyond community networking and CTCs: Access, development and public policy. *TPRC*, p. 24.

Sunnyvale. (2008). *Website.* Retrieved from http://sunnyvale.ca.gov/Departments/Library/Community+Profile.htm

Turner, S. D. (2005). *Broadband reality check.* Washington, DC: Free Press.

US Census Bureau Quick Facts. (2010). *Website.* Retrieved from http://quickfacts.census.gov/qfd/states/06/06075.html

View, M. (2008). *Google provides free wi-fi.* Retrieved from http://www.mountainview.gov/services/learn_about_our_city/free_Wi-Fi.asp

Vision 2 Mobile. (2008). *Portland seizes MetroFi muni wi-fi gear.* Retrieved from http://www.vision2mobile.com/news/2008/10/portland-seizes-metrofi-muni-wi-fi-gear.aspx

Vos, E. (2005). *Reports on municipal wireless and broadband projects.* Retrieved from http://www.muniwireless.com

Vos, E. (2008). *Oklahoma City rolls out world's largest muni wi-fi mesh network.* Retrieved from http://www.muniwireless.com/2008/06/03/oklahoma-city-deploys-largest-muni-Wi-Fimesh-network/

Wilson, M., & Kraatz, J. (2008). *Wireless internet company serving Cupertino, Sunnyvale shutting down.* Retrieved from http://www.siliconvalley.com

Wireless Silicon Valley. (2006). *Requests for proposals for a regional broadband wireless network for Silicon Valley.* Retrieved from http://www.jointventure.org/programs-initiatives/smartvalley/projects/wirelesssv/documents

Wireless Silicon Valley. (2008). *Wireless Silicon Valley adds Covad Communications to the team and plans test in San Carlos.* Retrieved from http://www.jointventure.org/programs-initiatives/wirelesssiliconvalley/updates.html

Chapter 3
Legal and Political Barriers to Municipal Networks in the United States

Eric Null
Juris Doctor, USA

ABSTRACT

The United States has been one of the most active countries in the deployment of municipal broadband networks. In America, many remote areas have no Internet access or are served by a single provider that might not meet local needs. Increasingly, access to the Internet is vital for social and economic development as well as prosperity. Without access, rural areas lose economic competitiveness and have lower quality of life standard of living. An attractive solution for such localities is to provide Internet access themselves, provided they believe that their area will realize significant benefits from it. However, the States' complex legal frameworks are a significant barrier to local network success. Each state makes its own laws governing its municipalities, and states have almost unfettered ability to constrain, either by ban or by lesser restraint, the emergence of local networks. These states are further influenced by the lobbying of incumbent service providers. Moreover, judicial remedies can be used strategically by incumbents to hinder, delay, or prevent local networks from succeeding. Despite all this, local networks can be and have been successful. This chapter discusses various legal and political barriers to municipal networks and explores case studies with the goal of learning from past successes and failures.

DOI: 10.4018/978-1-4666-2997-4.ch003

INTRODUCTION

The Internet, since inception, has become a bastion of commerce, democracy, and community. Access to the Internet has become vital for improving standard of living as well as economic and social prosperity. Yet, there are some areas of the U.S. that are not served by private service providers. Others are underserved as they have slow Internet speed or broadband subscription prices are prohibitive. Crawford (2011) termed this problem the "communications crisis in America." Assessments of this nature are further corroborated by recent report by the Organization for Economic Cooperation and Development (OECD) that shows that the U.S. ranked eighteenth in advertised Internet speeds (OECD, 2011) and fourteenth in total fixed broadband penetration rate (OECD, 2010).[1] A viable and effective solution is to allow those underserved communities or their local government to build networks for themselves through a community or municipal project (Abdelaal & Ali Khazanchi, 2009). Municipal networks have the potential to disrupt the telecommunication industry. Therefore, municipal networks are under vigorous attack by incumbent service providers that stand to protect their market share, revenue, and monopoly power. Some municipalities could overcome these attacks. However, municipalities should be aware of the legal and political barriers they may face and this is the focus of this chapter.

Unfortunately, the U.S. legal framework regarding municipal networks is inconsistent and constantly changing. The lack of substantive federal policy relevant to municipal networks, a tapestry of state laws forms the framework within which municipalities must maneuver. The majority of states have not enacted relevant legislations, but about twenty states have (Baller, 2012a). Some of these legislations address a specific technology such as wireless (Wi-Fi or WiMAX), fiber-optic lines, Digital Subscriber Line (DSL), cable services, broadband over power lines, telephone, or any medium that transfers voice, video, or data. Others are based on the entity such as the municipality itself, an electric or power utility, or political subdivisions. For simplicity, this chapter will use the term "municipal networks" to refer to all types of networks, unless otherwise specified.

As part of the planning stage of creating a municipal network, a municipality should consider the likelihood of being sued, the likelihood of a lobbying battle in the state legislature, as well as how costly it will be (in both time and money) to overcome any regulatory barriers. Combined, these barriers may seem substantial, but they can be planned for in advance and adequately handled. The costs they incur may be slight in comparison to the vast social and economic benefits that the network will provide. The positive externalities or "spillovers" brought about by increased Internet access, while they cannot be accounted for monetarily, are abundant and should not be overlooked or underestimated (Frischmann, 2012, pp. 317-318).

This chapter discusses the legal and political challenges facing municipal networks. It will introduce the topic by discussing the current federal regulatory framework. It will then provide a detailed account of legislative developments at both the federal and state level. Then, it will detail the strategies used to halt, delay, and hinder municipal networks. The chapter concludes with lessons learned from case studies discussed therein and a summary checklist for municipalities.

BACKGROUND

In 1934, the Communications Act of 1934 was passed ushering in an era where communications technology would be available to more people. Unfortunately, the law brought about monopolistic behavior from telecommunications (mainly telephone) providers by the early 1980s (Travis, 2006, pp. 1707-1710). Overall, the industry lacked effective government oversight. By 1984, this brought about the divestiture of the primary long-distance provider, AT&T, on antitrust grounds (Benjamin, 2006, pp. 723-724). In an effort to remedy the regulatory situation that culminated in AT&T's divestiture, the Telecommunications Act of 1996 was passed (mostly encompassing amendments to the 1934 Act [Benjamin, 2006, p. 1171]). It was designed to remove barriers to entry for new entrants into the telecommunications ("telecom") industry and to encourage competition so that, according to Senator Trent Lott (R-MS), "everybody [could] compete everywhere in everything" (Petitioner's Brief, *City of Abilene v. FCC*, p. 1).

In exchange for increased local competition, the telecom providers demanded that they be deregulated to allow them to enter other, lucrative, markets (Carlson, 1999, p. 46). Theoretically, this was sound. In a typical market, the theoretical increase to local telecom competition would keep prices and behavior of the incumbents in check. Competition in the other, lucrative, markets would keep behavior in those markets in check as well. In reality, local telecom competition largely failed to materialize. As discussed below, legal roadblocks to municipal networks contributed to the suppression of local competition.

The 1996 Act included a peculiar section over which dispute arose quickly. Congress wrote "[n] o state or local statute or regulation . . . may prohibit or have the effect of prohibiting the ability of any entity to provide any interstate or intrastate telecommunications service" (§ 253(a)). If "any entity" included municipalities, states would not be allowed to pass state laws designed to prevent municipal telecom competition. However, this interpretation ultimately did not succeed (Christensen, 2006, pp. 689-695). The story of this controversy, which took place in the courts, follows.

Before the passage of the 1996 Act, Texas enacted legislation precluding municipalities and municipal electric utilities from providing their own telecom services. The City of Abilene petitioned the Federal Communications Commission (FCC) to preempt the Texas law, alleging that Section 253(a) protected "any entity," including

any public entity, from state barriers to entry. The FCC disagreed and allowed the Texas law to stand as applied to municipalities, such as Abilene, that did not operate their own electric utilities. The FCC did not rule on whether the Texas law was valid as applied to municipal electric utilities. On appeal, in *City of Abilene v. FCC*, the District of Columbia Circuit agreed. The court said states can regulate their municipalities however they see fit, and if the federal government wants to interfere, it must include a very clear statement to that effect. The court felt the words "any entity" were not sufficient to infer congressional intent to interfere (Christensen, 2006, pp. 689-91).

The second case, *In re Missouri Municipal League*, involved a challenge to a similar Missouri law, but this case focused on whether Section 253(a) protected municipal electric utilities from state barriers to entry. The FCC, as in *Abilene*, allowed the law to stand, thus precluding both municipalities and their municipal electric utilities from providing telecom services. The FCC thought it was nearly impossible for a municipally owned entity to be sufficiently separate from the municipality to constitute a completely separate entity. Importantly, however, the FCC did state that municipalities "have the potential to become major competitors in the telecommunications industry" and that they "further the goal of the 1996 Act to bring benefits of competition to all Americans, particularly those who live in small or rural communities" (Christensen, 2006, pp. 692-693). On appeal, the Eighth Circuit overturned the FCC's holding by relying on the Supreme Court's previous decision in *Salinas v. U.S.*, wherein the Court stated that if Congress uses the word "any," and it is not restricted in any way, courts must interpret that term as broadly as possible. The court ultimately concluded that Congress did intend to include municipalities in the prohibition in Section 253(a) (Christensen, 2006, p. 693).[2] The Supreme Court took the appeal to resolve the circuit-split.

In *Nixon v. Missouri Municipal League*, the Supreme Court agreed with the District of Columbia Circuit in *Abilene*. The Court held that, to preempt a traditional state power, such as telling its municipalities what services they can and cannot provide, Congress must express its intent to preempt clearly and unambiguously on the face of the statute. Because the term "any entity" did not go far enough to meet this standard, the Court ruled against the municipalities, even while finding that they had a respectable position on policy grounds. To reach that decision, the Court relied on its federalism standard from *Gregory v. Ashcroft*: if Congress intends to abrogate a state's sovereign powers, it must make its intention unmistakably clear in the language of the statute (Travis, 2006, pp. 1730-1731). Without unmistakably clear language, there was no federal preemption. Thus, states were free to pass laws restricting municipal telecom networks. With this decision, the Court held steadfast to the reign of "new federalism" whereby states' rights are protected against inappropriate federal aggrandizement (Dunne, 2007, p. 1149).

As a municipality, it is important to understand the distinction between *information services* and *telecommunication services*. Some state laws make use of this technical difference, and federal laws apply differently to each. A *telecommunications service* means almost exclusively telephone services (or "local exchange services"). Those services are required to follow the "common carriage" rules under Title II of the 1996 Act, requiring just, reasonable, and nondiscriminatory customer rates, and requiring the network to interconnect with other networks ("open access") (*National Cable v. Brand X*). Only *telecommunications services* were at issue in *Nixon,* though the Court's reasoning in *Nixon* could apply to other services too. *Information services*, on the other hand, encompass essentially everything else: cable modem services, fiber-optic connections, wireless services, and even DSL, to the extent it "facilitates changes in the content of data," are considered information services (Crawford, 2008; Travis, 2006, p. 1755). These are all considered

"broadband" networks. They are subject to Title I of the 1996 Act, which allows for much less FCC regulation and federal oversight. Information services are not subject to Title II's more restrictive open access requirements (Crawford, 2008). Therefore, if a state bans or restrains the municipal offering of "telecommunications services" as defined by federal law, then this does not, on its own, preclude municipal provision of information or broadband services. However, some states do mention broadband technologies by name.

Nixon set the stage for subsequent state and federal developments regarding municipal networks. After *Nixon*, state governments could impose barriers unless and until the federal government "unmistakably clear[ly]" preempted them—an event that has yet to occur. Without uniform federal policy, municipalities continue to be subject to state-by-state roadblocks in courts and legislatures.

UNENACTED FEDERAL BILLS

In 2005, three bills were introduced, but not enacted, in Congress to address municipal broadband. The first was the "Preserving Innovation in Telecom Act of 2005." This curiously titled bill was designed to "prohibit municipal governments from offering telecommunications, information, or cable services except to remedy market failures by private enterprise to provide such services" (O'Loughlin, 2006, p. 494). In other words, if passed, it would have prohibited municipalities from providing their own telecommunications, information, or cable service when a private entity provides a "substantially similar service," an undefined term. This was an exceptionally broad prohibition because it prohibited essentially all technologies (telecom and information services)—many states do not go that far. As some have pointed out "it is difficult to see a basis for this type of measure" other than industry protectionism (Botein, 2006, p. 985).

The second bill was the "Broadband Investment and Consumer Choice Act" ("BICCA"). While taking middle ground, this bill would also have enacted formidable barriers to municipal networks. The bill required municipalities intending to provide a "communications" service (which includes essentially every technology except broadcast TV) to provide "conspicuous notice" to the community, including costs, services to be provided, areas to be covered, terms, and architecture; municipalities would also be required to perform a detailed accounting of any special considerations of its proposed service, including beneficial tax treatment or free- or below-cost use of rights-of-way (O'Loughlin, 2006, p. 495). Moreover, the bill required municipalities to hold "open bids" to allow non-governmental entities the chance to provide the service first. It is not difficult to see that this bill was the result of the telecom industry "carefully organiz[ing] a lobbying and letter writing campaign, encouraging local industry groups . . . to lobby their Senators extolling the virtues of the bill" (O'Loughlin, 2006, p. 497). Thus, in 2005, there were two pro-industry bills in Congress, but this bill had the most support (O'Loughlin, 2006, p. 494).[3]

The third bill, the "Community Broadband Act of 2005" ("CBA 2005"), aimed to foster greater competition in local Internet access by prohibiting states from precluding publicly provided "advanced" telecommunications services (advanced telecom services include data, video, and voice). This bill was the answer to *Nixon* and *Gregory*. It also prevented municipalities from favoring themselves in franchising deals, and by requiring municipalities to comply with other federal and state laws, such as those prohibiting cross-subsidization of services and those preventing predatory pricing. Intel, a supporter of municipal networks, said this bill struck a more appropriate balance between preempting state prohibitions of municipal networks and requiring municipalities to compete in a neutral manner (Travis, 2006, p. 1764).

Widely perceived as fair and balanced, CBA 2005 received broad support from both parties in both houses of Congress. It eventually passed the full House and the Senate Commerce Committee, but the 109th Congress ended without passing any telecommunications legislation. CBA 2005 was reintroduced, with modest changes, as the "Community Broadband Act of 2007," but it too failed to pass.

There have been many other bills that impact municipal networks, though they did not deal with state barriers to public entry.[4] The federal government has been much more concerned, recently, with funding next-generation networks. The American Recovery and Reinvestment Act of 2009 specifically allowed public entities to apply for grants, but it also required all stimulus recipients, including those public entities, to comply with all relevant federal, state, and local laws, thereby sidestepping the state barrier issue. For now, deciding unitary federal policy for municipal broadband is simply too political.

THE IMPACT OF THE FEDERAL COMMUNICATIONS COMMISSION

The FCC is the primary independent agency in charge of communications regulation. The FCC, originally named the Federal Radio Commission in the Radio Act of 1927, was renamed to the FCC in the 1934 Act (Benjamin, 2006, pp. 52-53). Its power to regulate the industry is vast—it covers everything from spectrum policy to licensing frequencies to indecent language on broadcast networks (Benjamin, 2006, p. 59; *FCC v. Pacifica Foundation*). The FCC's involvement in municipal networks was detailed above, and it continues to be involved in a limited way. The National Broadband Plan (NBP), promulgated by the FCC in 2010, encourages municipal networks. The FCC adopted as recommendation 8.19 that "Congress should make clear that state, regional and local governments can build broadband networks" (Connecting America, 2010, p. 153). In essence, the FCC wanted Congress to make the statement required by the *Nixon* case, and the recommendation is consistent with the FCC's pro-municipal network statement in *In Re Missouri Municipal League*. However, this is just a recommendation. Congress is not bound by it, and Congress is unlikely in the near future to provide such straightforward guidance on this issue. Their inability to reach consensus during the 2005-2007 period is indicative.

The FCC has a limited role as a funder, but the primary funding program is heavily biased toward private companies.[5] The FCC recently enacted the Connect America Fund (CAF) designed to be an "engine for rural broadband deployment" (Gillett, 2012). In the plan, the FCC gives funding recipients $775 per home that the entity agrees to build to (in a rural, "unserved" area defined by census block). Its first phase, which ended in July 2012, invested $115 million in rural broadband deployment in thirty-seven states and hopefully will vastly increase the number of homes with access to broadband connections (FCC Connect, 2012). However, most of this money went to private companies, including CenturyLink, Frontier, and FairPoint (Buckley, 2012). Those private companies were also required to provide only 4 mbps download speeds and 1 mbps upload speeds, nowhere near the gigabit speeds of municipal networks such as Chattanooga, TN, or Lafayette, LA. While municipalities are not precluded from receiving this funding, Phase I benefitted few, if any, municipalities directly.

FCC Regulation of White Spaces for Wireless Networks

Wireless (or "Wi-Fi") is a very attractive and popular type of network because it is easier to deploy than wired alternatives, provides increased mobility for users, and is, in general, cheaper to deploy and cheaper for users to use (Wong, 2007). Some municipalities have attempted to provide

this service for free, including the now-defunct St. Cloud, Florida network. However, free wireless networks can often be successful with adequate planning, such as in Cleveland, Ohio (Vos, 2011b). Others have charged for wireless networks, albeit very cheaply. The decision to charge a subscription fee could depend on the regulations in the state, which, like Pennsylvania, may not allow a fee-based service (Baller, 2011, p. 3). Costs imposed by procedural requirements, as detailed below, may also require revenue to recoup. On the other hand, wireless networks tend to have security problems and may be lower quality than wired service (Ganapati & Schoepp, 2009, p. 556). However, if the FCC opens white spaces for unlicensed use, deployment of wireless networks will likely become easier.

Wireless networks use spectrum frequencies that are available for unlicensed use. Many believe that spectrum is a natural but scarce resource (Benjamin, 2006, p. 31). The FCC licenses particular frequencies on the spectrum to users that apply for those licenses. The FCC has left certain bands "unlicensed" that can be used by anyone, including municipalities or private users. In-home wireless routers operate on unlicensed spectrum (Benjamin, 2006, p. 83). The recent boom in Wi-Fi services, networks, and equipment clearly shows that unlicensed use of the spectrum can be tremendously beneficial to society. Even wireless carriers such as Verizon and AT&T rely to a significant extent on Wi-Fi signals to carry their excess wireless traffic (Cooper, 2012, pp. 11-12, 19). Many wireless municipal networks exist based on wireless technology, including Corpus Christi, Texas, and Newton, North Carolina (ConnectCC, 2012; Vos, 2011a).

A current topic of debate is whether the FCC should open up "white spaces" to unlicensed use. White spaces are the frequencies between those dedicated to TV channels (which also use spectrum). The 2009 digital TV transition allowed more efficient use of spectrum for TV signals, increasing the white space between channels

(Benjamin, 2011, p. 17). One of the greatest benefits of white spaces over regular Wi-Fi signals is that they are lower in frequency and thus travel longer distances than typical Wi-Fi signals (which are typically 2.4GHz) (Rich, 2012). The primary counterargument to opening the white spaces to wireless network use is interference: those wireless networks could interfere with TV signals, defeating the purpose of both. However, it is generally believed that use of those extra white spaces is unlikely to interfere with TV channels (Benjamin, 2011, p. 17; Meinrath & Calabrese, 2008). Recently, the FCC has released some white spaces for unlicensed use (Benjamin, 2011, pp. 17-25; FCC Press Release, 2010). Additionally, the FCC plans to further liberate white spaces in order to encourage innovative and experimental use of the spectrum (FCC White Space, 2012). Municipalities and their network equipment vendors should pay close attention to future developments in this area because wireless over white spaces can allow the municipality to provide high quality, cheap wireless Internet access to its citizens.

New Hanover County, North Carolina, has deployed the first and only (thus far) successful Wi-Fi network using white spaces (Vos, 2010). The network was deployed in 2010, not long after the city transitioned to digital TV. The wireless signal is used to supplement the city's existing fiber-optic infrastructure by providing wireless access where the wires do not reach. The network is used to monitor water quality, lights in sports fields, medical telemetry, and has saved the city of Wilmington nearly $800,000 per year in energy costs alone (Vos, 2010). More recently, the county deployed a "Super Wi-Fi" system that is to be faster, and reach farther than the typical wireless network (Mitchell, 2012a; Rich, 2012). The network was deployed, in part, to deal with increased use of smartphones, tablets, and various other wireless devices, which have increased demand for wireless signal. The new network does not require a line-of-site to towers, thus increasing signal strength, reach, and utility to users. It also

reaches speeds of two megabits per second, a very fast connection for an area with "dense foliage" (Tynan, 2012).

BURDENS OF STATE LEGISLATION

State governments are the primary players in regulating municipal networks. Generally, states give their municipalities the authority to act by enacting general home rule measures or more specific authorizations or limitations. Thus, state legislatures generally have unfettered power to preclude, allow, or erect barriers to municipal networks, and can erect broad restrictions or restrictions that apply only to certain municipalities (Vermont Office of Secretary of State, 2009, p. 3). This ultimate state authority, unchecked by federal law or other states' laws, has allowed states to pass whatever restrictions they prefer, and many have done so (Baller, 2011; Community Broadband Networks, 2012a).

Many states have a morass of laws that are difficult to navigate. Detailing every state law would be onerous. This section will proceed by discussing the more popular mechanisms by which states increase barriers for municipal networks. It is strongly suggested that any interested municipality research their state's legal requirements independently; this section is simply a brief overview. A state-by-state table is included as Appendix.

Full or Partial Bans

Currently, no state has a law completely prohibiting every technology by every public provider. Many regulations prohibit only certain types of networks (cable, telecom services), or prohibit certain providers from providing the networks (the municipality itself, power utilities). I call this the "partial ban." This is different from the de facto ban (discussed later), which means a state enacted onerous administrative and/or financial restrictions, rendering municipal networks nearly impossible.

Texas bans municipalities and municipal electric utilities from providing certain telecommunications services to the public, but it does not ban municipal provision of cable service or broadband Internet access service. Nebraska bans municipal provision of "wholesale or retail broadband, Internet, telecommunications or cable service," and public power utilities can only provide these services on a wholesale basis, subject to various onerous terms and conditions (Baller, 2011, pp. 3-4). In Nevada, municipalities with populations greater than 25,000 and counties with populations greater than 55,000 are prohibited from providing telecommunications services as defined by federal law (Baller, 2011, p. 3). However, as discussed above, state laws affecting only "telecommunications services" will not necessarily preclude networks based on information services.

The University of Nebraska at Omaha was able to avoid the harsh Nebraska prohibition against municipal networks by simply providing the network itself, as the University (Omaha Wireless, 2012). Because the University is not the municipality, it is not precluded by law from providing free "broadband, Internet, telecommunications or cable service." With a relatively simple work-around, the city still gains the economic and social benefits of increased Internet access while the network is in compliance with state law.

In Utah, a municipality is prevented from providing direct or indirect *retail* cable or telecommunications services, meaning the municipality cannot provide service to the end-user. Therefore, any network can only provide wholesale service (lease access to private retail service providers). That is exactly what the Utah Telecommunications Open Infrastructure Agency (UTOPIA) has been doing. UTOPIA encompasses a group of cities in Utah that pooled resources to build fiber-optic broadband services in their area. Because of the state laws, it provides only wholesale service

(Coleman, Behunin, & Harvey, 2012, p. 3). While it cannot provide retail services, it still built a fiber-optic network where one did not exist before, and it controls the network and can impose its own rules on retailers if it wishes. Thus, by working within state law, the towns that make up UTOPIA still benefit economically and socially.

Administrative Requirements

States that impose administrative burdens most often impose (1) a feasibility study, (2) public hearings, and/or (3) a referendum. A municipality could lose its battle at any step, and many have, thereby negating any previous time and money investment. Clearly, these requirements make it difficult and less likely that municipalities will want to build a network. This many obligations (sometimes requiring large expenditures of money) can be difficult to justify for nearly-immeasurable future economic and social development. For example, Utah requires all three of the above-listed requirements (Utah Municipal Code, § 10-18-202, -204). In addition, under Section 10-18-203, the feasibility study must determine the following: whether the municipality's cable or telecom provision "will hinder or advance competition," other entities would have offered the same service; the fiscal impact of the network including capital investment and expenditures on labor, financing, administering the network; and many more subsections worth of requirements. Municipalities are at a disadvantage because private entities are not subject to these requirements by law, though may be internally.[6] However, even if not required, municipalities should undertake a feasibility study anyway, because it is important to know what to expect; the feasibility study can provide at least some of that information. Virginia and Wisconsin also require feasibility studies by law (Baller, 2011, p. 4).

Requiring hearings is another way for states to disincent municipal networks. If hearings are required, it is likely that the state will also have to make the feasibility report publicly available a few days before the hearing. Hearings are often the first time citizens collectively voice their opinions, so they could provide a way for government to meaningfully connect to its citizens on this issue. Many states require two public hearings, including North Carolina and Florida; others require only one, including Tennessee and Wisconsin (Baller, 2011, pp. 2-4).

Referenda require the municipality's constituents to vote in favor of the network before construction begins. This ensures that those who would pay for and primarily benefit from the network are in favor of it. States that impose this requirement include Alabama (for cable services), Colorado (unless the community is unserved and incumbents have refused to meet demand), Louisiana, Minnesota (requiring a super-majority of 65% voter approval), North Carolina, South Carolina, and Virginia (for cable services) (Baller, 2011, pp. 1-4). During the construction of a municipal network in Lafayette, Louisiana, a Louisiana bill imposing restrictions was passed. It did not require a referendum. However, incumbents were able to force a referendum by "challenging the City's authority in other areas." The referendum passed, but the new bill was later amended to require a referendum for future networks (Mitchell, 2012c, p. 20). Thus, even if a state's municipal broadband law does not impose a particular requirement, there may be other ways for incumbents to impose one.

In theory, referendum requirements are not unreasonable: constituents should agree to the network before it is built with (potentially) their tax money. However, the reality is that a referendum (and perhaps hearings too) is merely another way for incumbents to influence the results. First, these referenda are time-consuming and burdensome on municipalities. That, on its own, is not so bad. However, in addition, industry often vastly outspends municipalities during these campaigns because municipalities, in most cases, are not allowed to spend money lobbying the public. Even

if they could, they would still be greatly outspent by multi-million or billion dollar incumbents (Baller, 2011, p. 1). For example, in a referendum on whether Batavia, Geneva, and St. Charles, Illinois, should develop fiber-optic networks, the cities' supporters had a budget of less than $5000, whereas Comcast and AT&T admitted to spending more than $300,000 in opposition to the referendum (Ordower, 2005). Similarly, in Chattanooga, Tennessee, Comcast and the Tennessee Cable Telecommunications Association (comprising multiple telecom companies) aired 2,600 TV ads urging citizens to oppose the network (Mitchell, 2012c, p. 38).

Lastly, some states require municipalities to accept bids by private companies to provide the network before the city itself can provide the service. In North Carolina, before a municipality can undertake building a network, it must "solicit proposals from private businesses to partner with the government to provide the communications services . . ." (Millonzi, 2011). Michigan, furthermore, allows the municipality to set the terms it wants for the network, must solicit private bids, and can only then build the network itself if it receives fewer than three bids (Baller, 2011, p. 2). This law actually favors municipalities because they set the terms of the network (build-out requirements, speed, and technology). If the municipality receives three viable private bids, then there is no need for the municipality to build the network. If it does not, then the municipality builds the network itself as planned. It is a win-win situation for any municipality.

Two success stories illustrate that, while these requirements are burdensome, compliance is possible even despite strong opposition. First, Missouri state law does not allow municipalities to provide cable television services without a referendum. The City Council of North Kansas City, Missouri, had voted to adopt a plan to provide a fiber-optic (broadband) network that could provide data, voice, and video. Though the City's engineering firm recommended that it provide cable

TV services in order to create a self-supporting network, the City made no determination regarding cable TV services. Despite the City's refusal to answer that question, Time Warner Cable filed a lawsuit against it, claiming that the City was required to hold a referendum before providing its planned service. The Eighth Circuit dismissed the lawsuit on ripeness grounds, meaning there was no actual controversy yet. The court said there was no indication that the City planned to provide cable TV services. Just because a self-supporting network required providing cable TV services, said the court, does not mean that the City is going to provide cable TV—in other words, the City does not need to provide a self-supporting network in the first place. This plan meant that in the future, if the City chooses to upgrade the network through the simple and cheap addition of a connection to a cable facility, the citizens will have little reason to oppose it given the benefits of the municipally-provided cable service and the relatively slight cost (Tech Law Journal, 2005).

Second, in Longmont, Colorado, it took the city many years and multiple referenda to garner enough support for the network. The city built a fiber-optic network in the mid-1990s, but in 2005 the legislature (with the help of telecom lobbyists) passed a law prohibiting municipalities from providing telecom services directly, at least without first complying with a variety of procedural requirements (Vos, 2005). In 2009, Longmont held a referendum, and 56% voted against the network. After two years, during which time the citizens became upset with the incumbent and another information campaign was waged, a second referendum was held. This time, 60% voted for the network (Badger, 2011). Because many telecommunications companies are in disrepute, municipalities should embrace that and use it to their advantage to garner support.[7] Municipalities can be the "local" alternative to distant corporations with scattered shareholders. This strategy worked well for Longmont.

Restrictions on Funding

Financing is extremely important; without funding, municipalities cannot construct the network. State legislatures and lobbyists are aware of this, and some legislatures impose financial burdens for municipal networks. The following is a discussion of a few of those burdens.

The most obvious place from which to fund a network is tax revenue. A municipality has the power to tax its citizens, and it could potentially use those funds to build a network. Therefore, precluding funding of networks from general tax revenues is a very effective way to cut-off any network. States such as Alabama impose this requirement until the network itself is (essentially) self-sustaining (Baller, 2011, p. 1). Given the capital-intensive nature of building a broadband network, revenues are likely not to exceed costs for many years—this kind of law would require the municipality to charge a subscription fee to bring in revenue. Without tax revenues at their disposal, municipalities will likely have to issue municipal bonds, which can be regulated as well. Municipal bonds are simply bonds issued by a municipality to raise funds for certain projects. Some municipalities have successfully funded networks by issuing bonds, including Dunnellon, Florida (Mitchell, 2010).

Municipalities should be sure that the type of funding they use is not a form of cross-subsidization. This means that the municipality cannot use funds from another project to subsidize (to reduce its costs and thus its subscription rate) network services. Cross-subsidization has been the target of legislation in many states, including Tennessee (Mitchell, 2012c, p. 38), Virginia (Baller, 2011, p. 4), Florida, Iowa, and Wisconsin (Travis, 2006, p. 1768). The logic is that this would provide too much of a competitive advantage for municipalities and would drive private providers out of business because of an inability to compete with a below-cost municipal broadband network. The ban on cross-subsidization is extremely important for municipalities to be aware of. First, this claim has been the basis of lawsuits against cities, including Lafayette, Louisiana, which was sued *twice*, unsuccessfully, because its bond structure looked like, but did not constitute, cross-subsidization (Mitchell, 2012c, p. 22). Second, the ability to cross-subsidize the service is potentially a significant benefit for the area because it could reduce subscription rates drastically, but this would raise prices for another municipal project.

Another way for incumbents and states to prevent so called "predatory" pricing (subscription rates below cost) is to require municipalities to impute the costs of a private provider onto themselves. What exactly must be imputed is not certain, though states usually outline what is required (South Carolina Code Ann. 58-9-2620(4)). This burden negates any potential competitive advantage brought about by lower costs that come about *because* the network is provided by the municipality rather than a private provider, such as marketing costs or discounted capital costs. It also acts as a disincentive because any monetary benefit will be slight, as this mechanism forces the municipality to charge artificially higher prices. These requirements can be seen as a "form [of] legislatively-sanctioned price fixing" (Baller, 2011, p. 2). These costs can also be subjective, especially if the municipality is unserved by a telecom provider, because there will be no private entity to which to compare the municipality's costs. Thus, *any* amount of cost imputing could be the subject of a lawsuit claiming the costs imputed were not high enough. States that require imputed costs include Louisiana, North Carolina, South Carolina, and Virginia (Baller, 2011, pp. 2-4).

"De Facto Bans" Created by Numerous and Onerous Barriers to Entry

Some states have such onerous requirements that it is virtually impossible to successfully create a network. For instance, a state imposing a "break-even" requirement within a certain time limit will often altogether prevent successful networks. Florida requires that any municipality planning a network must ensure that revenues will exceed operating expenses and payments on debt within *four years*. Even more onerous is Virginia's requirement that any municipally provided cable service must be predicted to break even in the *first year* of operation, a virtual impossibility given the high up-front investment required to build a broadband network (Baller, 2011, pp. 2-4). Whether these states needed to impose such restrictions is not clear, because both Florida and Virginia have enacted so many other barriers that any future municipal project is unlikely to be built.

In Pennsylvania, if a municipality would like to provide broadband service, it must first give the local telephone company the chance to provide a comparable service (meaning same data speed *only*, Baller, 2011, p. 3) within fourteen months of the municipality's request. This gives the incumbent telephone provider a right of first refusal before any Pennsylvania municipality can provide its own broadband service (Christensen, 2006, p. 700). The incumbent's response is relatively simple because it only needs to match the speed the municipality was planning to offer. On the other hand, if the incumbent allows the network, the incumbent can still challenge it in the legislature or the court. This leaves any municipality subject to the decisions of the incumbent. However, like Michigan, this could be seen as pro-municipality, because the municipality does, in the beginning, set the terms (at least related to speed).

South Carolina is perhaps one of the most difficult states in which to build a municipal network. Prior to 2012, when the most recent municipal broadband bill was passed, South Carolina already had many burdens in place for government-owned telecommunications (primarily telephone) service providers. These included compliance with state and local law, no cross-subsidization of the network, private provider cost imputation, and annual audits (South Carolina Code Ann. § 58-9-2620 enacted 2002). The amendment passed in 2012 expands these requirements to all broadband services, which includes anything that delivers video or provides Internet access at speeds in excess of 768 kbps download and 200 kbps upload speeds, meaning these high barriers apply to very slow connections.[8] The statute does exempt from these requirements certain "unserved" areas. To be considered an unserved area, the municipality must apply for such designation and assert either (1) that they are a "persistent poverty county" *and* 75% of the households within the municipality's area do not have access to broadband service, or (2) that 90% of the households within the municipality's area do not have access to broadband service. A persistent poverty county is a county with 20% of its population defined as "poor" over the past four decennial census surveys. The exemption only lasts for a short period of time, and residents and competitors can seek reclassification of the municipality if they feel the area no longer meets the definition of "unserved." Competitors and residents are free to file objections during the application process as well. South Carolina's new bill does not grandfather currently pending networks, meaning networks existing at the time of passage will have to follow these new rules retroactively. The bill only exempts certain networks, such as institutional networks (provided by and to the government alone) and networks that applied for or received funding under the Broadband Technology Opportunities Program (BTOP) through the National Telecommunications and Information Administration, so long as those areas meet certain criteria outlined in the statute.

For most municipalities, the new requirements will be too much. There is at least one municipality that is forging ahead: Orangeburg County, South Carolina. After South Carolina passed its law, the city council voted in favor of a contract to complete the third phase of their BTOP network deployment plan. Orangeburg is considered a persistent poverty county, and it has decided that it needs high-speed Internet access for its residents and businesses (Mahanta, 2012). It secured $18.65 million in BTOP funding in 2009, and because of

that grant the town is exempt from the new law within area covered by the BTOP grant. Other areas of the County will, however, be subject to the law.

TACTICS USED TO PREVENT THE SUCCESS OF MUNICIPAL NETWORKS

There are two primary mechanisms through which incumbent providers seek to hinder, delay, or prevent municipal networks: lobbying state legislatures and litigating in court (Mitchell, 2012c, p. 37). Both have their advantages: a lobbyist can influence legislators without the procedural and substantive constraints of a courtroom or laws against ex parte proceedings, and can take advantage of the significant imbalance between municipal and corporate lobbying budgets; a lawsuit can delay planning and increase costs for municipalities even if the challenger ultimately loses (as seen below, Bristol, Virginia was successful in the state legislature and in court, but incurred $2.5 million in unexpected legal fees (Mitchell, 2012c, p. 4). Municipal networks of all kinds have been subject to these tactics.

State-enacted legal barriers have largely been the result of heavy industry lobbying, which is a common technique for incumbents to convince state legislatures to block or erect barriers to municipal networks. Those laws then provide the basis for lawsuits that hinder and delay the success of new municipal networks when the municipality does not carefully study their state's laws. This section will first discuss how and why lobbying has become so effective and what municipalities can expect. Next, it will discuss a cross-section of lawsuits that have been filed against these networks.

Lobbying State Legislatures

Many political battles have resulted in relaxed or no passage of municipal broadband barriers (Swirbul, 2006). However, as the political tide turned in the 2010 midterm elections, more state legislatures took action and passed laws restricting municipal networks. This has given incumbents a new hope in certain states, and they have only become more tenacious.

Incumbents have a significant advantage over those arguing in favor of municipal networks. Incumbents are nearly unconstrained in lobbying government. In general, private entities are given antitrust immunity for lobbying, even when that lobbying results in an anticompetitive law. This is because of the *Noerr-Pennington* doctrine, which is justified by the first amendment right to petition governments (Travis, 2006, p. 1740; *Knology v. Insight Communications Co.*). As a result, private entities can lobby vigorously, and they do: incumbent could hire "more lobbyists than [the state] has representatives in the capitol" (referring to a Texas lobbying effort by Southern Bell Corporation) (Nolan, 2005).[9] Municipalities typically cannot rise to industry's level of political posturing. Municipal leagues, the entities that typically represent municipalities at the state level, often have full agendas and are not able to fully devote themselves to one issue (Carlson, 1999, pp. 52-53). Moreover, public interest groups that argue in favor of municipal networks are either busy helping municipalities that *can* create networks, are focused on lobbying at the federal level where there is much more to lose, or do not have a multi-state program capable of combating incumbents in all states considering relevant legislation. Cities themselves often cannot spend money to fight the lobbyists either (Baller, 2011, p. 1). For example, recent lobbying efforts in North Carolina were successful in passing a bill erecting many new barriers to municipal networks, wherein telecom donors gave thousands of dollars to key North Carolina officials (Mitchell, 2012b).

Lobbying state legislatures also does not require the formalities of a lawsuit. Lawsuits are procedural and formulaic; they require evidence, burdens of proof, and hundreds of hours of work honing legal arguments and going through dis-

covery, settlement agreements, and perhaps trial. Additionally, court proceedings often favor fairness, and ex parte communications are frowned upon. Lobbyists are subject to far fewer rules, greatly increasing their power to influence. In essence, it is much easier to convince a legislator of a particular point than a judge or jury. Thus, lobbying is favored over filing a lawsuit, but both will be used if necessary.

In Virginia, the legislature enacted a ban on municipal networks in 1998 (Virginia, 2004, p. 1). In 1999, Bristol successfully sued to hold Virginia's law preempted by the 1996 Telecom Act (*City of Bristol v. Earley*; this holding was abrogated by *Nixon*). Upon the *Earley* holding, Bristol secured passage of a Virginia law allowing municipalities to offer telecom services (Mitchell, 2012c, p. 4). However, the network was met with incumbent attack regarding phone service (from Sprint) and its cable service (*Marcus Cable Associates v. City of Bristol*). Sprint lost at the state agency level. Marcus Cable, also known as Charter Communications, won at the federal district court level on Dillon's rule grounds (discussed later). After the loss to Charter, Bristol went back to the state legislature and convinced them to grant Bristol the authority to provide cable and triple-play services. Unfortunately, that permission was granted within a law that provided barriers to entry for all subsequent municipal networks, excepting Bristol (Mitchell, 2012c, pp. 3-5). Today, Bristol is viewed as one of the great municipal network success stories (Mitchell, 2012c; Null, 2013).

In North Carolina, a multi-year battle ensued between municipal network proponents and incumbent attackers. Beginning in 2008, Time Warner Cable, the incumbent in the state, lobbied for the passage of an anti-municipal network law. Between 2008 and 2010, Time Warner lost. During this time, the legislature was under Democratic party control, and the issue was given a fair hearing. Each year the proposed bill, seeking to erect barriers, failed to pass as written. When the 2010 elections switched the state leadership to the

Republican party, the story changed. Then, fair hearings were not held. State leaders were more sympathetic to the woes of local incumbent service providers. Local (municipal) officials were heavily lobbied by incumbents. Much more was done behind closed doors. Catharine Rice, President of the SouthEast Association of Telecommunications Officers and Advisors (SEATOA) and strong advocate for municipal networks, stated that she did not see a version of the 2011 bill (H. 129, passed in May 2011) until the seventeenth draft—highly unusual given the previous three years. Despite SEATOA's attempts at talking to every member of every committee it could reach, state legislators let incumbents call the shots. For example, the bill co-sponsor called a meeting to discuss the bill. Instead of hearing from both sides, she called on Time Warner's chief lobbyist, one of the bill's authors, to *run* the meeting.

A different story unfolded in Georgia. A very similar bill to North Carolina's H. 129 was brought to Georgia's legislature, but it was ultimately rejected. The key difference, according to Ms. Rice, was that Republicans (the majority in Georgia's government) were convinced by the businesses within the state that this bill was bad for business. North Carolina's legislature only seemed to care about Time Warner's problems. Georgia's legislators cared more about the private industry that used, and required access to, the network for their business—a group largely ignored by North Carolina. However, nothing is stopping the bill from being re-introduced next term, and continually until it is passed.

Philadelphia, Pennsylvania planned to create a citywide wireless network in the mid-2000s designed to bridge the digital divide and provide near-universal access for all within its 135-square-mile city. However, the city encountered problems including technical and business model issues (Christensen, 2006, p. 696). It also ran into opposition on Pennsylvania House Bill 30. Verizon and Comcast, upon hearing of Philadelphia's plan, drafted the bill and then lobbied the legislature.

Municipal network advocates fought back. After the confrontation, a compromise was reached: Philadelphia could continue with its network, but all subsequent networks would be required to give Verizon (and any subsequent telephone provider) the right of first refusal (Christensen, 2006, p. 700; Shaffer, 2007, p. 205). Thus, while Philadelphia perhaps won one battle, the war against the incumbents was lost.

Perhaps the most important lesson to be learned from these stories is that early education of legislators is very important. From a political perspective, there are many compelling sound bites that are easy to believe, but are not necessarily true. Many incumbents will argue that government should not undertake such a risk with taxpayer's money, or that the private incumbent already serves the needs of the area, or that it is anticompetitive. But these are all weak arguments: these networks are rarely funded with tax revenues anymore (more often it is with bonds backed up by the infrastructure itself); private incumbents often do not provide the speeds or the reach that municipalities can guarantee; and adding a competitor to the incumbent monopoly or duopoly is the exact opposite of anti-competitive—in fact, the *incumbents* are the ones engaging in anticompetitive behavior by lobbying for anticompetitive laws. With early education of these important points, legislators will be less likely to fall victim to incorrect characterizations of municipal networks by incumbents.

The quid pro quo for the 1996 Act was that telecom providers were given more freedom to enter into other markets in exchange for increased local telecom competition. Today, heavy industry lobbying and the FCC's inability to force open access on information services has depressed local competition to the point of non-existence. However, the telecom companies took advantage of their half of the deal by entering other, lucrative markets. Thus, telecom companies have created a win-win situation for themselves: entrance into other lucrative markets *and* suppression of local telecom competition.

Lawsuits to Hinder, Delay, and Prevent Municipal Networks

Often, there are multiple ways for incumbents to challenge municipal networks judicially (through the court system) and administratively (through particular state or local agencies). Incumbents can challenge the same network through public service boards, bond commissions, or other administrative bodies while simultaneously challenging the network in a state or federal court. These challenges drastically increase costs and delay for those networks. Delay benefits the private providers by delaying or fully preventing competitive entry into the market. Municipalities should be prepared for these challenges.

This section will begin by discussing legal challenges based on the U.S. constitution and federal statutes. Then it will discuss challenges based on state law, which are generally more common.

Federal Constitutional and Legal Challenges

Federal challenges to municipal networks have largely failed, including antitrust claims[10] and claims based on the first and fourteenth amendments (freedom of speech and due process). The following municipalities have won lawsuits based on federal antitrust or constitutional grounds: Glasgow, Kentucky (first amendment and antitrust), Paragould, Arizona (first and fourteenth amendments and antitrust), Niceville, Florida (first amendment), and Morganton, North Carolina (first amendment) (Merton Group, 2003, pp. 69-72). First amendment challenges lose because mere economic injury does not rise to the level of first amendment injury and private enterprises do not have a first amendment right to use publicly-owned rights-of-way (Merton Group, 2003, pp. 70-71). Antitrust challenges lose because of the state action exemption from antitrust laws (*Paragould Cablevision*, 1991, p. 1312). Constitutional challenges in general lose because franchise agreements can

involve the bargaining away of certain constitutional rights (*Paragould Cablevision*, 1991, p. 1315). Despite these losses, challenges may still be brought under these theories, so a municipality should be ready to defend those claims. However, incumbents have increasingly resorted to lawsuits based on state law.

State Law Challenges

As a preliminary matter, the state statutes discussed in detail above (requiring specific procedures such as a referendum or hearings) could be the basis of a lawsuit. To avoid the most obvious lawsuits, municipalities should know their state's regulations and follow them closely.

State level legal challenges are often based on a municipality's authority to provide broadband services in the first instance. Incumbents have been very successful asserting authority claims when municipalities are not careful regarding their grants of authority (Travis, 2006, p. 1768). Whether a municipality has authority to provide broadband service depends on whether the state follows the "Dillon" rule or the "home" rule.

Dillon's rule, named after former Iowa Chief Judge John Dillon (who coined the rule), states that "a municipal corporation possesses and can exercise the following powers and no others: First, those granted in express words; second, those necessarily implied or necessarily incident to the powers expressly granted; third, those absolutely essential to the declared objects and purposes of the corporation—not simply convenient, but indispensable . . ." (*Merriam v. Moody's Executors*, 1868, p. 170). This is a strict interpretation of municipal power, but it has been followed in many states since that case in 1868.

Put another way, in a state that follows the Dillon rule, municipalities receive express grants of authority and can act pursuant only to those express grants (Dunne, 2007, p. 1148). Any municipal act that is not tied to an express grant of authority or express purpose is beyond the municipality's power. If, for instance, a state law says that localities can establish and operate "waterworks, sewerage, gas works (natural or manufactured), electric plans, public mass transportation systems, storm water management systems and other public utilities" (as the Virginia state law provided in *Marcus Cable Associates*), then municipalities are allowed to provide those services and no others. In that particular case, the judge decided that "other public utilities" did not include a cable TV system, and thus Bristol was prevented from providing cable services (Bristol secured the necessary legislative grant thereafter). On the other hand, in a California case, *Cequel III v. Truckee-Donner Public Utility District*, the state law at issue said the utility could build public works "for supplying its inhabitants with light, water, power, heat, transportation, telephone service, or other means of communication, or means for the disposition of garbage, sewage, or refuse matter . . ." (*Cequel III*, 2007, p. 317). The court in that case said that "other means of communications" included broadband services. This dichotomy shows that even within the Dillon rule, there is still opportunity for argument depending on the exact phrasing of the statute.

For municipalities in states that follow the Dillon rule, including Vermont and New Hampshire (Merton Group, 2003, p. 75), and that lack the express authority to provide telecommunications or broadband services, there is not much choice other than to ask the legislature for a law granting the municipality authority to provide the service explicitly, as in Bristol. In today's political climate, that might be difficult.

By contrast, in a "home-rule" state, municipalities are given plenary power and the state may carve out exceptions in areas where it does not want cities to act (Dunne, 2007, p. 1148). More states follow the home rule rather than Dillon rule based on a 1992 study (Dunne, 2007, pp. 1148-1149). Thus, in many states, including Massachusetts and Rhode Island, municipalities have the power to provide broadband services short of specific restrictions in the municipality's grant of authority (Merton Group, 2003, pp. 74-75). Municipalities subject to the home rule are unlikely to be subject to lawsuits based on authority. Express preclusion of these networks is required by statute and will, in general, be obvious. However, they are still susceptible to lobbying efforts in order to enact such a preclusive law.

POLICY IMPLICATIONS AND LESSONS LEARNED

This chapter has discussed the many ways in which incumbents can attack a municipal network either indirectly (lobbying) or directly (lawsuits, information campaigns, and others). The veracity with which these companies fight municipal networks is at least one indicator of how successful the networks can be. If they were not effective competitors, incumbents would not spend hundreds of thousands, or millions, of dollars fighting the referenda, lobbying legislators, donating to campaigns, filing complaints, and creating slanderous advertisements. Municipal networks are disruptive, and incumbents know it.

The following are key learned lessons from case studies discussed within:

1. The expected outcomes and ultimate goals of municipal networks often make building them worth the fight.
2. If your state has no relevant laws, check to see if your state is a home-rule state or a Dillon's rule state, and how that rule operates in your state. Either way, there could still be challenges: once a municipal network enters the planning state, incumbents will likely find or manufacture some reason to attack it either in court, in the state legislature, or at the polls. These incumbents cannot afford to set the precedent of allowing municipal networks.

3. If your state does have relevant laws, leaders of the project should study them carefully. There are often ways to work around or avoid restrictions, depending on their wording. For instance, if the state defines "telecommunications" as the federal government does, as in Nevada, or has important exceptions to the restriction, as in Missouri, there may be ways to provide the desired services without violating the law. In other words, a careful selection of technology or provider could avoid state law altogether.

4. It is extremely important to have a budget for potential legal disputes. Do not begin planning a network until you can afford to pay a lawyer to help you understand your state laws and to represent you in court or in state agencies, if necessary. Legal costs can grow quickly. In the extreme case, Bristol, Virginia spent $2.5 million in unanticipated legal costs (Mitchell, 2012c, p. 4). Municipalities should be ready for the worst case scenario, even if it does not happen.

5. If your municipality can afford to hire a lobbyist, it can be helpful to have one. Lafayette, Louisiana, and Chattanooga, Tennessee each hired one lobbyist, while incumbents hired "the rest" (Mitchell, 2012, p. 53). That being said, many municipalities can be successful without serious lobbying efforts.

6. Politically, it is important to know your leadership, and it is extremely important to educate them early in the process. Incumbents will use any tactic they can to convince legislatures that municipal networks are risky, and the municipality must be able to counter these arguments. If the municipality has data (e.g. percentage of underserved people) to bolster its claim of need, that is even better. Data from the municipality can be more accurate than federal government data (often based on census blocks, which can be misleading).

7. Always follow legal developments in your area and in the federal government (especially the FCC and the NTIA). Be sure to get involved in funding projects if possible.

8. Do not let daunting lobbying and legal action dissuade your municipality from providing Internet service if the municipality believes the area will benefit in some important way. Municipal networks can help solve the communications crisis in America. Incumbent attack is only serving to augment that crisis. Many municipalities have succeeded even in states with barriers and after attack from incumbents.

9. Know what you are getting into. The opposition can be daunting. Prior knowledge of what to expect can be invaluable, and it can allow the municipality to plan ahead to avoid the worst of the opposition. For instance, educating legislators as early as possible can potentially negate the incumbent's likely mischaracterization of municipal networks.

SUMMARY CHECKLIST

Once a municipality has determined that the social and economic benefits of the network will justify its construction, *and* that there is at least some indication that there is demand for the network, the municipality should make some initial determinations regarding the state's laws and how to comply with those laws. This is not a comprehensive checklist, but it offers some guidance to municipalities that want to ensure they are following their state's laws.

1. Learn your state's laws inside and out and follow those laws exactly; confer with others in the state, if any, who have previously developed a community broadband network; and hire a lawyer who specializes in this municipal work, as this can be exceedingly helpful and cost-effective.

2. Determine what type of financing is available to you keeping in mind your state's laws (can you use general tax revenue? Bonds? Can you cross-subsidize the service? If some or all of your project area is currently underserved, can you seek CAF funding?

3. Determine what type of technology you will use, keeping in mind your state's laws and your financial means.

4. Plan, with specificity, the steps to bring the idea to completion.

5. Acquire as much independent data as possible. The more data you have, the more compelling your case will be when you are trying to prove (to legislators, to citizens) the broadband needs of the area that you would like to serve.

6. Be sure to keep a timetable of the steps in your plan and any of the specific requirements that the state's law requires, including a feasibility study, hearing, referendum, and any potential information campaigns that may need to be waged against incumbents.

7. If incumbents lobby your state government for more restrictive rules, be prepared to mount as effective a response as possible; try to enlist support from state and national associations, public interest groups, and other allies; if possible, hire at least one lobbyist to counter the incumbent's arguments—it will not be an equal or fair fight, and it will require extensive activity for several months, but there is no alternative if you want to have a fighting chance to succeed.

8. Be prepared both monetarily and mentally to defend lawsuits that seek to prevent or delay your network, and be prepared with legal arguments as to why your network should continue—this is where having a lawyer will come in handy, as well as the documentation.

CONCLUSION

Municipal networks could be America's salvation from recent reports of poor broadband deployment and poor access speeds. It could also be the solution to America's communications crisis. In a world where online innovative "apps" consist more and more of streaming music, video, and other activities that require a high-speed connection, high-speed Internet access will soon be vital for contribution to and participation in society for decades to come. Many municipalities are recognizing a local need for broadband services (especially when local businesses require it) and are meeting those needs by providing the service themselves rather than relying on flaky, bottom-line-driven private providers. There is a real opportunity in the near future to boost America's standing in the world regarding high-speed Internet access; reliance on purely private providers has only served to drop America's ranking and now it has fallen behind countries such as Slovenia and Estonia in advertised broadband speeds. Legal barriers, as discussed in this Chapter, have created problems for many municipalities, but they should not prevent a municipality from building a network if the network would be effective. There are ways to comply, or ways around, those laws and the increased burdens could still be worth the investment if the social and economic development will be substantial. If more municipalities are successful in doing so, they could help put America back on top.

REFERENCES

Abdelaal, A., Ali, H., & Khazanchi, D. (2009). The role of social capital in the creation of community wireless networks. In *Proceedings of the 42nd Hawai'i International Conference on Systems Sciences*. Waikoloa, HI: IEEE Press.

Badger, E. (2011). *How the telecom lobby is killing municipal broadband.* Retrieved from http://www.theatlanticcities.com/technology/2011/11/telecom-lobby-killing-municipal-broadband/420

Baller, J. (2011). *State restrictions on community broadband services or other public communications initiatives.* Retrieved from http://baller.com/pdfs/BallerHerbstStateBarriers(7-1-12).pdf.

Benjamin, S., Lichtman, D. G., Shelanski, H., & Weiser, P. (2006). *Telecommunications law and policy.* Durham, NC: Carolina Academic Press.

Benjamin, S., Lichtman, D. G., Shelanski, H., & Weiser, P. (2011). *Telecommunications law and policy: 2011 supplement.* Durham, NC: Carolina Academic Press.

Botein, M. (2006). Regulation of municipal wi-fi. *New York Law School Law Review. New York Law School, 51,* 975–988.

Buckley, S. (2012). *CenturyLink gets $35m in FCC CAF funding for broadband expansion.* Retrieved from http://www.fiercetelecom.com/story/centurylink-gets-35m-fcc-caf-funding-broadband-expansion/2012-07-25

Carlson, S. C. (1999). A historical, economic, and legal analysis of municipal ownership of the information highway. *Rutgers Computer & Technology Law Journal, 25*(1), 3–60.

Cequel III Communications v. Local Agency Formation Commission of Nevada, 149 Cal. App.4th 310 (2007).

Christensen, A. D. (2006). 'Wi-Fi'ight them when you can join them? How the Philadelphia compromise may have saved municipally-owned telecommunications service. *Federal Communications Bar Journal, 58*(3), 683–704.

City of Abilene v. FCC, 164 F.3d 49 (D.C. Cir. 1999).

City of Bristol v. Earley, 145 F.Supp.2d 741 (W.D. Va. 2001).

Coleman, R., Behunin, J., & Harvey, M. (2012). *A performance audit of the Utah telecommunication open infrastructure agency.* Retrieved from http://le.utah.gov/audit/12_08rpt.pdf

Community Broadband Networks. (2012a). *Community broadband network map.* Retrieved from http://muninetworks.org/communitymap

Connect, C. C. (2012). *About.* Retrieved from http://www.connectcc.com/aboutconnect.html

Connecting America. (2010). *The national broadband plan.* Retrieved from http://download.broadband.gov/plan/national-broadband-plan.pdf

Cooper, M. (2012). *Efficiency gains and consumer benefits of unlicensed access to the public airwaves: The dramatic success of combining market principles and shared access.* Retrieved from http://www.consumerfed.org/pdfs/EFFICIENCYGAINS-1-31.pdf

Crawford, S. P. (2008). *Information/telecommunications services.* Retrieved from http://scrawford.net/blog/informationcommunications-services/1181

Crawford, S. P. (2011). The communications crisis in America. *Harvard Law & Policy Review, 5*(2), 245–263.

Dingwall, C. (2006). Municipal broadband: Challenges and perspectives. *Federal Communications Bar Journal, 59*(1), 67–103.

Dunne, M. (2007). Let my people go (online): The power of the FCC to preempt state laws that prohibit municipal broadband. *Columbia Law Review, 107*(5), 1126–1164.

FCC. (2012). *Connect America fund (CAF) phase I.* Retrieved from http://www.fcc.gov/maps/connect-america-fund-caf-phase-i

FCC Press Release. (2010). *FCC frees up vacant TV airwaves for "super wi-fi" technologies.* Retrieved from http://hraunfoss.fcc.gov/edocs_public/attachmatch/DOC-301650A1.pdf

FCC v. Pacifica Foundation, 438 U.S. 726 (1978).

Frischmann, B. M. (2012). *Infrastructure: The social value of shared resources.* Oxford, UK: Oxford University Press. doi:10.1093/acprof:oso/9780199895656.001.0001

Ganapati, S., & Schoepp, C. F. (2009). The wireless city. In Reddick, C. G. (Ed.), *Handbook of Research on Strategies for Local E-Government Adoption and Implementation* (pp. 554–568). Hershey, PA: IGI Global. doi:10.4018/978-1-60566-282-4.ch029

Gillett, S. (2012). *FCC launches connect America fund.* Retrieved from http://www.fcc.gov/blog/fcc-launches-connect-america-fund

Grayson, T., Crawford, S., & Baller, J. (2012). *The present and future of alternative fiber networks.* Unpublished.

Greeley, B., & Fitzgerald, A. (2011). *Pssst ... Wanna buy a law?* Retrieved from http://www.businessweek.com/magazine/pssst-wanna-buy-a-law-12012011.html

Gregory v. Ashcroft, 501 U.S. 452 (1991).

In Re Missouri Municipal League, 299 F.3d 949 (8th Cir. 2002).

Knology, Inc. v. Insight Communications Company, 393 F.3d 656 (6th Cir. 2004).

Mahanta, S. (2012). *Why are telecom companies blocking rural America from getting high-speed internet?* Retrieved from http://www.tnr.com/article/politics/102699/rural-broadband-internet-wifi-access

Marcus Cable Associates v. City of Bristol, 237 F.Supp.2d 675 (W.D. Va. 2002)

McGarty, T. P., & Bhagavan, R. (2002). *Municipal broadband networks: A revised paradigm of ownership.* Retrieved from http://www.lus.org/uploads/MunicipalBroadbandNetworksStudy.pdf

Meinrath, S. D., & Calabrese, M. (2008). White space devices & the myths of harmful interference. *New York University Journal of Legislation and Public Policy, 11*(3), 495–518.

Merriam v. Moody's Executors, 25 Iowa 163 (1868).

Merton Group. (2003). *Municipal broadband report: Feasibility study report for Hanover, NH.* Retrieved from http://www.telmarc.com/Feasibility/Hanover%20Feasibility%20Study.pdf

Millonzi, K. (2011). *New municipal broadband limitations.* Retrieved from http://canons.sog.unc.edu/?p=4967

Mitchell, C. (2010). *Florida muni dunnellon building FTTH network.* Retrieved from http://www.muninetworks.org/content/florida-muni-dunnellon-building-ftth-network

Mitchell, C. (2012a). *North Carolina county turns on first white spaces wireless network in nation.* Retrieved from http://www.muninetworks.org/content/north-carolina-county-turns-first-white-spaces-wireless-network-nation

Mitchell, C. (2012b). *Big bucks: Why North Carolina outlawed community networks.* Retrieved from http://www.muninetworks.org/content/big-bucks-why-north-carolina-outlawed-community-networks

Mitchell, C. (2012c). *Broadband at the speed of light: How three communities built next-generation networks.* Retrieved from http://download.broadband.gov/plan/national-broadband-plan.pdf

National Cable and Telecommunications Association v. Brand X Internet Services, 545 U.S. 967 (2005).

National Telecommunications and Information Administration. (2012). *About BroadbandUSA.* Retrieved from http://www2.ntia.doc.gov/about

Nixon v. Missouri Municipal League, 541 U.S. 125 (2004).

Nolan, C. (2005). *Taking sides in the municipal wireless showdown*. Retrieved from http://www.eweek.com/c/a/Government-IT/Taking-Sides-for-the-Municipal-Wireless-Showdown

Null, E. (2013). Municipal broadband: History's guide. *I/S: A Journal of Law and Policy for the Information Society*.

O'Loughlin, D. S. (2006). Preemption or bust: Fear and loathing in the battle over broadband. *Cardozo Law Review, 28*(1), 479–510.

OECD. (2010). *Fixed broadband penetration and density*. Retrieved from http://www.oecd.org/sti/broadbandandtelecom/39574903.xls

OECD. (2011). *Average advertised download speeds*. Retrieved from http://www.oecd.org/dataoecd/10/54/39575095.xls

OECD. (2012). *OECD broadband portal*. Retrieved from http://www.oecd.org/document/54/0,3746,en_2649_34225_38690102_1_1_1_1,00.html

Omaha Wireless. (2012). *About us*. Retrieved from http://omahawireless.unomaha.edu/index.html

Ordower, G. (2005). Broadband quest cost $300,000. *Daily Herald*. Retrieved from http://lafayetteprofiber.com/imagesNRef/Docs/TriCitiesCosts.html

Paragould Cablevision v. City of Paragould, 930 F.2d 1310 (8th Cir. 1991).

Petitioner's Brief, City of Abilene v. FCC, 164 F.3d 49 (D.C. Cir. 1998)

Rich, S. (2012). *New Hanover county, N.C., first in nation to deploy 'super wi-fi' network*. Retrieved from http://www.govtech.com/e-government/New-Hanover-County-NC-Super-Wi-Fi-Network.html?elq=cc575b8cbd0a44588d262afb34031bbb

Salinas v. United States, 522 U.S. 52 (1997).

Shaffer, G. (2007). Frame-up: An analysis of arguments for and against municipal wireless initiatives. *Public Works Management & Policy, 11*(3), 204–216. doi:10.1177/1087724X06297347

South Carolina. (2002). *South Carolina code Ann. § 58-9-2620*. Retrieved from http://www.scstatehouse.gov/code/t58c009.php

South Carolina. (2012). *South Carolina code Ann. § 58-9-2620*. Retrieved from http://scstatehouse.gov/sess119_2011-2012/prever/3508_20120607.htm

Spector, D., Lubin, G., & Giang, V. (2012). *The 15 most disliked companies in America*. Retrieved from http://www.businessinsider.com/the-most-hated-companies-in-america-2012-6?op=1

Swirbul, C. (2006). *State broadband battles*. Retrieved from http://www.baller.com/pdfs/APPA_Broadband_Battles.pdf

Tech Law Journal. (2005). *8th circuit rules in north Kansas City municipal broadband case*. Retrieved from http://www.techlawjournal.com/topstories/2005/20051229.asp

Travis, H. (2006). Wi-fi everywhere: Universal broadband access as antitrust and telecommunications policy. *The American University Law Review, 55*(6), 1697–1800.

Tynan, D. (2012). White spaces: The next generation of wireless broadband has landed. *PCWorld*. Retrieved from http://www.pcworld.com/article/248847/white_spaces_the_next_generation_of_wireless_broadband_has_landed.html

Utah Code Ann. §§ 10-18-202, -203, -204, -302 (West, Westlaw current through 2012).

Vermont Office of Secretary of State. (2009). *Municipal law basics*. Retrieved from http://www.sec.state.vt.us/municipal/pubs/municipal_law_basics.pdf

Virginia Regulatory Issues Pertaining to Municipal Broadband. (2004). *Website*. Retrieved from http://top.bev.net/tamp/7-Common_Appendices/ Main_Project_Papers/Virginia_Regulatory_Issues_Pertaining_to_Municipal_Broadband.pdf

Vos, E. (2005). *Colorado anti-muni bill passed by house local govt committee*. Retrieved from http://www.muniwireless.com/2005/04/07/ colorado-anti-muni-bill-passed-by-house-local-govt-committee

Vos, E. (2010). *Wilmington, NC uses white for smart city, eco-friendly wireless applications*. Retrieved from http://www.muniwireless. com/2010/02/24/wilmington-uses-white-spaces-for-smart-city-ecofriendly-wireless-applications

Vos, E. (2011a). *Newton, NC deploys free downtown wi-fi service on the cheap*. Retrieved from http://www.muniwireless.com/2011/09/27/ newton-nc-deploys-free-downtown-wifi-service-on-the-cheap

Vos, E. (2011b). *Cleveland, Ohio neighborhood deploys large outdoor free wi-fi network*. Retrieved from http://www.muniwireless.com/2011/11/10/ cleveland-ohio-neighborhood-deploys-large-outdoor-free-wi-fi-network

White Space, F. C. C. (2012). *Website*. Retrieved from http://www.fcc.gov/topic/white-space

Wong, M. A. (2007). Community wireless: Policy and regulation perspectives. *Journal of Community Informatics, 3*(4). Retrieved from http://owl.english.purdue.edu/owl/resource/560/10

ADDITIONAL READING

Baller, J. (2005, May). Deceptive myths about municipal broadband: Disinformation about public ownership is impeding progress. *Broadband Properties*, 14-17.

Barkow, R. E. (2010). Insulating agencies: Avoiding capture through institutional design. *Texas Law Review, 89*(15), 15–79.

Benkler, Y. (2006). *The wealth of networks: How social production transforms markets and freedom*. New Haven, CT: Yale University Press.

Breitbart, J. (2007). *The Philadelphia story: Learning from a municipal wireless pioneer*. Retrieved from http://www.newamerica.net/files/ nafmigration/NAF_PhilWireless_report.pdf

Crawford, S. P. (2007). The internet and the project of communications law. *UCLA Law Review. University of California, Los Angeles. School of Law, 55*(2), 359–407.

Crawford, S. P. (2009). Transporting communications. *Boston University Law Review. Boston University. School of Law, 89*(3), 871–937.

Crawford, S. P. (2013). *Captive audience: The telecom industry and monopoly power in the gilded age*. New Haven, CT: Yale University Press.

Ellison, C. (2008). Municipal broadband: A potential twenty-first century utility. *New York University Journal of Legislation and Public Policy, 11*(3), 453–465.

Farivar, C. (2012). *South Carolina passes bill against municipal broadband*. Retrieved from http://arstechnica.com/tech-policy/2012/06/ south-carolina-passes-bill-against-municipal-broadband

Freiden, R. (2012). Assessing the need for more incentives to stimulate next generation network investment. *I/S: A Journal of Law and Policy for the Information Society, 7*(2), 1-50.

Frischmann, B. M., & Lemley, M. (2006). Spillovers. *Columbia Law Review, 100*(2), 101–143.

Frischmann, B. M., & van Schewick, B. (2007). Network neutrality and the economics of an information superhighway: A reply to Professor Yoo. *Jurimetrics Journal, 47*, 383–428.

Hussain, H., Kehl, D., Lennett, B., Li, C., & Lucey, P. (2012). *The cost of connectivity*. Retrieved from http://newamerica.net/sites/newamerica.net/files/policydocs/The_Cost_of_Connectivity.pdf

Lehr, W., Sirbu, M., & Gillett, S. (2012). *Broadband open access: Lessons from municipal network case studies*. Retrieved from http://people.csail.mit.edu/wlehr/Lehr-Papers_files/Lehr%20Sirbu%20Gillett%20Broadband%20Open%20Access.pdf

Lide, E. C. (2008). Balancing the benefits and privacy concerns of municipal broadband applications. *New York University Journal of Legislation and Public Policy, 11*(3), 467–493.

Mitchell, C. (2010). *Breaking the broadband monopoly: How communities are building the networks they need*. Retrieved from http://www.muninetworks.org/sites/www.muninetworks.org/files/breaking-bb-monopoly.pdf

Mitchell, C. (2011). *Publicly owned broadband networks: Averting the looming broadband monopoly*. Retrieved from http://www.newrules.org/sites/newrules.org/files/cmty-bb-map.pdf

Mitchell, C., & Meinrath, S. (2012). *Want to pay less and get more?* Retrieved from http://www.slate.com/articles/technology/future_tense/2012/08/community_based_projects_make_broadband_internet_access_high_speed_and_affordable_.single.html

Null, E. (2011). The difficulty with regulating network neutrality. *Cardozo Arts & Entertainment Law Journal, 29*(2), 459–493.

Null, E. (2012). *Public ownership of networks can solve broadband policy fights*. Retrieved from http://muninetworks.org/content/public-ownership-networks-can-solve-broadband-policy-fights

Ozer, N. A. (2008). No such thing as "free" internet: Safeguarding privacy and free speech in municipal wireless systems. *New York University Journal of Legislation and Public Policy, 11*(3), 519–566.

Paul, R. (2006). *Evaluating the pros and cons of municipal wireless*. Retrieved from http://arstechnica.com/business/2006/12/8517

Santorelli, M. J. (2007). Rationalizing the municipal broadband debate. *I/S: A Journal of Law and Policy for the Information Society, 3*(1), 43-82.

Settles, C. J. (2007). *The economic development impact of municipal wireless*. Retrieved from http://cjspeaks.com/msp/snapshot-7-07.pdf

Shapiro, R. J., & Hassett, K. A. (2012). *The employment effects of advances in internet and wireless technology: Evaluating the transitions from 2G to 3G and from 3G to 4G*. Retrieved from http://ndn.org/sites/default/files/blog_files/The%20Employment%20Effects%20of%20Advances%20In%20Internet%20and%20Wireless%20Technology_1.pdf

Staff Report, F. T. C. (2006). *Municipal provision of wireless internet*. Retrieved from http://www.ftc.gov/os/2006/10/V060021municipalprovwirelessinternet.pdf

Van Schewick, B. (2010). *Internet architecture and innovation*. Cambridge, MA: MIT Press.

KEY TERMS AND DEFINITIONS

Agency (or Regulatory) Capture: When a regulatory agency is influenced by a powerful interest group.

Broadband: Describes the technology that provides users with high-speed Internet access. Currently, the FCC's "Tier 1" Broadband standard is 768 kilobytes per second to the user.

Broadband Over Power Lines: High-speed Internet access provided over a power utility's infrastructure. This is an unpopular technology because it is generally believed to be outdated.

Broadband Technology Opportunities Program: A funding program, implemented through the National Telecommunications and Information Administration, to support the deployment of broadband infrastructure, enhance and expand public computer centers, encourage sustainable adoption of broadband service, and develop and maintain a nationwide public map of broadband service capability and availability (according to the BTOP website). All grants have been awarded.

Connect America Fund: A funding program, implemented through the Federal Communications Commission, to encourage the build-out of high-speed Internet infrastructure to rural areas. Currently, phase one has been completed, giving out $115 million to rural areas in 37 states.

Communications Act of 1934: The principal statute governing telecommunications. It was heavily amended by the 1996 Act.

Cross-Subsidization: Using surplus revenues from other municipal projects to reduce the costs and subscription rates of the broadband service. This could perhaps undercutting private providers, or it could provide a much-needed service for a lower price, to the benefit of the subscribers.

Digital Subscriber Line (or DSL): High-speed data provided over telephone lines through use of hardware that converts the telephone line into a data-transfer line. DSL is an information service to the extent that data is transferred over the phone line.

End-User: The person using the Internet connection.

Externality (or Spillover): A benefit or detriment incurred by a third person, not internalized by the parties in the transaction. Toxic run-off is a negative externality, affecting surrounding areas and people.

Federal Communications Commission (or FCC): An independent federal agency primarily responsible for telecommunications regulation.

Fiber-Optic (or Optical Fiber): A transparent fiber made of glass that facilitates the transfer of light, and therefore, information travels at the speed of light across the fiber. This is generally considered to be the next-generation technology that will not become obsolete at least in the near future.

Incumbent Service Provider (or Incumbent Local Exchange Carrier): The established local high-speed Internet service provider, or, in the case of local exchange carrier, the established local telephone provider.

Information Service: The name given to services that use "telecommunications" to deliver a service, including a cable modem service. These are not subject to common carrier regulations, but are subject only to the FCC's "ancillary" jurisdiction, which essentially encompasses the FCC's mandate to protect broadcasters. The 1996 Act defines it as follows: "the offering of a capability for generating, acquiring, storing, transforming, processing, retrieving, utilizing, or making available information via telecommunications, and includes electronic publishing, but does not include any use of any such capability for the management, control, or operation of a telecommunications system or the management of a telecommunications service."

Internet Service Provider (or ISP): A firm that provides Internet access to end-users.

Municipal Bonds: Bonds issued by a local government, public school district, or other quasi-government entity to raise funds to finance certain projects, usually to meet the infrastructure needs of the municipality.

Municipality: Usually any political subdivision of a state with powers of government.

National Broadband Plan: A series of recommendations from the FCC as to how Congress and the FCC should move forward in regulating the communications industry. It was released March 16, 2010.

National Telecommunications and Information Administration (or NTIA): The executive agency principally responsible for advising the President on telecommunications and information policy issues.

Open Access: Requiring broadband providers to carry multiple Internet service providers across their infrastructure.

Predatory Pricing: Pricing goods or services at a very low level to force competition out of the market.

Retail Service: A broadband service provider that provides Internet connectivity directly to the end user.

Ripeness: A judicial doctrine used to dismiss a case when the controversy to be adjudicated has not yet occurred.

Telecommunications: Defined under the 1996 Act as "the transmission, between or among points specified by the user, of information of the user's choosing, without change in the form or content of the information as sent and received."

Telecommunications Act of 1996: The most recent federal overhaul of the communications industry, encompassing many amendments to the 1934 Act. It primarily applies to the telephone industry, as that remains one of the few services left still considered a "telecommunications service."

Telecommunications Provider (or Carrier): Defined under the 1996 Act as "any provider of telecommunications services, except that such term does not include aggregators of telecommunications services (as defined in section 226). A telecommunications carrier shall be treated as a common carrier under this Act only to the extent that it is engaged in providing telecommunications services, except that the Commission shall determine whether the provision of fixed and mobile satellite service shall be treated as common carriage."

Telecommunications Service: Defined under the 1996 Act as "the offering of telecommunications for a fee directly to the public, or to such classes of users as to be effectively available directly to the public, regardless of the facilities used."

Underserved Community: This term is typically used for a locality with only one Internet service provider.

Unlicensed Spectrum: Frequencies on the spectrum band that the FCC does not license to specific users, thereby opening the frequencies up to innovative and experimental uses by any member of the public. Interference is a common complaint when users use products that take advantage of unlicensed spectrum (such as wireless telephones and Wi-Fi routers).

Unserved Community: A locality with no incumbent local Internet service provider (or local exchange provider).

White Space: It either means (1) frequencies licensed to broadcasters but not used locally, or (2) frequencies between licensed frequencies that are not being used. Interference is a common counter-argument to use of white spaces, but increasingly smart technology is alleviating that problem.

Wholesale Service: A network owner that provides access to the network to other Internet service providers for resale to the end-user (thus, the ISP in this case is the retail provider).

Wireless Fidelity (or Wi-Fi): Wireless transmission of broadband data among devices. In-home wireless routers use this technology, and recently it has been used as a basis for ubiquitous wireless access across municipalities.

Worldwide Interoperability for Microwave Access (or WiMAX): Part of the fourth-generation ("4G") wireless technology, it is a much improved (faster, reaches farther) version of Wi-Fi[11].

ENDNOTES

[1] For more rankings, see OECD (2012).

[2] A Virginia District Court came to the same holding in *City of Bristol v. Earley*, but this was abrogated in *Nixon*.

[3] The constitutionality of this bill has been questioned as a violation of the Tenth Amendment (O'Loughlin, 2006).

[4] The following is a non-exclusive list of un-enacted bills (Dingwall, 2006, pp. 87-90): Communications, Opportunity, Promotion and Enhancement Act of 2006 (H.R. 5252): precluding state prohibitions of municipal networks, requiring non-discriminatory

application of laws to municipal utility, and disallowing favoritism for municipal providers.

Communications, Consumer's Choice, and Broadband Deployment Act (S. 2686): requiring municipalities to provide thirty-days notice and to solicit private bids, but no requirement that the municipality accept the private bid.

Internet and Universal Service Act of 2006 (S. 2256) and the Universal Service Reform Act of 2006 (H.R. 5072): both provide for funding of the Universal Service Fund that can help fund municipal networks.

The "White Space" Bill (H.R. 5085), the Wireless Innovation Act of 2006 (S. 2327), and the American Broadband for Communities Act (S. 2332): all call for freeing unused spectrum for use for wireless networks, potentially for municipally provided wireless networks.

5   The National Telecommunications and Information Administration (NTIA) has a small role limited to funding as well. The NTIA, in general, is the executive agency that makes telecom policy recommendations to the President. However, under the American Recovery Reinvestment Act of 2009, the NTIA was charged with granting $4.7 billion in local broadband deployment grants as part of the Broadband Technology Opportunities Program (BTOP) (National Telecommunications and Information Administration [NTIA]). The BTOP, part of the Rural Utilities Service's Broadband Initiatives Program, was designed to help communities provide anchor institutions, such as libraries and schools, with high-speed Internet access, and to bridge the digital divide by providing cheap or free Internet access through a variety of mechanisms. All grants were awarded as of 2010. Universities might benefit from Gig.U and Air.U, both designed to fund high-speed networks at Universities around the nation.

6   To illustrate the amount of work involved, the feasibility study for Hanover, NH was nearly ninety pages of dense legal and economic analysis (Merton Group, 2003).

7   Six of the top fifteen most disliked companies are telecommunications companies. In growing order of hatred: DirecTV, CenturyLink, Cox Communications, Time Warner Cable, Comcast, Charter (Spector, 2012).

8   In its Advanced Services Report issued in 2010, as required by Section 706 of the Telecommunications Act, the FCC found that a person cannot have a meaningful Internet experience with connection speeds of less than 4 Mbps downstream and 1 Mbps upstream. Afterward, in implementing its Connect America Funding initiative, requires funded entities to provide 4 mbps download speeds and 1 mbps upload speeds; even these speeds are considered very slow when compared to gigabit connections.

9   Many of these lobbying efforts are attributable to the American Legislative Exchange Council. This ideologically conservative group is known for drafting model legislation that is passed to friendly state representatives to be enacted into law (Greeley & Fitzgerald, 2011), including the "Cable and Video Competition Act" that precludes municipalities from competing in that sector and gives control to state governments (Grayson, 2012, p. 7).

10 This is despite the call of some (McGarty & Bhagavan, 2002, p. 24) to "aggressive[ly] prosecut[e]...the antitrust laws." For a more complete discussion of antitrust issues, see Travis (2006).

11 Many definitions contained herein come from the Glossary in Benjamin (2006, pp. 1171-1183).

APPENDIX

This is not a complete list of all regulations (Table 1); it is only intended to be a summary and to give municipalities a general understanding of what to expect depending on the state in which your municipality is situated. Please refer to the relevant state laws when planning your own municipal network for a complete, accurate, and up-to-date list of requirements.

I am grateful to Jim Baller for providing much of the information for this table, and Harold Feld for the suggestion that I include a table in this chapter.

Table 1.

State	Technology regulated	Administrative Burdens	Financial Burdens	Other Burdens or Requirements
Alabama (*Alabama Code 11-50B-1, -12*)	Cable, telecom services	For *cable*: prior hearings, referendum	Imputation of private provider costs; no use of local taxes for start-up expenses; service must be self-sustaining	
Arkansas (*Ark. Code 23-17-409*)	Local exchange (telephone)	If governmental entity operates through an *electric utility*, it may provide local exchange services if it gives reasonable notice and holds a public hearing. Otherwise, no municipal network is allowed.		
California (*Calif. Government Code 61100(af)*)	Advanced telecom services (data, voice, video, benchmark is a technology-neutral 4Mbps download, 1Mbps upload—from FCC)	A District may provide broadband services if it makes reasonable efforts, but ultimately fails, to find private provider. If a private provider subsequently comes forward, state law requires a forced sale or forced lease or the District's network at fair market value.		
Colorado (*Colo. Rev. Stat. Ann. 29-27-201, -304*)	Cable, telecom, advanced services (Internet access > 256kbps upstream or downstream)	Municipalities can provide cable, telecom, or advanced services if the area is unserved as defined by Colo. Rev. Stat. Ann 29-27-202. If a municipality wants to provide these services, a referendum is required.		
Florida (*Florida Statutes 350.81, 125.421*)	Advanced service (Internet access > 200kbps upstream or downstream), cable, telecom services	Two public hearings, no less than 30 days apart with specific notice requirements; must consider statutorily imposed issues at these hearings; must make available the business plan; must have referendum after hearings; must hold annual meetings to review progress.	No cross-subsidization; if revenues do not exceed operating costs and payment on debt after four years, hearing must be held to determine network's fate under §350.81(2)(*l*). Florida imposes ad valorem taxes on municipal *telecom* services.	Municipality must not favor itself or give itself favorable pricing on rights-of-way or pole attachments; municipalities must comply with all state and local rules.

continued on following page

Table 1. Continued

State	Technology regulated	Administrative Burdens	Financial Burdens	Other Burdens or Requirements
Louisiana (*La. Rev. Stat. Ann. 45:844:41, -56*)	Telecom, Advanced services and Cable Television services, regardless of type of technology	If a municipality wants to provide these services, it must first hold a "preliminary" public hearing, then must hire a feasibility consultant, and determine when annual revenues will exceed costs. It must complete a feasibility study (with statutory requirements), and have two more public hearings. Then, a referendum must be held.	No cross-subsidization; must impute costs of private provider.	Municipality must follow all state and local laws; must not unduly or unreasonably favor itself; municipality is allowed to bundle services as a private provider would. Municipal provision of service suspends all private franchises.
Michigan (*Mich. Comp. Laws Ann. 484.2252*)	Telecom services: "regulated and unregulated services offered to customers for the transmission of 2-way interactive communication and associated usage"	A public entity (county, city, village, township, agency, or any subdivision of these) cannot provide telecom services unless it first: (1) requests private bids, (2) after 60 days, receives fewer than three bids, and (3) offers the same service as requested under part (1).		
Minnesota (*Minn. Stat. Ann. 237.19*)	Telephone	A municipality must obtain a super-majority of 65% of voters in a referendum before providing telephone services.		
Missouri (*Mo. Rev. Stat. 386.020, 392.410(7)*)	Telecom services			No political subdivision of the state shall provide telecom services except (1) for its own use, (2) for 911, E-911 and other emergency services, (3) for medical or educational purposes, (4) to students at an academic institution, or (5) Internet-type services (undefined)
Nebraska (*Neb. Rev. Stat. Ann. 86-575, -594*)	Broadband, Internet, Telecom, or video services			Municipality not allowed to provide broadband, Internet, telecom or video service. It may lease excess capacity in the form of dark fiber. If network is for education purposes, it may qualify under Educational Service Units Act.
Nevada (*Nevada Statutes 268.086, 710.147*)	Telecom services			The following municipalities cannot sell telecom services to the public: (1) a city with a population > 25,000; (2) a county with a population > 55,000.
North Carolina (*NC Statutes 160A-311, -339*)	Cable, video, telecom, broadband, or high-speed Internet	If a city provides these services, directly or indirectly, to the public for a fee, then it must follow these requirements: (1) solicit proposals from private businesses to partner with government, and *must* negotiate with them (municipality cannot provide the network unless no viable private provider can); (2) hold two public hearings with notice; (3) conduct a feasibility study; (4) hold a referendum before funding can be authorized.	No cross-subsidization; must impute costs of private provider.	No special favor for municipality; network must be open access. **Exceptions:** these obligations do not apply if area is "unserved" defined as a census block in which at least 50% of households have no high-speed Internet access or can only access it through a satellite provider. Municipality must petition state to be deemed "unserved."

continued on following page

Table 1. Continued

State	Technology regulated	Administrative Burdens	Financial Burdens	Other Burdens or Requirements
Pennsylvania (*66 Pa. Cons. Stat. Ann. 3014(h)*)	Telecom, advanced, broadband services	No municipal networks can be provided to the public for a fee unless the municipality has requested the local telephone company to provide that service (same data speed only), and the company has rejected that request. If the company accepts the offer to provide the service, it has 14 months to do so.		
South Carolina (*SC Code Ann. 58-9-2600, -2670*)	Telecom (broad: includes Internet services as well as telephone), broadband (speed > 192kbps)		No cross-subsidization; must impute costs of private provider; must prepare annual audit.	No special favor for municipality. **Exceptions:** municipality is not subject to these rules when providing service in "unserved" area. "Unserved" area is defined as either (1) in a persistent poverty county, at least 75% of households have no broadband access or only satellite broadband access, or (2) in all other counties, at least 90% of households have no broadband access or only satellite broadband access. "Persistent poverty county" is defined by the USDA as 20% or more of residents were "poor" as measured by last four decennial censuses. South Carolina also is harsh in that it does not grandfather in current networks, only specific networks that already received federal funding. NOTE: if municipality provides free Wi-Fi service, none of these obligations apply.
Tennessee (*Tennessee Code Ann. 7-52-401, -410, 7-52-601, -611, 7-59-316*)	Telecom, cable, Internet, video services	Municipalities acting through *electric utilities* can provide cable, video, Internet, and telecom services if they file a detailed business plan, hold a public hearing, and hold a public referendum. Municipalities not acting through electric utilities are limited to providing services in "historically unserved areas," and only through joint ventures with private firms.	*Telecom* services cannot be cross-subsidized.	Municipalities must follow all laws and rules regarding rights-of-way, pole attachments, and allocations of costs for rates and insurance.
Texas (*Texas Utilities Code 54.201, -2025*)	Telephone service			Municipality or municipal electric system is not allowed to offer telephone/telecom service to the public directly or indirectly through a private telecom provider. Municipality can lease unused (dark) fiber.

continued on following page

Table 1. Continued

State	Technology regulated	Administrative Burdens	Financial Burdens	Other Burdens or Requirements
Utah (*Utah Code Ann. 10-18-201,-306*)	Cable television, telecom service	If a municipality wants to provide these services, it must first hold a "preliminary" public hearing, then must hire a feasibility consultant to conduct a feasibility study (with statutory requirements), and have two more public hearings. Then, a referendum must be held.	No cross-subsidization; must impute costs of a private provider; must publish rates in a newspaper; municipality cannot receive funding under Utah's Universal Public Telecommunications Service Support Fund, established in 54-8b-15.	Municipality must not unduly or unreasonably favor itself; must follow all laws and rules that would normally apply.
Virginia (*Virginia Code 56-265.4:4, 56-484.7:1, 15.2-2108.5, -.11, 15.2-2160*)	Cable, telephone, broadband service	For *cable services*: must hold a preliminary hearing, then hire a feasibility consultant to conduct a feasibility study evaluating numerous factors, including whether average annual revenues in the first year (and five years out) will exceed average annual costs; must hold public hearings on feasibility study, must then hold referendum. For *telecom services*: must obtain certificate under 56-265.4:4. For *other services*: if city not acting through electric utility, and has fewer than 30,000 people, can offer high-speed data service <u>but not</u> cable service, and it subject to the state's approval under 56-484.7:2.	Municipal *electric utilities* can provide telephone and other communications services (except cable) so long as they do not subsidize the service, and they impute private-sector costs, do not charge rates lower than incumbents, and must undergo other statutory requirements.	
Washington (*Wash. Rev. Code Ann. 54.16.330*)	Telecom services		For *wholesale services*: rates and terms are not to be discriminatory, and no cross-subsidization.	A public utility district may maintain a telecom network for two purposes: for internal needs, and for provision of *wholesale* telecom services.
Wisconsin (*Wis. Stat. Ann. 66.0420, .0422*)	Video, telecom, broadband services	No local government may provide these services unless it holds a public hearing with notice, and makes available a report wherein the local government predicts the financial situation of the network for at least the first three years. Exceptions for *broadband service*: if local government asks incumbents whether they are providing, or intend to provide within 9 months, service in the area, and those incumbents do not respond in writing within 60 days, these administrative requirements are not required; OR if the local government plans to provide *wholesale* access only and there is not more than one competitor.	For *cable*: no cross-subsidization. For *telecom*: must pay local treasury back as soon as practicable out of revenues from the network.	

Chapter 4
The Social and Economic Benefits of Nepal Wireless

Mahabir Pun
E-Networking Research and Development, Nepal

ABSTRACT

Information and Communication Technology (ICT) has become a vital instrument for delivering a number of services such as education, healthcare, and public services. Community wireless networks are community-centric telecommunication infrastructures developed to provide affordable communication for those who live in remote areas. This chapter discusses the role of Nepal Wireless in achieving socio-economic development of rural communities by facilitating affordable Internet access. In particular, the authors discuss the philosophy and objectives of the project, used network technology, financial resources, and management structure. In addition, the chapter discusses its key services including e-learning, tele-medicine, e-commerce, training, and research support. The authors also analyze the challenges Nepal Wireless faced and articulate on the approaches it took to address those challenges. These challenges include lack of technical skills, selecting appropriate technology, ensuring funding resources, difficult geographical terrains, unstable political situation, and expensive devices. They conclude the chapter with some suggestions for policy makers, community developers, and academicians.

BACKGROUND

Governments and individuals in developing countries usually give higher priority to fundamentals of living such as healthcare, education, agriculture, clean water, roads, and other public infrastructures. That is greatly overshadowing the need for developing Information and Communication Technologies (ICTs) in remote areas where the majority of population live (Aitkin, 2009). As a result, ICTs services are available mostly for the people living in big cities and their peripheries.

DOI: 10.4018/978-1-4666-2997-4.ch004

This group of people is known as 'haves.' On the contrary, access to information and online resources in most of rural areas in developing countries is limited (Heeks & Kanashiro, 2009). Therefore, people living in these areas cannot explore the opportunities. This group of people is known as 'have-nots.' This is one of the reasons why there is a huge digital divide between people living in rural and urban areas. The difference of socio-economic opportunities between 'haves' and 'have-nots,' due to the technology is pronounced as digital divide (Herselman & Britton, 2002; Cullen, 2001). The concept of digital divide is not limited to the physical access of technology, but it is extended to social, political, economic, human, and cognitive spheres (Heeks, 2008; Warschauer, 2003).

However, besides having major inequalities across the regions and continued gaps in access to ICT, such tools and techniques can have positive impacts on development of the poor communities (Hamel, 2010; Nair & Prasad, 2002; Unwin, 2009). The cheaper version of new technologies can create conducive opportunities in the rural and remote areas. Studies show that proper implementation of the ICT services can facilitate socioeconomic development (Jensen, 2007; Díaz, Andrade, & Urquhart, 2009; Heeks & Kanashiro, 2009), and can reduce the digital divide (Zheng & Walsham, 2008).

Following a similar approach, the Nepal Wireless was established to narrow the digital divide prevailing in the mountain regions of Nepal. Before describing the details of the conception, implementation, and achievements of the projects, the following section gives a brief background of the ICT scenario of Nepal. Thereafter, subsequent sections will describe how the Wireless project started from one village and extended to more than 150 villages. Furthermore, the chapter assessed the achievements of the Nepal Wireless using Unwin's (2005) model of successful ICT4D projects.

Nepal is a small country located between China and India. It has about thirty million people. It is mostly a mountainous country with majority of the highest mountains of the worlds. Politically, Nepal is the youngest republics of the world, which is officially called the Federal Democratic Republic of Nepal (CIA, 2012).

A brief overview of ICTs services that are presently available in Nepal has been given in order to make readers understand clearly the reason behind the evolution of Nepal Wireless in 2001. Even if the mobile penetration rate has increased significantly in recent years in Nepal, many people in rural areas are still deprived from communication services including the Internet. Table 1 shows that in 2011 about 55% of Nepali people do not have telephone services and almost 90% of them have no access to the Internet (NTA, 2011).

Nepal Wireless started in 2001 as a grassroots project to connect some of the mountain villages of Nepal. The projected population of Nepal for 2010 was 28,584,975. Table 2 shows some statistics of the growth of telephone and Internet users between 2003 and 2011 (NTA, 2003).

SETTING THE STAGE

Prior to the start of Nepal Wireless in 2001, rural residents and remote communities had very limited knowledge about computer and Internet. In addition, there was no telephone or electric infrastructure in the villages. People had to walk five to eight

Table 1. Percentage of people using telephone and internet services in 2011

Voice Services being provided by different operators	Users	Penetration Rate %
Fixed	837,705	2.93
Mobile	11,299,735	39.53
Others (LM, GMPCS)	695,893	2.43
Total	*12,833,333*	*44.90*

Table 2. Telephone and internet users in 2003 and 2011

Telephone and Internet Users	2003	2011
Fixed Telephone Services	379,235	837,705
Mobile Telephone Services	88,551	11,299,735
Others Voice Services (LM, GMPCS)	5470	695,893
Email Subscribers and Internet Users	180,000	2,940,623

hours to the nearest city just to make a telephone call. Nepal Wireless started as an extension of several community development projects that have been developed since 1993. These projects include building schools, starting programs for income improvement, building healthcare clinics, and protecting the nature and the green coverage. Examples of the income improvement programs are yak farming, cheese making, yak and cow crossbreeding farm, fishery, paper production, camping and lodging facilities for trekkers, jam making, and mushroom farming, etc.

In the absence of modern telecommunication and transportation means before 2001, people of Nangi had to use human messengers or face-to-face communications to exchange messages, run their business and practice their daily life affairs. Unfortunately, this inefficient and costly way of communication still the most widely adopted way in remote corners of the world.

Thus, Nepal Wireless was developed to fulfill the desperate need of remote and underserved communities for electronic communications and services as well as access to online resources. However, the political situation at that time was not favorable. Therefore, Wi-Fi equipment had to be smuggled with the help of international volunteers. In addition, team members had to develop homebuilt and low-cost antennas and test them at different ranges. Because of lack of knowledge and

limited funds, the project had to adopt such low-cost solutions. We also obtained donated Wi-Fi equipment, normal hubs and recycled television satellite dishes. As for the power needed at the repeating stations, solar panels and wind generators were used to charge the storage batteries.

During the first year of the project, Internet connectivity was provided through a dial-up proxy server based in the city of Pokhara. Five villages shared the dial-up Internet link. In 2004, the Internet link was upgraded from dial-up to 64/64 kbps dedicated wireless link provided by the Internet service provider based in Pokhara. Now Nepal Wireless subscribes to a 5 Mbps Internet link from the Internet Service Provider to its base stations in Pokhara and Kathmandu. The base stations are running a Web server, a proxy server, a network management server, an Internet bandwidth management server and an audio/video streaming server.

Over time, a broad range of electronic services and applications such as e-learning, tele-medicine, tele-teaching, tele-training, e-government, and e-commerce were introduced. Moreover, Nepal Wireless supports a number of research groups for monitoring glacial lakes and studying weather and climate change in the Himalayas. It is also working on developing a surveillance system based on IP cameras at Chitwan National Park for the purpose of monitoring hunters and save endangered animals like tigers, leopards, and rhinos

PROJECT DESCRIPTION

The idea of using wireless networks to provide affordable local communication services to some mountain villages evolved few years before the real start of Nepal Wireless. In particular, community developers informally implemented a pilot project for testing wireless technology in Nangi in 2001. Nangi is a small mountain village of 800 inhabitants. It is located in the foothills of the Annapurna and Dhaulagiri Himalayas of western

Nepal at 7,300 feet elevation. It takes pedestrians about 7 hours to climb the mountain from the nearest highway to Nangi. The low population density and harsh geography of Nangi make it less attractive for telecommunication companies to invest in it. In addition, Nangi is an isolated and remote village that lack infrastructures and economic opportunities other than farming. More specifically, its residents are subsistence farmers who use primitive farming tools such as wooden plows, iron spades, axes, and sickles, etc. Moreover, there are no roads in the area. Therefore, villagers carry all kinds of loads such as supplies, firewood, building materials, composts and others on their back. This inefficient way of living daily life has been used for centuries. For these reasons, providing affordable Internet access is crucial for these communities to benefit from e-learning, e-commerce, e-government, e-entertainment, and other online resources and services.

In 2001, when the plan was made to connect the Nangi village to Pokhara city to bring the Internet through Wi-Fi, the responses from several communication engineers and experts in the communication field were negative. Their main concern was the lack of proper equipment and the long distance that the radio signal had to cover. In spite of the negative feedback, the team members of Nepal Wireless had decided to go ahead for doing field experiments to find if a long-range wireless link can be made using simple Wi-Fi radios and homebuilt antennas. The experiment was successful to the surprise of many technical experts who had been skeptical of the project. As a result, Nepal Wireless has now expanded its network to more than one hundred twenty villages in thirteen districts of Nepal. It is expanding its network to more districts of Nepal in coming years.

The following is a brief discussion of the objectives of the project, network infrastructure, funding resources and business partners, technology used, and server setup has been given below to explain the technical part of the network:

Objectives of the Project

Our project was guided by the success factors for partnerships of Information and Communication Technology four Development (ICT4D) projects developed by Unwin (2005, pp. 64-67). These factors are trust, focus, champions, leaders, sustainability, balance between demand and supply, networking, and transparency and sound ethical basis.

The goals and objectives of the project have evolved over several years working on community development. At the initial stage of the project, the goal was to connect just a few neighboring villages of Nangi and provide communication services such as IP telephony and email communication. The founders of the project were not aware of the full capabilities and potential services of wireless networks. New objectives and services have emerged with experience and interactions with similar community wireless networks. For example, it was found that the minimum intranet bandwidth in local wireless network was around 2Mbps, which was enough to run simple video conferencing application for teaching and training purposes. Therefore, the current objective is not only to bring the Internet and computers in the rural areas but also to maximize the benefits of wireless and information technology for rural population in remote areas. This is done by introducing useful applications and services in the field of education, health, communication, e-commerce activities and climate change monitoring activities. We also hope that using the wireless technology would make the life of rural people easier and bring some socio-economic transformation. In other words, our overall long-term goal is to become one of the biggest rural Internet Service Providers in Nepal that help bridge the digital divide. The digital divide, in this context, expands beyond physical access to the Internet and computers to include the social, political and economic differences (Warschauer, 2003).

Nepal Wireless has identified the following objectives (E-Networking Research and Development, 2009, pp. 10-11):

1. **Education:** One of our key objectives is to offer better educational opportunities in the rural communities by creating tele-teaching and tele-training programs. In addition, we facilitate the delivery of e-learning materials in local language to students, teachers, and community members through e-libraries and online services.

2. **Health:** Another objective is to connect rural health clinics and health workers to city hospitals in order to provide quality medical assistance through tele-medicine programs. This includes providing healthcare to rural communities by virtually bringing medical doctors to villages.

3. **Communication:** Increasing communication facilities in rural areas by providing telephone services through Internet phone system (VoIP). We also work on making Internet available for email communication, and help villagers to communicate using Nepali language. Using bulletin boards for community discussion is also one of our tools to improve civic engagement and build the spirit of community.

4. **Local E-Commerce:** Nepal Wireless helps villagers to buy and sell their products in both the local and international markets through local intranet and Internet.

5. **Job Creation:** We aim at creating jobs in the rural areas for younger generation through ICTs related services such as communication centers, VoIP phone services, remittance services, and virtual ATM machine.

6. **Research and Field Testing:** Another emerged objective is to help researchers of climate change monitoring projects to collect data remotely and provide real time weather information about air routes in the Himalayan valleys. This real-time delivery

of information is crucial, particularly for airlines, during bad weather conditions and the monsoon season. Moreover, we also provide technical support to the Department of National Park of Nepal. Specifically, we installed surveillance system at the parks to monitor the movement of the poachers to save the endangered species in the park.

The Network Infrastructure

In the beginning, Nepal Wireless did not have any funding to buy high-end equipment for building wireless network. It had to utilize mostly used equipment including wireless radios, switches, power sources, and computers donated by individual donors. Now, the network has grown bigger and better equipment and applications have been added as mentioned below.

Nepal Wireless network was built step by step and expanded to many villages over a time span of about ten years and it is still growing. The villages that are connected to the network get Internet and intranet services via a server in Kathmandu or in Pokhara. The servers in Kathmandu and Pokhara are linked through a leased optical fiber line. We also installed a series of mountaintop repeater stations that relay the signal from Kathmandu and Pokhara up into the mountain villages. There are also access points at the mountaintop that play as relay stations to distribute Internet to end users in the neighboring villages. Villages are connected to the access points using point-to-point wireless links or by connecting to a wireless access point that serves many villages at once. Table 3 shows the growth of users until 2011.

The high-speed backhaul radios at the relay stations operate on a dedicated core Local-Area Network (LAN) that reaches from the base stations (Pokhara and Kathmandu) to different districts through relay stations. The longest point-to-point link the project has made is 42 km from a mountain top to another mountaintop. The distance between access points ranges from 2 km to 18

Table 3. Progress of Nepal wireless networking project from 2002 to 2011

Year	2002	2003	2004	2005	2006	2007	2008	2009	2010	2011
No. of Villages	1	2	5	13	22	35	60	82	102	122
No. of Rural Schools	1	2	5	8	19	24	40	58	74	80
No. of Rural Clinics	0	0	0	2	3	5	5	6	8	10
No. of Individual Users	0	0	0	0	0	0	0	15	40	65

km. We divided the districts that are connected to the network into different subnets in order to manage the network smoothly. Thus, served villages use relay stations that operate on separate local LANs through VLAN switches. Routers at each of the relay stations provide DHCP services to end users and serve as interfaces between the backbone LAN and the local distribution LANs.

Transport Technology

A considerable amount of resources have been dedicated over time to the network infrastructure and management. This includes wireless devices, network servers and associated software, and power generation systems at the relay stations. The transport devices that are being used in the network are high speed and long-range backhaul radios, access points and cheaper client radios that are available in the market, directional and Omni directional antennas with different capacities, routers, manageable switches, normal switches, and IP phones. Some of the Omni directional antennas were built by volunteers and staff. Thus, the wireless network has used a variety of wireless devices to maintain connectivity between the nodes. Conceptually, the network has been divided into two parts, as follows:

1. Network backbone, which connects base stations in Pokhara or Kathmandu to a series of mountaintop repeater stations.
2. Client connections that connect villages to the relay stations. In some cases, a connected village has also acted as a relay station.

Proprietary devices made by Alvarion and Motorola Canopy at 5.8 GHz bands were used for the backbone connection due to their high reliability, robustness, and low signal interference. However, several brands of 802.11b/g standards such as smartBridges, EnGenius, Ubiquity, Deliberant, Mikrotik, and TP Link were used for the last mile connections due to their lower cost and the compatibility between manufacturers. All the wireless equipment including backhaul radios, access points, client radios and wireless routers installed by Nepal Wireless use de-licensed frequencies of 5.8 GHz and 2.4 GHz.

Access Technology

Before the wireless network was built for Internet connection, most of the "computers at the schools were assembled in wooden boxes" (BBC, 2001). They have been built using used computers parts brought by volunteers coming to Nepal. Now access to IT services is provided through modern desktop computers and laptops. With the addition of an Internet telephony system, the project added a number of network telephones in the villages. This Internet telephony enables users to make international phone calls at cheaper rates.

Network Servers

We set up several servers for different purposes such as Web server, sip server, proxy server, network management server, Internet bandwidth management server, and video conferencing. For example, the network server in Pokhara facilitates network monitoring and management. The video conferencing server is dedicated for tele-teaching, tele-medicine, and tele-training purposes only. All of the server computers are branded computers. The servers run Ubuntu with additional third party software. They are configured for maximum redundancy to guard against failure. We choose Ubuntu because of a number of reasons such as compatibility, features, and availability for free. In addition, it is well tested and proven in production environments and thus made a perfect choice for the server.

Currently, the servers provide the following services:

1. Voice over IP Phone system for local use through SIP server and Internet telephony.
2. Community (Nepali) bulletin board service and locally hosted home pages for the villages through the Intranet.
3. File sharing service through the server as the central point of coordination for sharing teaching materials and network maintenance software for the network.
4. Database for additional projects on the network, like the Haat Bazaar trading forum, which is hosted on the server.
5. Network name and route monitoring through which the system can resolve IP addresses to names and also maintains a table of availability for each host on the network.
6. Video conferencing for tele-teaching and tele-medicine.

The server also contains dozens of custom scripts to tie all the functionality together in a usable interface. Administering such complex functionality, even from a GUI interface like WebMin, is too difficult for the average user. Thus, we developed customized scripts using Perl and PHP to provide a user friendly server management system.

Funding Resources

After the successful completion of testing phase of the long range Wi-Fi link in May 2002, Nepal Wireless started getting some financial assistance and some in-kind support from individual supporters and organizations from around the world. Some organizations supported the wireless project directly and some did it through the Himanchal Education Foundation in Nebraska, USA. The donors, who helped the project at the initial years, are the International Center for Applied Studies in Information Technology (ICASIT) in Virginia, the Huguenin Rallapalli Foundation in California, and the Poverty Alleviation Fund of Nepal. Moreover, the International Telecommunication Union (ITU), AMD, and Asia Pacific Tele-Community (APT) provided financial support for buying some more equipment in order to expand the network to reach more villages and for doing a pilot project for tele-medicine.

The International recognition for the team leader of Nepal Wireless helped to get some more support for the project from more individuals and organizations. For example, the project received a number of used computers and wireless network equipment from several individuals and international companies like KDDI Japan, Cisco U.S.A., National University of Singapore, Polycom Australia, and Swiss Re, Switzerland. In 2009, the project got financial support from ISIF Asia for building its human capacity. In addition, the project received financial support from the Juniper Foundation, U.S.A. for the purpose of providing training for network operators and computer teachers. Most recently, Asia Pacific Tele-community provided a research grant for developing an open-source multi-destination audio/video conferencing system for tele-teaching, tele-medicine, and tele-training purposes.

Therefore, Nepal Wireless started a campaign in July 2008 called "Donate One Dollar a Month to Help Build Wireless Broadband Information Highway in Nepal" with the help of some Non-Resident Nepalese (NRNs) residing overseas. The project has received more than US$50,000 so far as contribution from countries like the U.S.A., the U.K., Japan, Hong Kong, and Malaysia donated by NRNs. We also decrease our dependence upon external support by encouraging local stakeholders such as community members, local schools, not-for-profit organizations, and local governments to donate money and provide technical support.

One of the reasons the NRNs contribute to the project is that Internet success provides them the opportunity for to connect to relatives and family members in their home villages in Nepal. In particular, they can send emails, use low-cost VoIP phone services, and benefit from live audio/video conferencing systems such as that of Skype and Google. The donations of NRNs funded the expansion of the network in more than fifteen villages to the wireless network and build computer labs in affiliated schools.

Business Partners

Nepal Wireless was started entirely with the help of technical volunteers from abroad. It still lacks high-technical expertise and adequate human resources necessary to fully achieve the objectives of the project. Therefore, Nepal Wireless works with partners that have different expertise and resources.

Despite the generously of donors and supporters, the received financial support was not enough to build a wireless network across the entire country.

Moreover, Nepal Wireless is continuously getting technical support from many national and international volunteers. These volunteers either come to Nepal for a short stay to provide technical support or provide online support from abroad for managing the wireless network, and for developing applications that the serve the specific needs of the community.

The following are the key partners of Nepal Wireless:

1. **E-Networking Research and Development:** This organization has been helping as a technical partner since 2006 to build wireless networks in rural areas and to provide training for local wireless technicians, network operators and computer teachers. It is helping the tele-teaching and tele-training program to rural areas through audio/video conferencing (http://www.enrd.org). One of our key collaboration projects is developing video conferencing software for twenty five high schools in seven districts of Nepal.

2. **Open Learning Exchange – Nepal:** It develops contents in Nepali language for the elementary level student and makes the contents accessible online for the students in community schools. It has also developed e-library contents in Nepali language for the students and villagers. Nepal Wireless is working with this organization to connect the schools and make the contents available (http://olenepal.org) in the rural areas.

3. **Nepal Research and Education Network:** It uses the network as a platform for Research and Development (R&D) and is supporting for technical issues like network designing, equipment testing, routing, server building and maintenance (http://nren.net.np).

4. **Kathmandu Model Hospital:** Doctors in this hospital use the network as an electronic facility for delivering tele-medicine services and tele-training programs for rural health workers (http://www.phectnepal.org).

5. **Thamel.com:** This entity has created a virtual ATM system for implementing credit card transaction services for the trekkers in trekking routes of Annapurna region (http://www.thamel.com).

6. **Kathmandu University:** Engineering and computer science students volunteer to develop applications needed for the project

such as local e-commerce and phpBB applications, and customizing them in Nepali. The students are also helping to develop video conferencing software for tele-teaching and tele-training using open source.

7. **Japan International ICT Association (JIIA):** *JIIA* provides technical support by frequently sending their experts to Nepal. JIIA also collects used computers and networking gear for the project (http://www.jiiasec.com).

Project Management

The Himanchal High School collaborated with the E-Networking Research and Development group to develop and manage the network. It managed the Nepal Wireless Networking Project only until 2006 because the wireless network was limited to few villages and not many services were provided. The school did not have a license to become an Internet Service Provider. Therefore, it did not provide services directly to the end-users of community members. Instead, the services were provided to end-users through the village communication centers. Each of the centers was managed differently and independently from others. The project did not collect revenues directly from end users. Instead, it billed communication centers.

Management committees have been formed from community members to manage different programs and services provided by Nepal Wireless. These include caretaker organizations, mother's groups, social clubs, school management committees, and communication centers. By running the communication centers, the respective caretaker organizations also gained an opportunity to share in the revenue generated from usage fees. Additionally, the telephone users were charged a very low surcharge on telephone calls (Approx. $0.01 per minute more than is the surcharge of Nepal Telecom). This additional fee, profit, is used for covering the operation cost of the network.

In order to comply with the rules and regulations of the Nepal Government and to become a legal rural Internet service provider, Nepal Wireless was registered as a not-for-profit sharing company on July, 2009, from the Office of Company Registrar. The company consists of a board with five board members. As a nonprofit sharing company, it can provide IT services, technical support to build wireless networks and training for bringing ITCs related services.

Until it became a nonprofit sharing company, Nepal Wireless provided connectivity only to rural schools, healthcare clinics, and community communication centers. Currently, it provides services to individual home users in the villages if they can afford to have it. Still, villagers who cannot afford to have computers and Internet connection at home can get access to the Internet through communication centers in each village or school. Nepal Wireless charges a monthly connectivity fee directly to schools, communication centers, and local service providers. This fee is just to cover the operating and maintenance cost of the network. This includes monthly cost for the Internet bandwidth, rent for the server rooms, electricity bills, and salaries for the managers of repeater stations and the fulltime technical staff.

In addition, Nepal Wireless also works with local Internet service providers to provide Internet access to home users. A group of individuals or schools or organizations can form a committee to become a local ISP. Local representatives from the villages have been put in the management committee so that they can come up with their own ideas and make their own plans on how to use and maintain the wireless network.

Thus, an organizational structure has been developed in which public and private partners have taken responsibility in managing the network and providing related electronic services, and supplying technical support. The reason to make public partners involved in the management of the network is to develop Nepal Wireless as a business enterprise with public and private stakeholders. Now, there are many community stakeholders involved including local schools, local businesses, local governments and individual partners. This allows an avenue for democratic participation as well as sharing of risk and benefits.

PROVIDED SERVICES

Nepal Wireless provides a wide range of services. The following section discusses the most important ones:

E-Learning

Nepal Wireless facilitates the delivery of a wide range of e-learning content in schools connected to the wireless network. This content is produced by Open Learning Exchange Nepal (OLE-Nepal). The students of grade two through six use interactive learning materials based on their textbook. So far, OLE-Nepal has developed content for English, Math, Science, and Nepali language. Moreover, OLE-Nepal has also created an e-library (http://www.pushtakalaya.org) with content in both Nepali and English languages for students and community members as well. Nepal Wireless is teaching villagers how to use this content. Currently, teachers and student can access this content through intranet. Nepal Wireless has also created a Wiki in Nepali language with more than four hundred useful articles on different topics such as health, education, agriculture, and technology. In addition, an audio/video conferencing application is under development using open source software and content for e-learning purposes. These resources and services create a new paradigm in the educational system of Nepal. They also assist in overcoming education related problems such as lack of printed content and shortage in qualified teachers in rural areas.

Tele-Medicine

We also use Nepal Wireless to overcome the problem of the absence of clinics and doctors in rural areas. Specifically, we use audio-video conferencing to connect rural healthcare practitioners with their peers in city hospitals. We started this program in 2006 by connecting one clinic in a village to a hospital in Pokhara. Currently, eight rural clinics have been connected to a hospital in Kathmandu. As a result, healthcare providers can consult with medical specialists at a city hospital whenever they need help to treat patients. With this service, Nepal Wireless has assured a virtual presence of doctors in remote areas. Providing virtual healthcare is faster, cheaper, and more efficient than utilizing traditional clinics and hospitals. Building clinics in these remote and geographically harsh areas is very expensive and usually takes several years. In addition, the area is not attractive to doctors who prefer high populated areas whose residents receive high income.

E- Commerce

For the long-term sustainability of any ICT4D project, income generation activities are important (Kumar & Best, 2006; Sein, Ahmad, & Harindranath, 2008). In this context, connected people, particularly farmers, also use the Internet to communicate with other villagers to tell about their products. Villagers use an application called Haatbazar to build a virtual market where they can put contact information and description, including images, of their products. It is actually a virtual marketplace, accessible in the network through intranet. In addition, community members and local businesses have used the Internet to promote their local areas for tourism. The development team of Nepal Wireless collaborated with the community and developed a room booking system for local lodges. Nepal Wireless also served as an infrastructure for money transfer stations in some of the villages. Moreover, Nepal Wireless provides virtual ATM service to trekkers with the help of Thamel.com so that trekkers can get cash in the villages or pay their bill using their credit cards.

Technical Training

Human resources development is also highly needed for technical sustainability. Nepal Wireless frequently runs training programs to provide training in the design, setup, running, and maintenance of used servers, repeaters, routers, switches, and other related devices. The offered training session include setting up access points and clients, managing bandwidth, IP sub-netting, line of sight calculation, link budget calculation, firewall configuration, etc. We also run basic and intermediate level computer literacy programs for villagers, teachers, and operators of communication centers.

Research and Field-Testing

Nepal Wireless facilitates research on climate change. In particular, researchers can access real-time data on weather and climate change. They also connect sensors and field servers set up in remote mountain sites to the Internet through our network. Three weather stations have been installed to date in the remote Himalayan Mountains. We also work with Wild Land Security (http://wildlandsecurity.org) in the U.S.A. to set up an IP based surveillance system developed by them for monitoring the movement of poachers in one of the national parks of Nepal to help protect endangered animals. The developers of Nepal Wireless are also developing a cheaper Electro Cardio Graph (ECG) machine in its laboratory. Once it is developed, the machines will be put in rural clinics and will be used for telemedicine.

CURRENT CHALLENGES

Nepal Wireless really contributes significantly to the socioeconomic development of Nepal through a wide range of e-services and online resources. Given Unwin's (2005) model of successful partnership, Nepal Wireless lacks some components such as full sustainability, and balance between supply and demand. Our future plans focus on addressing these shortcomings. We also face other challenges including:

1. **Lack of Technical Expertise:** Lack of technical expertise and know-how was certainly the first key challenge. As with developing societies, the served communities have high illiteracy rate and lack of knowledge about emerging technology such as wireless networks.

2. **Lack of Appropriate Technology:** Wi-Fi was just an emerging technology in 2001 and it was primarily built for indoor use. Therefore, several affiliated communication engineers did not believe that it would be possible to build long-range wireless network using the indoor Wi-Fi equipment available in the market. The Wi-Fi equipment that was donated to the project to build the first 40 km long wireless link with a 3,300m tall mountain between Nangi village and the city had just 60 mW transmitter power without any amplifier and long-range antennas.

3. **Difficult Geographical Terrains:** Geographical barriers created by the high mountains also added difficulties and barriers towards accomplishing the project. Many of the villages were not in Line-of-Sight (LOS) because of the mountains. Moreover, the repeater stations that had to be installed on the top of the mountains were far from the villages. Therefore, powering the equipment, ensuring its security, and providing power supply were difficult.

4. **Unfavorable Political Situation:** When the project started, there was a severe political instability in Nepal. This instability and conflict reached rural Nepal. Therefore, the project was facing threats of closure. In addition, community members were reluctant to participate due to fears of violence from authorities and the Maoist rebels.

5. **Legal Issues:** The autocratic regime of the king had severely restricted import and use of wireless networking equipment. It was almost impossible to obtain wireless equipment from abroad because of the restrictive laws and license requirements. In addition, there was a huge license fee required from the government to become an Internet Service Provider requires paying expensive license fee.

6. **Expensive Internet Cost:** The monthly connectivity fee during the early days of the project was about USD 2,500 per month for a 1 Mbps link. Therefore, the project had to use a dial-up link at the beginning. Although the cost of Internet bandwidth has fallen to approximately USD 500 per Mbps per month in 2011, it is still expensive. The 5 Mbps Internet bandwidth Nepal Wireless has subscribed to provide to more than 120 villages is too little for the villagers. As a result, Internet bandwidth is a bottleneck for the project to provide IT services smoothly in an affordable and reliable way manners.

7. **Sustainability of Funding Resources:** This is the major challenge many of the organizations like Nepal Wireless have been facing (Thapa & Sein, 2010). The challenge is not only to make a project financially sustainable but also to make it technically sustainable through upgrading the technology.

After the restoration of democracy in 2006, we allied with the Association of Internet Service Providers in Nepal and formed a lobby that advocates rural ISPs. This lobby demands the government to de-license the 2.4 GHz and 5.8 GHz frequencies of the ISM band. The team leader of Nepal Wireless gave two presentations to the Members of Parliament on how Nepal Wireless had been set up and how the rural communities were benefit from its services. As a result, the Government de-licensed these frequencies in September 2006. Now the wireless equipment at those de-licensed bands can be legally imported and used.

Another regulatory obstacle to bring IT services to rural areas was also overcome. That was the requirement of license by paying high license fee to become an Internet Service Provider. This fee was almost $4,000 a year (NTA, 2010). As a result, no investor was motivated to become a rural Internet Service Provider because there was very limited return on investment due to the low population density. Nepal Wireless lobbied with the regulatory body of the government to reduce the license fee. Nepal Telecommunication Authority (NTA) issued a new law making the licensing procedure simple and decreasing the license fee to become only Nepali Rupee 100 (Approx. USD 1.50) a year for rural Internet Service Providers. Currently, Nepal Wireless has a license and it legally provides Internet access and other IT services in the rural areas.

The issue of balancing between demand and supply is still a big challenge for Nepal Wireless. Nepal Wireless receives requests from many enthusiastic villagers and local governments from around the country for help to connect their villages to the Internet. They are also able and willing to raise some funds for expansion. At present, Nepal Wireless cannot fund all of them, although it can definitely provide technical assistance. Thus, there is a huge demand that Nepal Wireless cannot fulfill.

SIGNIFICANCE FOR POLICY AND PRACTICE

Ruth and Schware (2008) identify some of the unique characteristics of Nepal Wireless in the following statement:

There are several aspects, which make the Nangi case unusual. First of all, it was not initiated through government, NGO or multilateral interventions. Second, it was managed entirely by Nepalese local leaders, with assistance provided through skillful selection of outside helpers, mostly volunteers, from developed nations. There was no role of the USAID, UNDP or World Bank in the project. Third, the hardware used was top-of-the-line, and not a compromise among cheap or aging technologies. Fourth, the cost of the project was relatively modest, since there were no "middle-man" charges and because several equipment suppliers were so impressed with the Nangi concept that they gave large price concessions. Perhaps the most unusual aspect of the case is that the entire effort was not driven by an application, like licensing, taxation, or finance. Instead, the idea for Nangi was, "if you build it they will come," like the theme from the film Field of Dreams.

Working over a decade on the project, we can identify the following key lesson learned:

Engaging Local Stakeholders

Project leaders should engage local stakeholders such as local businesses, local community organizations, local governments, and individual community members. Engaging local stakeholders (as operators, sponsors, developers) can make the project self-sustainable. In addition, they can host access points or repeaters and solve the right-of-way problem.

Engaging Volunteers

The issue of technical sustainability has been solved by hiring some full time and part time staff after Nepal Wireless has obtained a nonprofit status in 2009. Still the number of full-time technical staff is not adequate. Therefore, there are quite a

number of people volunteering or doing internship in the project, and Nepal Wireless seems to depend more on the support of the volunteers.

Financial Sustainability

We suggest that local stakeholders should collectively fund the project. In particular, each user, whether it is an individual, a school, a business, or a clinic or a government entity, has to pay the installation cost and monthly fee for Internet access. The monthly connectivity fee is based upon the Internet bandwidth used by a user, which is from USD 15 to USD 30 a month. The contributions received from outside donors is spent to build the backbone of the network and to install back up power generators for the relay stations. Support from local government and district government also goes to build local networks and backup power systems. Nepal Wireless also provides ongoing trainings to teachers, administrators, and local people to manage and use the network effectively. Since the community members have also invested money for the installation cost, there is a feeling of ownership towards the network. While support from donor organizations are helpful, Nepal Wireless always believes in the maximum level of local participation for achieving its goal.

Replicating the Project

Facing a number of challenges and interacting with community needs and circumstances, it has been realized that it is very difficult for a single organization to bridge the digital divide in the whole country. An organization or some organizations like Nepal Wireless can help to develop a model or system and connect only a few hundred or so villages for bridging the digital divide if it works in its full capacity. However, there are over 6,000 villages in Nepal and many organizations like Nepal Wireless need to be created to bring ICTs to the Nepali society at large.

Integrating Computer Literacy in Education

Another recommendation for bridging the digital divide is integrating ICTs in education programs in all public schools. The Ministry of Education of Nepal Government recently has started an initiative for integrating ICTs in education with the support of UNESCO. Its objective is to harness the power of ICTs for teaching and learning. This initiative is part of an ongoing major education programs that include the School Sector Reform Plan, the Education for All campaign and teachers education projects (UNESCO, 2010).

Government Intervention

The intervention of governments is crucial for the success and sustainability of community and municipal wireless networks. In the case of Nepal, about 80% of the population lives in the rural areas and these people will not have affordable broadband without government intervention or subsidy. Therefore, the government must take the initiative to build broadband information highway in rural areas. In addition, it should consider providing financial support to grass roots level organizations like Nepal Wireless. One of the reasons government intervention is recommended is that it is less likely for commercial companies to invest in the remote rural areas. It is unwise for commercial operators to invest huge amounts of money to build infrastructure to bring broadband information highway and to provide Internet services in sparsely populated and scattered rural areas of Nepal because there is a slim chance for them to get a return. In addition, the low income of rural residents and the mountainous nature of the area decrease the viability of any telecommunication infrastructures.

Given the experience Nepal Wireless has gained, it is recommended that the government intervention for bridging the huge digital divide in developing countries must come both in the form of open policies and favorable spectrum regulations for the expansion of wireless or optical fiber networks, whichever is feasible, in the rural areas. In addition, governments should also subsidize the service or provide financial support to rural operators that are interested to bring Internet and IT services in the remote rural areas for which it is recommended to provide subsidized financial support for building infrastructure and for paying the cost of the Internet access. Governments can use the Universal Service Obligation Fund (USOF), which the commercial operators pay to the government.

Some countries require incumbent operators to allocate part of their net earnings for providing information and communication services to rural population. For instance, India achieves this policy through a program called Universal Service Obligation Fund in India. Similarly, Chile has the Telecommunications Development Fund (FDT). Pakistan has the Universal Service Fund Company, and Taiwan has the Universal Service Fund (Wikipedia, 2012). In the case of Nepal, such a program is called Rural Telecommunication Development Fund. The Nepal government is planning to use the fund for connecting all the 75 district headquarters through optical fiber link and then the villages through wireless link. The amount of money collected for Universal Service Obligation Fund in each country is quite a large amount. Therefore, it will be a good resource for bridging digital divide in developing, as well as, in developed countries. According to the National Broadband Plan of America, approximately 100 million Americans do not have broadband at their home. Therefore, the U.S. government has set up the "Create the Connect America Fund (CAF) to support the provision of affordable broadband and voice with at least 4 Mbps actual download speeds and shift up to $15.5 billion over the next decade from the existing Universal Service Fund (USF) program to support broadband" (Connecting America, 2012, pp. xiii).

Even the governments of developed countries, like the United Kingdom, feel that they are obliged to do more for integrating ICTs in education. In particular, it allocated £363m to improve broadband connection in rural areas of England and Scotland. In addition, the British government has brought forth a scheme to give "free laptops to 270,000 low-income people in England" (BBC, 2011). In order to target digital divide, "low-cost computers are to be offered as part of a government scheme to encourage millions of people in the UK to get online for the first time" (BBC, 2010).

Another example of government intervention for bridging digital divide is through the "One laptop Per Child Program." Under this program, "over 1.7 million children and teachers in Latin America are currently part of an OLPC project, with another 400,000 in Africa and the rest of the world" (OLPC, 2011). Similar projects have been started by governments in Sultanate of Oman, Gaza, Afghanistan, Haiti, Ethiopia, and Mongolia. There is no doubt that such programs will be able to bridge not only the digital divide but make people connected to each other and create new opportunities for remote communities (Thapa, et al., 2012). Similarly, the Open Learning Exchange Nepal initiative has successfully adopted a "One Laptop per Child Program" as a pilot project with collaboration with the Department of Education in five districts of Nepal. This pilot program was done in two phases and it distributed more than 2,500 laptops to students and teachers (OLE Nepal, 2011).

CONCLUSION

In this chapter, we discussed how Nepal Wireless brought electronic education, healthcare, and public services to remote communities of Nepal through affordable Internet access. The chapter also demonstrated how the project turned challenges into opportunities such as providing training

courses, developing hardware and software, and building the sense of community. Furthermore, the chapter highlighted the key challenges facing the project such as limited technical expertise, sustainability of fund, harsh geography, legal and regulation barriers, and high Internet cots. Finally, the chapter suggested some practical implications for policy makers, community developers, and academicians. It is important to note that the Nepal Wireless Project could not be successful without the generous support of international volunteers.

ICTs have become one of the essentials of living and the basic human needs such as food, shelter, and clothes, etc. However, there are billions of people deprived of these important tools. Therefore, integrating computer literacy in primary education should be one of the priorities of governments. Learning basic computer skills has grown to be one of the fundamental rights of children.

REFERENCES

Aitkin, H. (2009). Bridging the mountainous divide: A case for ICTs for mountain women. *Mountain Research and Development*, 22(3), 225–229. doi:10.1659/0276-4741(2002)022[0225:BTMDAC]2.0.CO;2

BBC. (2001). *Village in the clouds embraces computers*. Retrieved July 22, 2011, from http://news.bbc.co.uk/2/hi/science/nature/1606580.stm

BBC. (2004). *Wi-fi lifeline for Nepal farmers*. Retrieved July 28, 2011, from http://news.bbc.co.uk/2/hi/technology/3744075.stm

BBC. (2010). *Poorer pupils to be given free laptops*. Retrieved July 25, 2011, from http://news.bbc.co.uk/2/hi/8449485.stm

BBC. (2011). *£98 PCs target UK digital divide*. Retrieved July 22, 2011, from http://www.bbc.co.uk/news/technology-12205412

Brinkerhoff, J. M. (2008). What does a goat have to do with development diasporas, IT, and the case of thamel.com. *Information Technologies and International Development, 4*(4), 9–14. doi:10.1162/itid.2008.00023

CIA. (2012). *Nepal - CIA worldfactbook*. Retrieved 12 July, 2012, from http://www.theodora.com/wfbcurrent/nepal/nepal_introduction.html

Connecting America. (2012). The national broadband plan. *Download the Plan*. Retrieved March 26, 2012 from http://www.broadband.gov/download-plan

Cullen, R. (2001). Addressing the digital divide. *Online Information Review, 25*(5), 311–320. doi:10.1108/14684520110410517

Díaz Andrade, A., & Urquhart, C. (2009). The value of extended networks: Social capital in an ICT intervention in rural Peru. *Information Technology for Development, 15*(2), 108–132. doi:10.1002/itdj.20116

E-Networking Research and Development. (2009). *Final report of APT ICT J3 project in Nepal*. Paper presented to Asia Pacific Telecommunity. Bangkok, Thailand.

Hamel, J.-Y. (2010). *ICT4D and the human development and capability approach: The potentials of information and communication technology*. Geneva, Switzerland: UNDP.

Harsany, J. (2003). *Taking wifi to new heights*. Retrieved July 25, 2011, from http://www.pcmag.com/article2/0,2817,1365140,00.asp

Harsany, J. (2004). *Wireless networks open up Nepal*. Retrieved July 25, 2011, from http://abcnews.go.com/Technology/ZDM/story?id=99622&page=1

Heeks, R. (2008). ICT4D 2. 0: The next phase of applying ICT for international development. *Computer, 41*(6), 26–33. doi:10.1109/MC.2008.192

Heeks, R., & Kanashiro, L. (2009). Telecentres in mountain regions - A Peruvian case study of the impact of information and communication technologies on remoteness and exclusion. *Journal of Mountain Science, 6*(4), 320–330. doi:10.1007/s11629-009-1070-y

Herselman, M., & Britton, G. K. (2002). Analyzing the role of ICT in bridging the digital divide amongst learners. *South African Journal of Education, 22*(4), 270–274.

ITU. (2000). *The internet from the top of the world: Nepal case study*. Retrieved from http://www.itu.int/ITU-D/ict/cs/nepal/material/nepal.pdf

Jensen, R. (2007). The digital provide: Information (technology), market performance, and welfare in the south Indian fisheries sector. *The Quarterly Journal of Economics, 122*(3), 879–924. doi:10.1162/qjec.122.3.879

Kumar, R., & Best, M. (2006). Social impact and diffusion of telecenter use: A study from the sustainable access in rural India project. *The Journal of Community Informatics, 2*(3), 1–21.

Levett, C. (2004). *Himalayan village joins wireless road*. Retrieved July 23, 2011, from http://www.smh.com.au/news/World/Himalayan-village-joins-wireless-world/2004/12/26/1103996439623.html?oneclick=true

Nair, K. G. K., & Prasad, P. N. (2002). Development through information technology in developing countries: Experiences from an Indian state. *Electronic Journal of Information Systems in Developming Countries, 8*(2), 1–13.

Nepal, O. L. E. (2011). Our projects. *OLPC Project*. Retrieved July 29, 2011 from http://www.olenepal.org/olpc_project.html

NTA. (2003). *Management information system*. Retrieved July 25, 2011, from http://www.nta.gov.np/articleimages/file/NTA_MIS_1.pdf

NTA. (2010a). *License fee.* Retrieved July 23, 2011, from http://www.nta.gov.np/en/content/index.php?task=articles&option=view&id=46

NTA. (2010b). *Ten year master plan (2011 - 2020 AD): For the development of telecommunication sector in Nepal.* Retrieved from http://www.nta.gov.np

NTA. (2011). *Management information system.* Retrieved July 27, 2011 from http://www.nta.gov.np/articleimages/file/NTA_MIS_51.PDF

OLPC. (2011). Over 2 million children and teachers in 42 countries are learning with XO laptops today. *About the Project/Countries.* Retrieved July 24, 2011, from http://one.laptop.org/about/countries

Ruth, S., & Schware, R. (2008). *Pursuing truly successful e-government projects: Mission impossible?* Information Technology in Developing Countries, 18*(3). Retrieved July 25, 2011 from* http://www.iimahd.ernet.in/egov/ifip/oct2008/stephen-ruth.htm

Sein, M. K., Ahmad, I., & Harindranath, G. (2008). Sustaining ICT for development projects: The case of grameenphone CIC. *Telektronikk, 104*(2), 16–24.

Thapa, D. (2011). The role of ICT actors and networks in development: The case study of a wireless project in Nepal. *The Electronic Journal of Information Systems in Developing Countries, 49*(1), 1–16.

Thapa, D., & Sæbø, Ø. (2011). Demystifying the possibilities of ICT4D in the mountain regions of Nepal. In *Proceedings of the 44th Hawaii International Conference on System Sciences (HICSS 44),* (pp. 1-10). Kuai, HI: IEEE Computer Society.

Thapa, D., & Sein, M. K. (2010). *ICT, Social Capital And Development: The case of a mountain region in Nepal.* Paper presented at third annual SIG GlobDev workshop. Saint Louis, MO.

Thapa, D., Sein, M. K., & Sæbø, Ø. (2012). Building collective capabilities through ICT in a mountain region of Nepal: Where social capital leads to collective action. *Information Technology for Development, 18*(1), 5–22. doi:10.1080/02681102.2011.643205

UNESCO. (2010). Nepal en route for introducing ICT in education. Communication and Information Sector's News Service. *Retrieved July 21, 2011, from* http://portal.unesco.org/ci/en/ev.php-URL_ID=30622&URL_DO=DO_TOPIC&URL_SECTION=201.html

Universal Service. (2012). *Wikipedia free encyclopedia.* Retrieved March 12, 2012 from http://en.wikipedia.org/wiki/Universal_service

Unwin, T. (2005). *Partnerships in development practice: Evidence from multi-stakeholder ICT4D partnership practice in Africa.* Paris, France: UNESCO.

Unwin, T. (2009). *ICT4D: Information and communication technologies for development.* Cambridge, UK: Cambridge University Press.

Warschauer, M. (2003). *Technology and social inclusion: Rethinking the digital divide.* Cambridge, MA: MIT Press.

Zheng, Y., & Walsham, G. (2008). Inequality of what social exclusion in the e - society as capability deprivation. *Information Technology & People, 21*(3), 222–243. doi:10.1108/09593840810896000

KEY TERMS AND DEFINITIONS

ICT4D: The full form of the abbreviation is Information and communication Technology for Development. It is usually used when information communication networks are used for the sake of development through e- services like e-learning, e-commerce, tele-medicine, e-governance, etc. to raise socio economic standard of life.

Wireless Networks: Interconnection between different networks of computer at different places without wired connection.

Wi-Fi: Wireless Fidelity.

VoIP: Voice over Internet protocol gives a platform for voice communication with the use of Internet as a medium.

Chapter 5
The Role of ICT in Empowering Rural Indians

Ashok Jhunjhunwala
IIT Madras's Rural Technology and Business Incubator (RTBI), India

Janani Rangarajan
IIT Madras's Rural Technology and Business Incubator (RTBI), India

N. Neeraja
IIT Madras's Rural Technology and Business Incubator (RTBI), India

ABSTRACT

This chapter discusses some attempts over the last decade in using Information and Communication Technology (ICT) to empower rural communities and those who are socially and economically left behind in India. It begins with discussing the drivers of telecommunication growth in India since the mid-nineties. Then, it addresses the role of the village Internet kiosks in bringing the Internet to remote villages and articulates on the challenges facing the kiosk model. It then touches upon the rapid growth of mobile telephony in rural India. Following this, it discusses a number of attempts that use mobile telephony to empower rural communities. The authors also use multiple case studies to explore the role of ICT in supporting agriculture, delivering healthcare, achieving financial inclusion and improving the overall livelihood of rural communities in India. The key lessons learned include that the "one-size fits all" model does not work for all communities. In addition, involving both local and federal governments is crucial for the success of community-focused initiatives. Moreover, engaging communities and educating them about the benefits of delivered services would help in sustaining such community-focused initiatives.

DOI: 10.4018/978-1-4666-2997-4.ch005

BACKGROUND

India is a developing country whose population exceeds 1.2 billion people (Census of India, 2011). Its economy has been growing rapidly, achieving seven to nine percent growth rate in recent years. Despite the decline of poverty rate, a careful look at the figures of its economy shows that there are a number of economic problems that have not been solved yet.

While the average income in India is low, it has been rising rapidly especially in the last five to six years, as shown in Figure 1 (National Council of Applied Economic Research- Centre for Macro Consumer Research (NCAER – CMCR, 2011). In 2009-10, the average annual income of a rural household is just about 2777 UDS at 2004-05 prices. In other words, the average rural income in rural areas is about 1.5 USD per day. This is just above 1.25 USD defined as threshold for poverty by United Nations (Gordon, 2005). The conditions of those who live in rural areas are even worse when we consider that the average family size is five members (Shukla, 2010). Given the fact that India has about 800 million people living in rural areas, serious efforts are needed to overcome the poverty problem.

This chapter studies a number of community-focused initiatives conducted over the last decade in using Information Communication Technology (ICT) to empower rural communities in India.

The following section discusses the situation of the telecommunication sector in India in the decade of nineties. It also shows how mobile telephony took off in the early part of the century and discusses how policy was focused on achieving affordability of ICT. Then, it discusses efforts to expand infrastructure to rural areas. Then, we discuss how the focus has been shifted from infrastructure provision to service and application development. After that, we present a multiple of case studies on how the TeNeT group and its partners used ICT to support farmers, deliver healthcare, achieve financial inclusion, and promote business outsourcing to rural areas.

TELECOMMUNICATION IN INDIA

The telecommunication sector in Indian had not taken off until 2000. In particular, India was struggling with only 28.55 million telephone lines in 2000, 1.80 million of which were mobile (Telecom Authority of India, 2004), as shown in Figure 2 (Cellular Authorities of India, 2004; Department of Telecommunications, 2004). The growth over the previous decade was very slow. Even though Indian telecom was liberalized in 1994 and private participation in operations started in 1994, the scenario did not change. The Indian telecommunication market was influenced by the hype of a very large middle-class market. Some

Figure 1. Yearly household incomes in INR 1000 (inflation adjusted to 2004 prices)

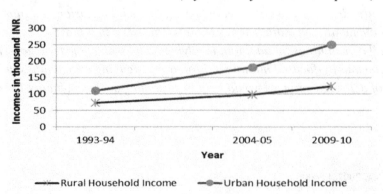

Figure 2. Number of telecom subscribers in India between 1993 and 2003

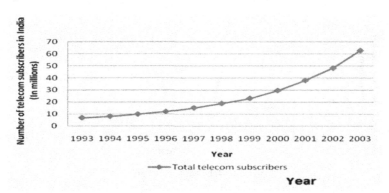

companies entered the market and invested astronomical amounts of money to obtain a license to become operators without considering the limited purchasing power of this middle class (Uppal, Nair, SKN, & Rao, 2006; Ministry of Finance, 2009).

It was realized that the installation and operation of telephones were too costly for the middle-class in India to afford. In 1994, the cost of telephone installation was dominated by the cost of the copper based local loop that connects the customer's premises to an exchange point (Jhunjhunwala, Ramamurthi, & Gonsalves, 1998).

In the early nineties, the idea of using "wireless technology in Local Loop" instead of copper wire had emerged (Jhunjhunwala, Jalihal, & Giridhar, 2000; Jhunjhunwala & Ramamurthi, 1995). However, wireless technology was immature and very expensive at that time. A research and development group in India, known as TeNeT group, realized the potential of wireless technology as an alternative of copper wire for Local Loop connections. While the cost of copper wire and digging has increased each year, the cost of electronics has come down driven by Moore's Law (Intel, 2005). Furthermore, the total cost has been driven by the cost of software rather than hardware as discreet radio components have been replaced by a programmable Digital Signal Processor. With bigger scale, the software cost gets allocated to larger number of systems, reducing per-line cost.

Armed with this argument, TeNeT group took an initiative of developing the corDECT Wireless network for Local Loop connections (Analog Devices, IIT Madras, & Midas Communication Technologies Pvt. Ltd., 1997; Ramachandran, 2002). The corDECT Wireless network did not provide only voice, but also data communications at a fraction of the cost of landline telephony. Therefore, the competition increased in the telecommunication sector in India. In other words, the capital and operational expenses of installing telephone lines started to decline. Other mobile technologies (e.g. GSM and CDMA [IS-95]) followed suit after a few years. Consequently, the stage was set for a wide wireless revolution in India.

Nonetheless, the large license fees continued to be a deterrent for the telecommunication sector to take off. The situation was dismal in 1999, with most of operators on the verge of bankruptcy. It is here that the government of India took a historic step in 1999, where it decided to forgo huge amounts of license fees that the operators had bid for and converted the license-based model to a revenue-sharing model. Accordingly, the telecom operators would pay the government a share of their revenue instead of license fees. Within a couple of years the wireless telephony took off and the Indian telecommunication sector achieved exponential growth, as shown in Figure 3 (TRAI, 2011).

Figure 3. Number of telecom subscribers in India between 2004 and 2011

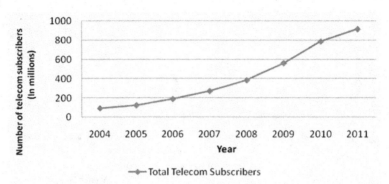

EXPANDING INFRASTRUCTURE

In 2000, wireless telephony in India started to grow exponentially. However, a look at the business plans of telecom operators showed that telecom expansion was limited only to urban areas. The license requirements necessitate operators to serve rural communities as well. However, service providers served limited rural areas to get away with this condition. This was because rural communities have limited capacity due to their low-income and low density.

The low income of rural communities is not sufficient for a family of five members to secure food, clothing, housing, education, and healthcare. In other words, it did not leave any surplus for telecommunications, which had a lower priority. At the same time, the cost of installing a telephone (landline or mobile) was far higher in rural India as compared to urban India. In addition, availability of power supply was a serious constraint. Furthermore, the market was ripe and more profitable in urban areas with huge expansion capabilities. Therefore, operators had focused on urban areas, leaving rural ones left behind.

It is important to note that telephones and the Internet are not just about communications. They are effective drivers for socioeconomic development. As data communications and Internet emerged, telecommunications were considered a powerful tool that could liberate rural com-

munities from deprivation and backwardness, which was thrust upon them during colonialism. Equipped with telecommunication infrastructure and an Internet access, a village could improve healthcare delivery and education, so that people could compete with their peers who live in urban areas and the rest of the world.

For telecommunication companies, expanding their services to villages did make any business sense at the time. Therefore, operators did not have plans to expand their networks to rural areas. In such a situation, innovation should step in and play a major role. Innovation often provides new approaches when traditional methods do not work. The following section presents key initiatives adopted to expand telecommunication infrastructure to rural areas in India.

Deploying Internet Kiosks in Rural India

The Public Coin-Operated Phones

In late '80s, it was recognized that it did not make business sense to install telephony in every urban home in urban India. This is because of the low-paying capacity of middle and lower-income households. However, telephony was known to not just enable communication, but bring in huge efficiency in all aspects of life. As India was getting more integrated and young people were migrating from one part of India to another for education and work, communications had become essential. This growing demand, but unaffordable, gave birth to a new "shared-telephony" model. Public Coin-Operated (PCO) telephones, installed on streets and public places, were common in the West. India first tried to adopt this model, but it did not work. One reason was that such PCO were not reliable. In addition, coins did not make them work all the time. Then there was vandalism. Further, they were not serviced well and maintained enough. Some broken telephones were often remained unfixed for weeks. As a result, the PCOs model failed in India. In such a situation, a new shared-telephone was conceived of and another model has emerged. At every street corner in India, there was a "paan-shop," a shop which sells beetle-nuts, leaves, cigarettes, cold-drinks, etc. These shops were open about 16 hours a day and serve local communities 365 days a year (Jhunjhunwala, Ramachandran, & Bandhopadyay, 2004, p. 34). The new shared telephone, initially called "paan-wala phone," was to be a regular phone with an attached meter kept at the shop. The neighbourhood folks usually come and use the phone and pay the shop-keeper the charges. This form of PCO quickly took off and had been enhanced to include Standard Trunk Dialing (STD-PCOs). These phones had the ability to make long-distance calls within the country. The STD-PCOs proliferated to every street-corner of urban and semi-urban areas to serve more than a million individuals. Telephony was then available to all in urban India.

The success of STD-PCOs in urban areas gave a clue to the solution of the rural telephony tangle. When a telephone was not affordable to individual homes in urban India, STD-PCO shops enabled sharing and became successful. As telephony (and now Internet) was not affordable for most rural households, a shared model on the lines of STD-PCO could possibly work. The Rural Internet kiosks (later referred to as CSC or common service center) providing telephony as well as Internet services was to be set up in each village. Such Internet kiosks were providing a whole variety of services to people living in the village. A typical village host from one to two thousand people. Aggregating the needs of such a number would drive the business viability of the established kiosk.

The n-Logue Business Model

The early attempts, sponsored by the government, to set up Village Internet kiosks showed the potential services and benefits of this model (Jhunjhunwala, Narayanan, Prashant, & Arjun, 2011). One of these early attempts was in the Dhar district in M.P. However, these kiosks were limited to areas where telecommunication infrastructure exists. In addition, they were dependent on government funds needed for setting up and operating the kiosks.

Meanwhile, n-Logue, a for profit company incubated at IIT Madras, went a step forward (Jhunjhunwala, Ramachandran, & Bandhopadyay, 2004). It decided to set up the telecom connectivity as well as Internet kiosks in villages. Some form of broadband connectivity (fibre or microwave) was available at the *taluka*-towns (sub-district) headquarters in most of India. Most villages will be within 25 Km radius of such *taluka*-towns. n-Logue used the already mentioned innovative wireless for Local Loop technology, corDECT, to provide connectivity to each village kiosk. n-Logue created a three-tier business model (Jhunjhunwala & Ramachandran, 2004). It had a Local Service Provider (LSP) in each taluka, who will provide connectivity to around 500 villages in 25 Km radius. At each village, there will be a Kiosk-Operator (KO), which serves as a sub-franchisee. This kiosk-operator invests in the Internet kiosk and provides services to the villagers, while taking connectivity from the LSP. The KO would have to invest about Rs 50,000 (in 2001) to get a computer, a printer, an Internet connection, an uninterrupted power supply (power back up), a camera, a microphone etc. The villagers used the

Internet kiosk for browsing the Internet, sending emails, obtaining computer training, in addition to telephone service. A typical KO needed to earn about Rs 3000 to break-even. For a village population of about 1000, this would indeed be doable.

Gradually, more services were added to the kiosk. These services included low-bit rate video-conferencing, electronic-payments, and banking and e-government services. Other kiosks explored computer-based education, health delivery (tele-medicine), job-portals, agricultural support services, mobile-charging and e-commerce services. The services rapidly grew to reach about 25 talukas in six states serving about 3500 villages. However, KO failed to sustain itself and started to fold in 2007.

Common Service Centres

In 2007, when most of the Internet Kiosk programs were down to their heels, the central government realized their potential and capabilities. Specifically, the government funded programs to set up Common Service Centres (CSCs), essentially Internet kiosks, in 100,000 villages (IL & FS, 2006). The key purpose of these CSCs is to offer e-government services to rural people. The government decided to go about this differently. They planned to invite large business to participate in a private-public initiative, where established large companies would set up the CSCs and provide services in the villages. There was enthusiasm and a significant number of corporations jumped into the fray. However, neither the government nor these corporations were willing to learn and build up from the experience of companies like n-Logue, which had already done this before. They believed that this public-private partnership would be successful as it secures fund, expertise, and political support. In addition, operators invested more on expanding the infrastructure to serve more rural areas and technical staff needed to maintain PCs has become available. However, power supply continued to be a problem. In addition, the govern-

ment was planning to provide subsidy; but there was so much hype that service providers would bid for zero or even negative subsidy in the reverse auction that was carried out.

The CSCs had started, but they operated in much smaller number of places than it was advertised. Very few of them have achieved break-even. Currently, most other CSCs are virtually closed or are losing money heavily. e-Governance services are still too patchy, bringing in revenue to only a few CSCs. The corporations which had invested have either closed the division, writing off the investment or are just waiting for e-government services to come to make an impact. Similarly, the Internet kiosk model continues to struggle.

There were several other similar attempts (though at smaller scale). They all started to decline in this period. However, the question is *why such sound and logical initiatives fail?*

The reasons for failures could be grouped into three categories (Jhunjhunwala, Kumar, Prashant, & Singh, 2009):

1. **Infrastructural Issues:** The biggest problem was the lack of reliable power supply. Most villages used to get power supply for only two to ten hours per day in summer months. Therefore, Internet kiosks needed battery back-up or diesel generators. Energy costs, with these alternatives, increased three to four times. When a village receives electric power only for two hours, charging the battery itself becomes another problem. In addition, the majority of rural areas did not have telecommunication infrastructure to support Internet kiosks. The n-Logue's model of bringing in wireless connectivity helped, but was somewhat expensive. Other operators depended on available telephone connectivity and could rarely provide affordable Internet connection in villages. They operated in villages near towns, where the problem was less severe. The third problem was the difficulty of maintaining PCs and

lack of spare units. Furthermore, rural areas did not have the technical staff needed to support such devices. Fortunately, such technical staff did develop as Internet kiosks were set up. Similarly, lack of adequately trained persons needed for operations was also largely overcome with the efforts of service providers. However, power and nonexistence of infrastructure continued to be key obstacles facing Internet kiosks.

2. **Inadequate Government Support and Encouragement:** Most programs were started involving some local government officials (mostly district collectors), who helped in setting up the Internet kiosks for target populations. Almost all these government officials promised to bring in government services to citizens through these kiosks. In addition, they helped in developing new services, particularly in education and healthcare. However, government services never became part of committed government programs and remained just initiatives of individual officials. As soon as the official was transferred to another district, the government support decreased and services disappeared. Therefore, the expected revenue generated from the e-government services for the kiosk operators remained unfulfilled, affecting the viability and sustainability of the kiosk. Furthermore, the government did pretty little in either helping overcome the infrastructural-bottlenecks or make available any subsidy. In fact, the purpose of n-Logue kiosks was to make telephony available (in villages where there were no telephones), but the state-owned telecom operator BSNL blocked this. Moreover, some government departments attempted to put tax on provided services, hurting the business model further.

3. **Lack of Services and Applications:** The biggest challenge is failure to create services needed by local communities and for which they would pay. The focus remained on infrastructure deployment, training, and operations rather than on services and content that could generate revenue and attract a wide range of users. There were a large number of initiatives for service creation. However, most of these initiatives were conducted by universities, which demonstrated the viability. They were rarely taken up by businesses and packaged as service offering. Operators were too focused on operations and did not invest enough on services or applications. These resulted in incomplete service-offerings and inconsistent revenues. The revenue model was, therefore, severely constrained.

There were some very successful kiosk operators. However, this success was mainly due to their effective entrepreneurship, rather than due to the strength of the services. Currently, none of the efforts that emerged in the period between 1999 and 2004 has continued to be in operation. However, wherever they were operated well, there was no doubt that the Internet kiosks were making a difference to the daily life of rural residents. There were a large number of stories of how provided services benefitted people (Utz & Dahlman, 2007). There was significant attention of government, media and intelligentsia. However the benefits at some individual villages could not be replicated everywhere. In addition, sustainability could not be achieved.

SERVICE DEVELOPMENT

By 2007, the situation in rural India started to change with the rapid growth of mobile phones (Jhunjhunwala, et al., 2011, pp. 11-18). In particular, the density of telecommunications had crossed 50% in urban India and operators started to realize that the urban market had saturated. At this point, they started seriously examining options to expand services to rural areas. Again, over 800 million people lived there and the density of telecommunications was still very limited, providing huge growth-potential. The key problem was the high operation cost in rural areas, which endures low revenue-potential (as rural incomes were still very low). It was clear that rural India will not provide Average Revenue Per User (ARPU) higher than Rs 100 per month. The question that arose was how operators could serve rural customers and make profits with such a low ARPU?

This was the biggest dilemma that operators started to ponder about. The switching cost was then negligible and already accounted for in the urban market. In addition, growth of subscribers would not add much cost. The key costs were, therefore, in the rural towers, the base-stations, power and cooling equipment at the towers and the back-haul connectivity from the existing backbone network to these towers. Fortunately, the GSM and 3G-1X (CDMA) base-stations costs had tumbled down, driven by huge volumes that India used. Similarly, the cost of radio-back-haul had declined substantially over the years. The main costs were those involved for towers and for back-up power generators at the towers. As the electricity-grid did reach most towers, the power could be interrupted for as much as twenty hours a day. The operators innovated to overcome this difficulty. So far, each operator owned its towers and thought of

it as a strategic advantage over competitors. But currently, aided by new regulations, they started to leave the installation and maintenance of towers to an infrastructure company. Currently, this company would not limit its service to a single operator, but it gets several operators to share a tower. This had two implications. First of all the cost of the tower was now shared amongst several operators, bringing down the cost of each operator. Further, the operator now did not have any upfront capital expenditure, and simply paid rent to the tower companies. In other words, this model has converted the CAPEX into OPEX. The only issue which did not find a satisfactory answer was the power supply, and it continued to contribute significantly to the costs for the operators as they ventured into Rural India. In a short time, mobile service was expanded to most of Rural India and started to grow at a rapid pace.

As shown in the Figure 4, the density of telecommunication in rural India has grown exponentially after 2007. This is mainly because the telecommunication companies started to direct their attentions towards the rural market (TRAI, 2011). It is important to note that most of rural families already have access to mobile service today.

Voice communications is the primary application of mobile phones. Yet, youngsters in Indian

Figure 4. Rural teledensity

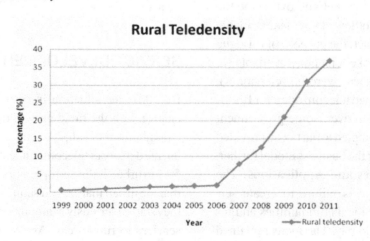

villages widely use SMS. As of today, SMS is available only in English and most villagers are not comfortable with the language; but that has not been a bottleneck. Rural India has innovated and have come up with new languages popularly known as "Hinglish" (Hinglish is a Hindi language written in roman script with a mix of Hindi and English words) and Tanglish. Ten other Indian languages have been developed to fit SMS. So the SMS flows widely. Further, the mobile connectivity in Rural India supports GPRS and 3G-1X data communication—even though a low-bit rate, it is possible to have data-connectivity in rural areas. The 3G data-connectivity is limited only to urban areas. It is expected that the 3G service will be available in most villages within few years.

With the growing advancement and penetration of mobile phones, they have become powerful tools for empowering rural population. First of all, most rural homes now have access to such phones. This is because the price of phones have declined to as low as Rs 1000. Furthermore, the tariffs have become very low, with voice call costing less than Re 0.50 per minute. Similarly, the costs of SMS have become less than Re 0.50 per minute and tariff plans allow unlimited free voice calls and SMS to some selected telephone numbers. Data tariffs are still a bit high, but are coming down. In addition, most phones have cameras, consequently allowing pictures to be taken and sent. Data communication is, however, limited only to about 50% of rural population who are mostly either illiterate or semi-literate and are not comfortable with any written communications. However, they can still send and receive pictures and video clips.

CASE STUDIES

While Mobile operators started to a mark in rural areas in 2007, the focus shifted from infrastructure deployment to the development of services and applications. Some serious attempts started in several locations, including those by the TeNeT group at IIT Madras. The TeNeT group created "Rural Technology and Business Incubator (RTBI) (www.rtbi.in)" focused on rural India. They focused on using ICT for delivering healthcare, achieving financial inclusion, improving output of agriculture, advancing education, promoting social networking and improving the overall livelihood of rural population in India. The group believe that developing new community-focused services and applications would significantly leverage the value of voice and SMS, and Web browsing. The following is a brief discussion of the key initiatives and programs implemented by the TeNeT group and its partners:

Case 1: Farming Support

One of the projects of RTBI focused on using mobile phones to deliver agricultural related support services to farmers. It primarily relies on voice communication, as farmers are more comfortable with it. Farmers interact with a computer using voice-recognition technology to add their (and that of their farm's) profile on a page created for each farmer. The data is captured, using simple question-answer format, collecting information about the farm size, type of soil, sources, and extent of irrigation, crops grown over the last few years, inputs used, yields, diseases encountered, etc. At the same time, pictures of the farm are uploaded on the farmer's page. Once the farmer is registered, he/she can access the agriculture support service, which is provided by a call-centre attendee. As soon as the farmer calls, the farmer's Web page pops up on the screen of the call center operator (agri-support person), who provides answers to farmer's queries taking into account the farmer's page and past history of diseases. If necessary, the agri-support person conferences in an agricultural expert to provide the right answer. Even while the queries are being answered, the support person enters the summary of the current interaction on the farmer's page, for future reference.

Similarly, ICT is leveraged to provide livelihood opportunity in villages. One of the interesting efforts involves setting up a "Business Processing Outsourcing (BPO)" in a village. This will be discussed below. Another example is where companies are outsourcing their manufacturing processes to rural areas. They use ICT to remotely manage and oversee the production facilities. This is happening, especially in areas, where manufacturing does not require heavy machinery or large power requirements and where some of the raw materials come from villages. It creates employment opportunities in the villages, slowing down rural to urban migration. RTBI has incubated companies like ROPE and Vastra which have been using this outsourcing model.

Similar initiatives are taking place in other areas. The following section presents some of these initiatives, a few of which had started in 2000.

Case 2: Healthcare Delivery

Over 20 million people in rural India (Jhunjhunwala, Prashant, & Kittusami, 2011) are pushed below the poverty line because of the high expenditure on healthcare which is stemmed from poor but expensive healthcare facilities available there. These facilities are provided by two sources namely the public system which is under the supervision of the government and the private system which consists of private clinics and nursing homes.

The public healthcare delivery system works at two levels. At the first level, the Government has set up Primary Health Centres (PHCs) where every PHC has a medical officer, two staff health nurse, a paramedical staff, one pharmacist, one sanitation and hygiene officer and a lab-technician (NRHM, 2010). India has just 23,485 PHCs, which means that each PHC has to serve 30 villages and people have to travel a long distance to access the service of one of the PHCs (Rural Health Statistics Bulletin, 2010). The second level of the rural public healthcare delivery system comprises of 147,069 health sub centres, each of which has a

Village Health Nurse who serves a population of 2000 people. The public healthcare system covers just 19 percent of rural population (NHRM, 2010) and has unfilled vacancies and non-attendance in both PHCs and the health sub-centres (Azad Indian Foundation). Most of private clinics and nursing homes are run by NGOs and charities. They provide healthcare services to about 15 percent of the rural population. The major drawback of the private healthcare system is that it is aimed at providing specialized healthcare and not basic healthcare. The remaining rural population (more than 50 percent) go to private and indigenous healthcare practitioners most of whom are ex-pharmacists, compounders, mid-wives, paramedics or traditional healthcare practitioners. They are referred to as Rural Healthcare Providers and are not recognized as medical practitioners by the Medical Council of India (World Health Organization, 2007). These RHPs are usually the ones delivering healthcare services in villages for years but are usually referred to as "quacks" (Kanjlal, Mandal, Samanta, Mandal, &Singh, 2007). Many rural healthcare initiatives which have been taken over the last 50 years have met with limited success but the arrival of ICT has opened the door for the delivery of health care in remote areas where doctors are not present.

The following section discusses how RTBI used ICT for healthcare delivery.

Remote Diagnostics

NCPL (http://www.neurosynaptic.com/) collaborated with IIT-M and designed a telemedicine product called ReMeDi, which could measure 6 parameters namely, non-invasive Blood Pressure, pulse-count, temperature, 12-lead ECG, Oxygen saturation and heart and breath sounds through the Internet (Bai, Prashant, & Jhunjhunwala, 2011). It also uses video conferencing as a feature to enable doctors to view patients remotely and serve them using an audio-video link via the Internet. ReMeDi works on a rechargeable battery. It could be connected to the kiosk PC on an infrared link. It consumes just two watts of power and facilitates the transfer of data communications and video-conferencing, even on low Internet bandwidth. These features of ReMeDi instantly overcame poor infrastructural facilities in rural India such as poor power supply and the lack of seamless bandwidth connectivity. Furthermore, ReMeDi is safe to use because the kit was not connected to electrical mains while in use and could be placed in village kiosks where a trained operator could help the patient get connected to a doctor. It was installed in many villages and it was used by many doctors between 2003 and 2006.

However, a product in itself does not provide a business model to deliver a service. In order to evolve a business model and understand the socioeconomic and implementation challenges of ReMeDi, a one-year pilot project was taken up in 2007 by RTBI (IIT Madras), NCPL and Nicholas Piramal Industries, a manufacturer and supplier of medicines. The pilot project was conducted in selected rural villages of Tamil Nadu using Re-MeDi (Jhunjhunwala, 2011). Post the pilot study, ReMeDi has been installed in more than 300 rural areas in several states. Sponsored by a program under Bill and Melinda Gates Foundation, 400 new installations of ReMeDi are in progress in Bihar (Jhunjhunwala, 2006). At the same time, IIT Madras has taken an initiative to come up with an upgraded ReMeDi kit that compatible with Bluetooth, Wi-Fi and USB in order to enable rural patients to connect to a doctor using a 3G mobile phones.

Apart from the telemedicine kits, a number of exploratory initiatives were undertaken by RTBI with the aim of making healthcare more accessible to rural communities. Though the Internet Kiosks set up by n-Logue were not targeted at healthcare services, the possibility of such a provision through n-Logue was serendipitous. The kiosk operators started using the video conferencing facility to connect patients to doctors, who had volunteered few hours every week for treating rural patients

after being contacted by n-Logue. The team also developed an online patient database to enable patients to book an appointment in advance and hospitals to schedule and track the appointments to ensure a hassle free service delivery. Another program brought in a computer to test for colour-blindness and eye-testing (Jhunjhunwala, 2006). ICT also enabled video-consultation with general healthcare practitioners and specialists in fields like cardiology, gynaecology, and paediatrics (Jhunjhunwala, 2006). In addition, kiosk opera-tors used this technique to connect with veterinary doctors to treat livestock diseases (Jhunjhunwala, 2004).

Empowering RHPs by First Care

The RHPs, as mentioned before, are the primary healthcare providers in rural areas. Therefore, training them to provide high quality healthcare would result in the improvement of the standards of healthcare facilities available in rural India. Keeping this in mind, RTBI, IIT Madras launched a two year pilot initiative for rural healthcare called First Care in 2007 (Jhunjhunwala, 2006). Under this initiative, 22 RHPs in Sivaganga district of Tamilnadu were provided basic training through computers by linking them to the existing Internet kiosks in their villages. In addition, RTBI started an e-learning program aimed at improving medi-cal knowledge of community members about the symptoms of various common ailments and dis-eases was conducted. The RHPs were also given short classroom based education on conducting simple lab tests for testing blood and urine for anemia, malaria, blood sugar, etc. They also provided training sessions on how to maintain online records for treated patients. Using the Internet, the RHPs were connected to doctors for guidance and supervision, so that irrational drug dispensing could be controlled, pregnant women could be monitored and anaemia in children could be prevented. The RHPs also had the facility of ordering and procuring medical supplies using the

Internet. However, continuous pressure from the medical council opposing the contact with RHPs forced the program to be abandoned.

Children's Healthcare

The government of India has adopted several initiatives to combat malnutrition and maternal deaths but has met with just limited success because of the lack of a proper healthcare delivery mechanisms. To ensure that the service reaches the targeted population, developing systems for monitoring healthcare delivery becomes essential. RTBI along with Uniphore Software Systems have developed a system for collecting children's health related data using voice telephony (Jhunjhunwala, Anandan, Prashant, & Sachdev, 2011; Soman, 2010). The system authenticates children using voice recognition to weed out false identities. This voice-based system was initially used in parallel with the paper based reporting to check its validity. After validation, the system was extended to most centers. The technology which was developed was simple yet powerful. It required just a telephone at each center. The voice authentication, the voice recognition technology, and the application resided only on the health-server, located at the state-capital. The conversation between the mother and the computer took place entirely in local language voice and could, therefore, be easily handled by even an illiterate mother. The records were collected instantaneously and the whole project could be monitored easily. The use of voice-telephony for providing healthcare services and monitoring is now being considered for implementation of other government health-projects in the state of Tamil Nadu, India.

Real-Time Bio-Surveillance

Mobile phones have become a useful tool to monitor remote conditions and situations. RTBI has adopted the mobile phone technology to undertake

a real time bio-surveillance (http://rs.rtbi-iitm.in/BioSurvey/). Detecting communicable diseases before it reaches an epidemic state is vital. The challenge lies in receiving health information in a timely manner in order to prevent diseases from reaching epidemic states. The current surveillance system requires the village health nurse to report disease incidence through a paper form process. The forms from all villages are collected and analyzed to detect the onset of any epidemic. The process is cumbersome and consumes time. Consequently, the epidemics are often detected too late. The need of the hour is a "real-time" information flow and analysis so as to detect events as well as any increase, even in scattered cases. The Real-Time Bio surveillance Program was started for the purpose of using ICT to detect diseases in real-time (RTBI, et al., 2010). The project was piloted in two countries: India and Sri Lanka. It used mobile phones to capture health related information, which was then communicated to a central database in real time for analysis and monitoring. The ICT system used General Packet Radio Service (GPRS) and Short Message Service (SMS) as transport technologies. The results obtained from the pilot phase conducted in two countries for over one year, showed the validity of the process.

Using ICT for delivering healthcare in rural India continues to be a struggle. The programs described above, shows the potential of ICT in healthcare delivery, but also points to the large amount of work ahead.

Case 3: Financial Inclusion

Access to basic financial services continues to be a major problem for people in rural India. A recent book about the financial behaviour of poor households in Bangladesh, India, and South Africa, *Portfolios of the Poor* (Collins, Morduch, Rutherford, & Ruthven, 2009), says that there are three big challenges facing low-income households as they seek to manage their money better: (1) Dealing with money-management on a day-to-day basis; (2) Coping with emergencies; and (3) Building useful large sums of money. With these challenges, managing day to day cash flow for the rural mass becomes a bottleneck with no access to financial services. Rural India has very limited banking services. Even-though almost half of the national income is generated (Indian Knowledge@Wharton, 2010) from rural areas. About 85 percent of India's workers are employed in the "unorganized sector"—where wages are paid in cash—and at least half of workers do not have bank accounts. With the development of Indian rural economy, people in the rural areas have cash and are willing to spend, unlike before. Therefore, there is an urgent need to take banking and financial services to Rural India.

The Gramateller Rural ATM Machine

In 2004, few banks prophesied a great need for installing ATM machines in rural India in order to provide people with the facility of cashing out their money for emergency situations or day-to-day needs. However, installing conventional ATM machines would mean spending about 0.8 to 1 million rupees per machine. In addition, ATMs need power backup and an air conditioned closet which increases costs significantly. Therefore, it is difficult to scale up the system to all villages, particularly if we consider the micro-transactions and low income of rural residents. In order to solve this problem, Vortex India (http://vortexindia.co.in/), an IIT Madras incubated a company that invented a low cost ATM machines called *Gramateller* ATM. These machines are powered by solar power and have the capabilities to work in harsh rural environment. They cost one-fifth of the price of the conventional ATMs and is cheap enough to deploy in rural areas where the average transaction cost is only a few hundred rupees. The system is built around a pared-down software

platform and connected to a bank via the Web. It also includes a fingerprint scanner to provide identification. This fingerprint-based identification system is suitable for a country where 70 percent of the population are illiterate. It runs on just 30 watts of electricity, a fraction of the 3,000 watts required by a conventional ATM machine. This low power requirement decreases the cost of supplying backup power in areas where blackouts are common. The new generation ATM also emits less heat. In other words, unlike a traditional cash machine, it does not need to be housed in an air-conditioned closet. Moreover, the ATM machine does not require brand new notes and hence re-cycled notes can be delivered which is helpful, since rural users are often suspicious that crisp new cash is forged (Blakely, 2008). The ATMs are now connected to the bank's switch using GPRS or 3G1X (CDMA) mobile technology. Therefore, a high bit-rate and leased line connectivity are no longer required.

The Low cost ATMs has been used by the (National Rural Employment Guarantee Scheme [NREGS]) in India to make payments on a weekly basis, which was previously being distributed by the village panchayat. These ATM machines increase the transparency of the system, ensure security, since they use a robust fingerprint biometric system, and also provide the ease and convenience of withdrawing money anytime. State Bank of India has deployed over 500 such ATMs in India and more are getting deployed (Rediff Business, 2010).

Nevertheless, access to financial services does not stop here with only being able to withdraw money. It means a huge realm of services like micro savings, fixed deposits, microloans, funds transfer, and handling cash-in and cash-out. New ATMs are being envisaged where a person can also deposit money, in addition to withdrawing cash. A few other services like bill-payment and bank-transfers have been added. Yet, the new ATM machine is not a replacement for traditional banks.

Business Correspondents

To access financial services, rural people have to travel to banks located in towns or cities nearby. When in need for small ticket loans, banks have stringent Know Your Customer (KYC) norms that need to be fulfilled. In addition, moneylenders charge an exorbitant interest rate (RBI KYC Norms, 2002). This pushes a common rural man either into a vicious cycle of debts with no proper facility to save his money. Therefore, the Reserve Bank of India (RBI) allowed banks to launch the Business Correspondent model in 2006 (RBI, 2006). The purpose of this model was to expand the ambit of the formal banking sector and to extend valuable financial services to rural populace. Accordingly, banks appoint third party banking agents, termed as Business Correspondents (BCs), who open no frill accounts with zero minimum balance and flexible KYC norms. These BCs provide basic banking services on behalf of the bank. The BC model is ideally the alternate viable business model in order to ensure greater coverage of rural clients (at the order of at least one CSP-customer service point per village) at low costs. The BCs work with small terminals, which are referred to as Financial Inclusion–Point of sale terminals. These terminals are connected to the bank's computers using mobile data-connectivity alone. This new channel works through a process of collaboration between banks, technology vendors, and BCs. Technology vendors provide the necessary hardware, software and processing facilities that link clients to the CSP of BCs and BCs to the bank. BCs are organizations or individuals that manage and offer points of transaction outside bank branches in partnership with the bank.

Over the years, several initiatives have been taken by banks and technology providers (e.g., ALW, FINO) succeeded in opening many bank accounts. However, most of these accounts would quickly become inoperative. The reasons for that include long distance to the nearest BC, negligible range of financial products, lack of awareness, etc. An additional problem was the high capital cost needed for the operation BCs. Unless the number of transactions is significant enough, the average capital cost would be too high and this would make the operation unviable. Fortunately, the exponential growth of mobile phones in India is changing the paradigm of people-enterprise interaction and increases commerce-driven telephony. Looking at the growth and penetration of mobiles in the country, the time is right to envision and create an environment that can leverage this medium with multilingual voice based information and dissemination together with telephony based transactions. Since text-based messages and transactions would only serve a few (because of lower literacy levels) using once own voice in carrying out banking transactions is the need of the hour. Uniphore also develops voice-based solutions for banking, health, agriculture and many more voice based applications that add value to a rural man. One of the key technologies being explored is the Uniphore's voice solutions which uses voice biometrics to identify and authenticate customers (in addition to call line identification from the customer's mobile). This solution enables customers to carry out banking transactions entirely based on voice using their own vernacular language. The application developed by Uniphore has the ability to understand what the caller speaks and it processes the request accordingly. Each caller is assigned a pre-registered voice-based password with the system and the caller will be authenticated based on his/her voice. This can be coupled with a BC model suggested by RBI. Using this system, rural customers can make transactions like cash-in, cash-out with his/her CSP available in the village centre. In addition, customer can use the voice based banking module installed in his handset to carry out other services such as funds transfer and balance enquiry. From the CSP's point of view, using end-user mobile phone for money transactions ensures *low cost and better business viability*.

Thus, having a CSP in every village and arming every villager with a banking enabled mobile phone would help them transact conveniently and safely. Thereby, this system makes a huge difference to the life of those who live in remote rural areas. In addition, such a negligible capital and operation cost would make this system viability even for small villages. It also offers a wide range of financial products such as savings, payments, small-ticket loan, and mobile top-ups with convenient access to the nearest customer service point. This new banking channel promises to provide financial services in every village by the end of 2015.

Case 4: Outsourcing Businesses

Recently, India had gained tremendously from carrying out IT based services for the West. The gains had been both in financial terms (as it generated huge wealth for the country) as well as in boosting self-confidence of youngsters. Carrying out the work for the Western companies, these youngsters feel equal to their counterparts in the West. It stemmed the large flow of educated Indians from India to the West.

The question was whether what was done in urban India could be repeated in Rural India by getting the rural youngsters to carry out IT based work for urban India. RTBI helped set up a company called "Desi-Crew" (http://desicrew. in/) in 2007, precisely to answer this question. Desi Crew was established to leverage the opportunities emerging in the ICT-based services by setting up rural BPOs in villages and provide employment opportunities for their unemployed youth. There was hope that it would create wealth in rural India and help in decreasing the flow of youngsters from villages to cities in search for jobs. Desi-Crew set up rural ICT delivery units to overcome outsourcing related problems such as lack of infrastructure in rural areas, assuring quality of services, security, etc.

Desi-Crew has developed a rural BPO delivery business model (Rediff News, 2009; Nasscom).

The model consists of a network of micro-centres, which have been set up in a wide range of rural locations. These micro-centres are situated in territories whose rural population ranges between 3,000 and 20,000 individuals. Each centre is professionally run with a 25- to 50-seat facility and works in two shifts. These centres provide back-end services to Indian as well as some global clients. The rural youth are provided necessary training on computer skills, basic office management, HR practices, data entry, and Internet usage. After training, they are given projects/works from clients. A central operation hub situated in the city acts as a one-point contact for the client and takes responsibility for assuring quality and delivering the requested services on a timely manner.

There were three major challenges, which were faced during setting up the Desi Crew centres. The first one was the creation of the requisite infrastructure including communication and power infrastructure in the villages to match the facilities available in the offices of the clients, which were based in the cities. Another roadblock was changing the mindset of people in urban India and convincing them that outsourcing work could be done from rural India without compromising on the quality of the service delivered. The third hurdle was establishing the concept of a BPO in remote villages. In addition, the need for meeting delivery parameters had to be emphasized.

After the initial phase of trials and errors, Desi-Crew has successfully provided non-agricultural job opportunities to the educated youth in rural areas. It has also made a positive impact in the lives of youngsters, especially young rural women. Cases of parents letting a woman continue her education at least upto high school and postponing very early marriage reflect the success of the model. Desi-Crew has also, to a certain extent, decreased the migration of youth from the villages to the cities in search of lucrative job opportunities. Employees of Desi-Crew also save money more than their urban counterparts because they do not have to spend as much on living expenses and maintenance. Exposure to the needs of the corporate world and the training to meet the demands adds to the self-esteem and self-worth of the employees.

This scalable Desi-Crew model employees a large number of workforce, facilitates access to an untapped and well-trained workforce, lowers operation and fixed costs (40 percent lesser than an urban BPO), and provides access to multiple Indian language skills. In addition, this model creates computer and knowledge base jobs for rural communities where there are no similar jobs. Moreover, employers of Desi-Crew became less inclined to leave their jobs given the improved quality of life and option of staying with their families. The major services provided by Desi-Crew include publishing, research organization, e-Governance, and back-end support services for the financial sector. Desi Crew has won many awards such as BiDs India, Sankalp 2009 (Highly Scalable Social Model), CII Seal of Appreciation and FICCI FLO (Best Woman Entrepreneur).

Since, the wealth is created in local communities by the inhabitants of communities and is used for their betterment. This business model has proven viability for the entrepreneur and has shown evidence on sustaining the development of rural India.

CONCLUSION

The potential of ICT in transforming rural lives has been well recognised in India over the last 15 years. A number of initiatives were undertaken by several groups in this direction. Most early efforts created a lot of enthusiasm and many interesting and insightful stories. The common focus was on affordability, sustainability, scalability, and innovation. The key learned lesson is the need for iterative processes to explore the socioeconomic and technical issues of the model in order to refine it make it more robust. In addition, the participation of local communities is crucial for the success of community-centred initiatives. However, none of these early efforts achieved significant success.

In other words, none of them were able to sustain or scale. In particular, the services that have been built are still rudimentary and immature. In addition, they have not attracted a wide range of beneficiaries and clients, considering the large population of rural India.

One of the reasons of these sustainability and scalability issues could be the inadequate infrastructure. Another reason is the lack of government support. In addition, the technology was not mature enough. In addition, rural communities were not aware of the benefits of these innovative initiatives. Furthermore, the harsh conditions of rural communities (e.g. low income, lack of technical skills, poor English language) were another obstacle that prevented them from significantly benefitting from these new technologies. Moreover, it is important to remember that India is a very diverse country. Villages in one state in India are quite different from that of another. Therefore, developers have to refrain from the "one size fits all" model. In other words, services need to be customized or may be even re-developed for each region. Patience and persistence are another success factors that needs to be considered while serving rural markets. There is also a need for programs to educate people about the benefits of the delivered services. In other words, benefits to the villagers have to be very clear. These are some of reasons that might have contributed to failure of these efforts.

We hope that future initiatives would be more successful for many reasons. Telecommunication infrastructure is rapidly growing in rural India. Today most villages have mobile telecommunication infrastructure and most households have access to affordable mobile services. At the same time, rural the average income in rural areas has increased and programs like NREGS are putting some disposable incomes in the hands of villagers. Coinciding with this, there is a new round of efforts, which leverage mobile and Internet connectivity and build community-cantered services and applications for rural communities. Many of the emerging services use voice communication.

Voice communication is the only mode, which is usable by all and works for all languages and local dialects. Though early, these efforts have met with far larger initial success than those, which were taken before. If any business model has to succeed, they need to become robust.

India's initiatives in using ICT to empower rural communities and integrate them in the information society are very useful for the rest of the world. Simply, the large size and the wide range of their socioeconomic issues make them very interesting. In addition, both successful and the failure stories provide many insightful lessons.

REFERENCES

Azad Indian Foundation. (2012). *Website*. Retrieved on January 14, 2012 from http://azadindia.org/social-issues/rural-healthcare-in-india.html

Bai, V. T., Prashant, S., & Jhunjhunwala, A. (2011). Rural health care delivery- Experiences with ReMeDi telemedicine solution in southern Tamilnadu. In *Proceedings of the International eHealth Telemedicine and Health ICT Forum for Educational, Networking and Business*. IEEE.

Blakely, R. (2008, March 12). State bank of India takes no-frills ATMs to masses. *Times (London, England)*.

Cellular Authorities of India. (2004). *GSM mobile statistics*. New Delhi, India: Cellular Authorities of India.

Census of India. (2011a). *Government of India, ministry of home affairs, census India*. Retrieved on January 14, 2012, from http://censusindia.gov.in/

Census of India. (2011b). *Provisional population totals: Rural urban distribution*. Retrieved on January 14, 2012 from http://www.censusindia.gov.in/2011-prov-results/paper2/data_files/india/paper2_at_a_glance.pdf

Collins, D., Morduch, J., Rutherford, S., & Ruthven, O. (2009). *Portfolio of the poor*. Princeton, NJ: Princeton University Press.

Gordon, D. (2005). Indicators of poverty and hunger. *Expert Group Meeting on Youth Development Indicators*. Retrieved on January 14, 2012 from http://www.un.org/esa/socdev/unyin/documents/ydiDavidGordon_poverty.pdf

Government of India. (2004). *Annual Report*. New Delhi, India: Government of India.

IL & FS. (2006). *The common service entre scheme*. Retrieved on January 14, 2012 from http://www.ilfsindia.com/downloads/bus_concept/CSC_ILFS_website.pdf

Indian Knowledge@Wharton. (2010). *MNCs in rural India: At a turning point*. Retrieved on January 14, 2012 from http://knowledge.wharton.upenn.edu/india/article.cfm?articleid=4472

Intel. (2005). *Website*. Retrieved on January 14, 2012 from ftp://download.intel.com/museum/Moores_Law/Printed_Materials/Moores_Law_2pg.pdf

Jhunjhunwala, A. (2004). *Applications and services through e-kiosks towards enabling rural India*. New Delhi, India: Asian Development Bank.

Jhunjhunwala, A. (2006). Rural connectivity towards enhancing health care. In *Proceedings of the Pugwash Conference on HIV/AIDS*. MSSRF.

Jhunjhunwala, A. (2011). *Can ICT make a difference? Case study: Rural India*. Copenhagen, Denmark: Infrastructures for Health Care.

Jhunjhunwala, A., Anandan, V., Prashant, S., & Sachdev, U. (2011). Experiences on using voice based data entry system with mobile phone in rural India. In *Proceedings of Infrastructures for Health Care* (pp. 59–67). Copenhagen, Denmark: Infrastructures for Health Care.

Jhunjhunwala, A., Jalihal, D., & Giridhar, K. (2000). Wireless in local loop, some fundamentals. *Journal of IETE, 46*(6).

Jhunjhunwala, A., Kumar, R., Prashant, S., & Singh, S. (2009). *Technologies, services and new approaches to universal access and rural telecoms.* International Telecommunication Union.

Jhunjhunwala, A., Narayanan, B., Prashant, S., & Arjun, N. N. (2011). *Innovative services & mobile access to drive broadband in rural India- Learning from recent successes and failures.* Unpublished.

Jhunjhunwala, A., Prashant, S., & Kittusami, S. P. (2011). *Can information and communication technology initiatives make a difference in health-care? Case study: Rural India.* Unpublished.

Jhunjhunwala, A., & Ramachandran, A. (2004). n-Logue: Building a sustainable rural services organization, case study on communication for rural and remote areas. *International Telecommunication Union.*

Jhunjhunwala, A., Ramachandran, A., & Bandhopadyay, A. (2004). n-Logue: The story of a rural service provider. *The Journal of Community Informatics, 1*(1), 30–38.

Jhunjhunwala, A., & Ramamurthi, B. (1995). Wireless in local loop: Some key issue. *Journal of IETE, 12*(5&6), 309–314.

Jhunjhunwala, A., Ramamurthi, B., & Gonsalves, T. A. (1998). The role of technology in telecom expansion in India. *IEEE Communications Magazine, 36*(11). doi:10.1109/35.733480

Kanjlal, B., Mandal, S., Samanta, T., Mandal, A., & Singh, S. (2007). *A parallel healthcare market: Rural medical practitioners in west Bengal.* India: Institute of Health Management Research.

Ministry of Finance. (2009). *Position paper on telecom sector in India.* Retrieved on January 14, 2012 from http://www.pppinindia.com/pdf/ppp_position_paper_telecom_122k9.pdf

Nasscom. (2012). *Women take the lead in rural BPOs.* Retrieved on January 14, 2012 from http://www.nasscom.in/women-take-lead-rural-bpos

National Rural Health Mission. (2010). *Rural health statistics in India.* Retrieved on January 14, 2012 from http://mohfw.nic.in/BULLETIN%20ON.htm

NREGA. (2012). *Smart cards: Biometrics scanning for signatures and handheld devices.* Retrieved on January 14, 2012 from http://www.nrega.net/ict/ongoing-ict-projects

Ramachandran, T. V. (2002). *Building on the initial success of cellular: The Indian experience.* ITU.

RBI. KYC Norms. (2002). *Circular: DBOD.AML. BC.18/14.01.001/2002-03.* Retrieved on January 14, 2012 from http://www.rbi.org.in/scripts/NotificationUser.aspx?Id=819&Mode=0

RBI. (2006). *Circular RBI/2005-06/288, January 25, 2006.* Retrieved on January 14, 2012 from http://rbi.org.in/scripts/BS_CircularIndexDisplay.aspx?Id=2718

Rediff Business. (2010). *Low cost ATMs: Vortex's pioneering gift to India.* Retrieved on January 14, 2012 from http://business.rediff.com/slideshow/2010/sep/16/slide-show-1-innovation-vortex-gift-to-rural-india.htm

Rediff News. (2007, November 29). *DesiCrew: A girl's pioneering vision for rural BPOs.* Retrieved on January 14, 2012 from http://specials.rediff.com/money/2007/nov/29sld1.htm

RTBI. et al. (2010). *Evaluating a real time bio-surveillance program – A pilot project report of interim findings and discussion workshop.* Retrieved on January 14, 2012 from http://rs.rtbi-iitm.in/BioSurvey/doc/Workshop/RTBP-FindingsWorkshop-REPORT.pdf

RTBI. IIT Madras. (2012). Technology innovations. *Gramateller.* Retrieved on January 14, 2012 from http://www.rtbi.in/innovate.html

Rural Health Statistics. (2010). *RHS bulletin*. New Delhi, India: Government of India.

Shukla, R. (2010). The official poor in India summed up. *Indian Journal of Human Development, 4*(2). Retrieved on January 14, 2012 from http://xa.yimg.com/kq/groups/12632651/1180166002/name/The+Official+Poor+in+India+Summed+Up_Rajesh+Shukal.pdf.pdf

Shukla, R. K. (2011). *NCR analysis (limited circulation)*. New Delhi, India: National Council of Applied Economic Research.

Soman, S. (2010, November 8). Talking their way to success. *The Times of India Chennai*, p. 3.

Telecom Regulatory Authority of India. (2011). *Report*. Retrieved on January 14, 2012 from http://www.trai.gov.in/Default.asp

TRAI. (2004). *Press release 47/2004*. Retrieved on January 14, 2012 from http://www.trai.gov.in/WriteReadData/trai/upload/PressReleases/160/Press%20Release -%2030%20July-04-Final.pdf

Uppal, M., Nair, S. K. N., & Rao, C. S. R. (2006). India's telecom reform: A chronological account. *NCAER*. Retrieved on January 14, 2012 from http://www.iipa.org.in/common/pdf/PAPER%20 15%20- %20TELECOM%20REFORM.pdf

Utz, A., & Dahlman, C. (2007). *Promoting inclusive innovation: Unleashing India's innovation: Toward sustainable and inclusive growth*. Retrieved on January 14, 2012 from http://siteresources.worldbank.org/SOUTHASIAEXT/Resources/223546 -1181699473021/3876782-1191373775504/indiainnovationchapter4.pdf

World Health Organization. (2007). *Not enough here, too many there: Health workforce in India*. Geneva, Switzerland: World Health Organization.

KEY TERMS AND DEFINITIONS

ATM: An automated teller machine or Automatic Teller Machine (ATM) is an automated telecommunications device that provides the clients of a financial institution access to financial transactions in public spaces without the need for any human help.

Average Revenue Per User: Average Revenue Per User (ARPU) is a measure used by telecommunication companies which helps one understand how much money the company makes for every customer that it has. It is the total revenue generated divided by the total number of subscribers.

Business Process Outsourcing: Business Process Outsourcing (BPO) is a part of outsourcing that involves the contracting of the operations and responsibilities of specific business functions to a third-party service provider.

Financial Inclusion: Financial inclusion otherwise called inclusive financing is the delivery of financial services to the low-income segments of society at affordable costs.

Tele-Density: The number of landline telephones in use for every 100 individuals living within an area/state/province/country is called the tele-density of the area/state/province/country.

Village Internet Kiosk: A Village Internet kiosk is a terminal that provides public Internet access in the village.

Chapter 6
Digital El Paso:
A Public–Private Business Model for Community Wireless Networks

Barbara Walker
Cisco Systems, Inc., USA

Evelyn Posey
University of Texas at El Paso, USA

ABSTRACT

The Digital El Paso (DEP) community wireless network was deployed as a public-private business model to achieve digital inclusion, sustain economic development, and enhance government and public services. The design, implementation, and funding of DEP were achieved through the collaboration of local businesses, core community members, non-profit organizations, academic institutions, and government entities. In particular, Cisco Systems, Inc. provided design and planning support to complement Intel Corporation's seed funding for the site survey. El Paso County, the City of El Paso, El Paso Independent School District, the Housing Authority of the City of El Paso, and El Paso Electric provided equipment and services. The purpose of DEP is to provide wireless Internet access to achieve social inclusion and economic development. DEP's main challenges include lack of funds, limited user acceptance, and insufficient user training. The policy implication is that leveraging public/private partnerships enhances collaboration and increases the chances of success of community wireless networks. A family-centric approach to drive the adoption of these emerging networks and increase bandwidth utilization, particularly in rural and underserved communities is also recommended.

DOI: 10.4018/978-1-4666-2997-4.ch006

BACKGROUND

The Digital El Paso (DEP) community wireless network was deployed as a Proof Of Concept (POC) to test wireless technology and explore the possibilities of using wireless networks to improve the delivery of state and local government services. Other objectives of DEP were to increase social inclusion and to drive economic development. The project was led by a collaborative, self-directed core group of state and local governments, education institutions, and private sector partners who brought the commitment and support of each of their sponsoring organizations to the project. The overall goal was to identify local needs and develop a solid business case to plan and build out a large-scale municipal wireless network. The lessons learned by this team provide a roadmap that could be replicated by other communities with similar broadband visions and digital needs.

El Paso, Texas is the largest U.S. city on the 2,000-mile U.S.-Mexico border. El Paso is the 6th largest city in Texas and the 23rd largest city in the United States. The closest U.S. metropolitan city is about 200 miles away. In other words, El Paso community lives in a relatively isolated and remote location. It has a fiber cable that serves as an Internet backbone expanding through El Paso from east to west. However, the county has minimal middle and last mile service to pockets of the community. El Paso County is about the size of the U.S. state of Rhode Island. The county includes communities with no access to basic utilities such as water or sanitary (Rubio, 2009). Because of the limited capacity and low population density, it may not make sense for traditional Internet Service Providers (ISPs) to build out to serve these areas. However, access to broadband service is essential for this underserved community to thrive and fully use the digital opportunities and resources of the current information age.

Ciudad Juárez is El Paso's sister city in Mexico. It is the largest city in the state of Chihuahua, the 5th largest city in Mexico, and the largest Mexican city on the U.S.-Mexico border. The city halls of El Paso and Juárez are about ten blocks apart. The international metro area is home to about two million people and is the largest community in the world where a developed country and a newly industrialized country join to form a regional economy (Rubio, 2009).

SETTING THE STAGE

In 2006, the frenzy of "free" municipal wireless networks was taking off, sparking a debate between community activists and telecommunications companies (Abdelaal & Ali, 2008; Abdelaal, Ali, & Khazanchi, 2009). Many people thought that municipal wireless networks were a viable solution for affordable Internet access, but the question was how to pay for them (Cisco, 2006a, 2006b; Rideout & Reddick, 2005; Simpson, 2005). In other words, what is the suitable business model?

The prevailing wisdom among community leaders seemed to fall into two camps. One group believed that they could negotiate free wireless Internet for all citizens by contracting with large ISPs (Brietbart, 2008). In this model, ISPs would make money from monthly subscriptions and sales of advertisements that could be delivered to connected citizens. Some envisioned a recurring revenue stream from other accompanying services. The other group believed that cities should own the network and fund the build-out by adopting an anchor productivity or municipal application such as automated parking meter reading, code enforcement, utility meter reading, emergency responder, or traffic management (Quinn, 2006). Both groups advocated for a participatory design model, where end users are involved in the design and outcomes of the project as a way to ensure the sustainability of such community-centered projects (Carroll & Rosson, 2007; Simpson, Wood, & Daws, 2003).

Cities such as Chicago, Los Angeles, and San Francisco all announced municipal networks during the middle of the last decade. City governments expected these networks to spur economic development and bridge the digital divide. However, many cities found that the implementation was rather difficult, if not outright impossible. Costs to set up the network, signal barriers, and political disagreements stymied efforts, especially for free access models. These projects failed or were delayed because of high costs, technical issues, and problems due to public-private partnerships. As Rolla Huff, chief executive of EarthLink, Inc. put it, "It was a great idea. It was not a great business" (Lavallee, 2008). Oklahoma City was a typical example. A pilot program was launched in 2006, but there were technical problems and not as many people subscribed to the network as expected. City officials abandoned the program, deciding that if they were to resurrect it they would focus on certain hotspots rather than the entire city.

Philadelphia and San Jose are planning for the second attempts after the failure of the initial efforts to establish municipal wireless networks. In Philadelphia, the City Council recommended that the city purchase the network for government use only (Vos, 2010). The city of San Jose announced in March 2012 that it is upgrading its network for public Wi-Fi access and municipal applications, but it has not yet announced a timetable for implementation of these services (Lawson, 2012). Minneapolis, one of the first cities in the country to successfully develop wireless coverage, continues to be the exception. "Wireless Minneapolis" offers free connectivity in public spaces and additional access to residents for only $20.00 per month. The city targets about 30,000 subscribers by 2013. Its success may stem from the decision to award a ten-year contract to a local company, U.S. Internet, using an anchor-tenant model, rather than going with a major provider such as EarthLink (Gorney, 2010).

In the backdrop of this debate, a small group of El Paso community members and technology leaders started to consider the potential of wireless technology. In addition, civic and business leaders announced some key local initiatives:

1. City of El Paso launched the El Paso Region Creative Cities Leadership Project (Florida, 2000; City of El Paso, 2006). The team worked together for one year to understand how the three pillars: talent, technology, and tolerance could be used to shape the city and drive economic development.
2. El Paso's Mayor formed the El Paso Lyceum committee (City of El Paso, 2006) to identify policy recommendations for programs to fuel a modern, innovation-driven, and high-value-added economy that could grow regional wealth, wages, and prosperity.
3. A group comprised of regional businesses and community leaders announced support for a major downtown revitalization process. This group saw free wireless access as an essential component as a way to increase tourism.

The concept of DEP was aligned to these three initiatives. In 2007 funding was secured and planning and implementation of the DEP wireless network began (City of El Paso, 2006). DEP was a project that had neither a service provider to fund it nor an identified anchor productivity application that the city could use to fund a build-out. DEP was started by a group of people who had the courage to say that they were willing to work together to understand what would make sense for El Paso and bring the community together.

CASE DESCRIPTION

The Proof Of Concept (POC) of DEP was funded and supported by a public-private sector consortium. Cisco Systems, Inc. provided design and planning support to complement Intel Corporation's $10,000 seed funding for the site survey. The technology was not bleeding edge and the initial objective was not a citywide network.

The approximately 2 square mile coverage area provided free outdoor wireless Internet access to about 5,000 underserved residents in Downtown El Paso. This area is one of the lowest per capita income populations in the U.S. Despite the poverty level of local residents, the area is rich in terms of culture diversity and community spirit. It houses a diverse community of homes, apartments, shops, Fortune 500 companies, K-12 schools, libraries, community centers, community-based healthcare centers, senior citizen centers, City Hall, the El Paso County Courthouse, arts and entertainment venues, museums and historic sites, hotels, bus depots, parks, recreation centers, fire, police and sheriff's departments, and two U.S.-Mexico border ports of entry. These entities would significantly benefit from free or affordable wireless Internet access.

The POC was the first step toward understanding the feasibility of providing countywide affordable Internet access. The overall goal of the project was to gather proof points to enable the city and county to understand the critical usage and adoption rate of Internet and phone. Such information is required to determine the feasibility of a countywide wireless network. Expected outcomes included:

1. A plan to use wireless Internet access as the foundation to achieve social inclusion and economic development and to develop specific programs to support these goals and use mobility to extend delivery of services into the community while they are on the go. Given the size of El Paso County, wireless was considered to be the most cost effective method to support service delivery.
2. Programs to provide computers and technology literacy to enhance the quality of life for the community and support social inclusion and the overall goal of economic development.
3. A test bed to identify, evaluate, and field test wireless applications that could improve government efficiencies and increase

productivity prior to purchase and provide funding sources to scale the network.
4. A business model to support a large-scale wireless network that serves El Paso County at large. This business model would clearly define plans for using municipal services, funding sources and service pricing, anchor tenant applications, shared services, and a strategy for infrastructure management and maintenance.

These proof points were critical to constructing a viable business model that would cost-justify and define the roadmap for the future build out of a large-scale wireless network. The business case would also identify municipal applications that could serve as anchor tenants to justify the viability of the build out with plans to provide free or low-cost Internet access in municipal entities, public spaces and underserved areas.

The DEP team recognized the unique opportunity to differentiate El Paso as a leader in planning and implementing community-centered telecommunication infrastructures as the fourth utility behind gas, electricity, and water. Wireless Internet access, in particular, would create opportunities for integrating new mobile technologies in all aspects of daily life affairs and business processes. In addition, wireless technology could extend and leverage community services, healthcare, social services, homeland security, justice, and public safety. This approach would minimize risk and reinforce prudent management of public resources and expenditure of public funds to implement ubiquitous, affordable community access to broadband Internet.

The following timeline and activities were set to enable the team to manage expectations and to ensure success:

Phase 1: 1st Quarter 2007
1. Identifying an ideal site for the POC.
2. Completing preliminary work to scope the POC.

Phase 2: 2nd Quarter 2007

1. Implementing the POC.
2. Completing the build out of pilot wireless network in the 79901 zip code to include downtown and an underserved community in the 2nd poorest zip code in the U.S.
3. Launching the DEP website in English and Spanish, with disclaimer policy, acceptable use policy, and local cultural content. E-mail service was not provided. However, links to popular e-mail sites, such as Google and Yahoo, were to be listed on the portal.
4. Monitoring bandwidth usage and adoption rates and report results to the development team (3-6 month process).
5. Testing wireless mesh technologies, planning for network management and security, and testing municipal and utility services and applications in production fields such as E-Citations and Integrated Records Management System.

Phase 3: 4th Quarter 2007

1. Determining the best business model based on data gathered during POC, formalizing organizational structure, and estimating costs to extend deployment countywide based on information gathered during the pilot.
2. Posting request for information for a countywide build out.
3. Evaluating responses and selecting the best proposal.

Phase 4: 1st Quarter 2008

1. Awarding project, acquiring equipment, and starting the build out of countywide wireless network.
2. Scaling social inclusion programs and service offerings defined during the POC to support the countywide build out.

El Paso County did not deviate much from the implementation plan for the first six months. All activities of phase one and two were completed in the first year, except for item five in phase two.

Organizational Structure

The DEP was a grassroots movement with a core team led by the technical staff of El Paso County. This team was supported by other partners including Cisco and Intel. In addition, the team grew to include other key players such as the city, K-12 education associations, higher education institutions, workforce development, the housing authority, economic development agencies, healthcare providers, Non-Governmental Organizations (NGOs), private sector companies, ISPs, and the local electric utility companies. Initially the project faced resistance from the local ISPs and local exchange carrier communities. They were concerned that the county would compete with them even though none of them were providing Internet access in the POC service area. Their initial decision not to provide service was due to the lack of density, or disposable income in the POC residences to support the cost to the build out. Once they understood the plan to provide affordable PCs and training, they supported DEP because they felt that DEP could become a new market for them. Although they were never wildly supportive, they did come to meetings to stay informed.

Self-directed committees formed to identify tasks that were critical to construct the POC and reported back to the team. The team met weekly at the start and then monthly. Each member checked their egos at the door and directed their passion and energy to making their shared vision a reality. As issues arose, the team worked together to identify and implement a solution.

DEP served as a catalyst to bring together many existing programs as the foundation for the work to be done. Many hours were spent discussing the need to have a formal team and a paid executive director. Given that there was no budget, the discussion was tabled, and the team focused on the work at hand. It was known, however, that this issue would be critical in order to scale the project.

Technology Components

Cisco was selected as the provider for the wireless network equipment. This made implementation relatively easy since the city and county were using Cisco networks and had the experienced and staff to help with the installation, management, and technical support. The network, still in place today, uses a standards-based architecture and is built on a Cisco centralized wireless controller that provides centralized management for the entire infrastructure. Three Cisco 1400 bridge access points (RAPS) mounted on rooftops to direct traffic to a 4th Cisco 1400 RAP on the roof of the County Courthouse. This RAP connects to the county Point Of Presence (POP) and serves as a backhaul to 30 Cisco 1510 access points mounted on light poles at street level, providing 802.11g to end users. The POP provides a 15 Mbps Internet backbone, 256 Kbps, guaranteed to each user, bursting up to 768 Kbps. Peak utilization has been about 8-9 Mbps. SolarWinds was used to monitor bandwidth utilization. The network is provisioned on a separated secure VLAN and firewall managed by the county. The DEP website (www.digitalelpaso.com) was developed and is hosted by the County. It was populated with locally created and socially relevant content, offered in both Spanish and English. After much discussion about security, the team adopted an acceptable use policy that is presented on the DEP splash page and must be accepted prior to using the portal. The county provisioned the public use portion of their POP to go directly to the Internet without traversing their network.

The team felt that software standardization would make troubleshooting a bit simpler once the buyer took the PC home. Refurbished PC specifications were defined by the local community college. They were a minimum of 900 Mhz or higher processor (Pentium III or later) and 128 MB RAM or greater. A generic 802,11(a) wireless access PCI adaptor card PC was added to the refurbished PCs. Microsoft OS, Office productivity Suite and Web browser software was configured on each PC. No anti-virus software included because of budget constraints.

As access points began to be deployed, it was realized that the city streets department and local electric utility should have been engaged. The City owns the street lights, but the electric utility company had the contract to maintain them. Wireless Access Points (APs) needed 25 watts of electricity to power.

We were able to engage the city and the local utility company. In fact, they have become key partners. The utility company did not charge for turn up or electricity used for the POC. However, full-blown expansion would require a funding stream to pay for electrical construction and consumption. The ideal solution would be to include this in the renewal of the maintenance contract.

Technology Concerns

Among technology concerns there was the question of additional technical personnel needed to manage deployment, monitor usage and provide ongoing support for the wireless network. The personnel concern was alleviated as the team worked together to manage the deployment of the wireless network, monitor the usage of the network, and coordinate with the city streets department and electric utility to design, install, and manage the wireless network. The team identified public sector-owned vertical assets that were used for the deployment. The plan was to use the lessons learned to construct a comprehensive infrastructure plan including costs and a timeline to build out a wireless infrastructure. Although outdoor wireless was a relatively new technology, it was not considered a bleeding edge technology. The county served as the network operations center because it had a strong technical staff, a standards-based reliable network, and sufficient Internet capacity. The centralized management model and mesh design did not add a significant burden to the load of technical support. Thus, no additional personnel were required.

Another major concern was whether there would be overwhelming load on the free wireless network that it would not be able support. This turned out not to be the case. Since the community did not have PCs, free wireless was relatively useless and the bandwidth sat idle. This dilemma was taken back to the team and was the catalyst for developing the social inclusion programs.

Additionally, there was a concern about signal interference from the Mexican side of the border. Both 802.11a and 802.11g standards use the 2.4 GHZ ISM band. It operates in the U.S. under Part 15 of the US Federal Communications Commission Rules and Regulations. The segment of the radio frequency spectrum used by 802.11 standards varies between countries. In the U.S., the 802.11a and 802.11g devices could be used without a license, as allowed in Part 15 of the FCC Rules and Regulations. These standards are not regulated in Mexico, but are not allowed for cross-border use. The Mexican rules were not being enforced by some companies who were using the spectrum and causing interference in the POC. This issue was resolved with some technical tweaking, but mostly was resolved by a phone call to the neighboring companies causing the interference. For the most part, it was agreed on which spectrum in the non-regulated frequencies would be used by the wireless network and they adjusted to accommodate.

Finally, costs also became a major concern as the POC progressed and the team began to extrapolate what it would take to fund a formal organization, sustain the project, and scale to a countywide network.

Social Concerns

The initial intent of the POC was to address the concerns of social inclusion. When the project started, we had very little in terms of baseline data on the community we were planning to serve. If we were able to secure funding to build out a countywide network. As we began the process of applying for grants, we realized that we needed more specific baseline data to support our applications. A survey was developed by the team for the purpose of supporting grant requests with reliable current survey data. We also felt that our populations would be missed by most traditional polling methods that relied on landline phone calls. Our most vulnerable populations had no phones or were using cell phones. We felt that we could get more inclusive data by going directly to community sites and doing face to face polls, since our populations were rapidly abandoning land lines for cell phones. We set a goal of surveying 500 people in areas with significant populations living below the poverty level, with the median income of $25,275, well below the U.S. national median household income (2006-2019) of $51,914 as reported by http://quickfacts.census.gov/qfd/states/00000.html.

We completed about 2,268 face-to-face surveys at a dozen locations including libraries, community college campuses, workforce centers, community centers, and community-based health clinics. The results were posted by community college students. About thirty students participated in tabulating the survey results. The participating students were part of a community college program called Student Technology Services (STS). STS is a student friendly program that allows students to concentrate on school and earn an hourly wage, while they learn technology skills that are valuable in the workplace. The survey results showed that:

- 80% of the most vulnerable populations did not have PCs connected to the Internet at home
- 63% wanted an Internet connected PC at home
- 56% were interested in a refurbished PC
- 57% were interested in free training to earn a refurbished PC

We did not poll income levels, but focused on pockets of the community with the lowest per capita incomes. Those polled, who were interested in having an Internet connected PC in their home, were willing to pay an average of $15 per month for Internet access and $15 per month for a PC. This clearly showed that access, cost of Internet service and PCs, and lack of training were significant barriers to entry for many community members. The implications of the digital divide in El Paso are dramatic. Computer literacy was a growing requirement for many jobs, including occupations traditionally not requiring computer skills. Without collective or government intervention, we would be an entire city lost in the digital divide.

The DEP family-centric approach was also critical to keeping families engaged in the learning process. The POC used free wireless Internet and social inclusion programs to increase computer and financial literacy. This in turn, set the basic building blocks for a more employable workforce and encouraged students to continue education by engaging them early. The fund obtained BTOP funding enabled DEP to expand its proven programs throughout the county and reach diverse un-served and underserved populations. However, the planning and progress must continue in order to make our vision a reality.

Finances and Business Partners

There was no formal budget for the POC. When Intel offered the initial $10,000 seed funding, we literally did not know how to proceed. Working through government processes is not easy, especially when it deals with technology. Fortunately, we had some seasoned team members who helped. While they worked with the city and county, others started calling clients who had presence in the proposed POC coverage area to gauge their interest and to secure their partnership.

All anchor stakeholders were public entities. Therefore, the project had to be presented to each of their respective boards in order to gain support and agreement to sign the memorandum. The project received overwhelming support and was approved unanimously by the boards. This process took about three months, compared to the two weeks it took to construct the network. Below is breakdown of the approximately $200,000 USD it cost to implement the network as well as commitment and contribution of partners:

1. El Paso County committed services that were absorbed by their normal operating budget including:
    a. $25,000 installation cost and engineering support to help construct the network.
    b. Ongoing engineering support, management troubleshooting, and maintenance of the network.
    c. Access to their vertical assets such as electrical poles.
    d. Access to their 150 Mbps Internet POP as the aggregation point/onramp to the Internet.
    e. Construction, hosting, and maintenance of the Web portal.
    f. Providing help desk services from 7:00 am to 7:00 pm, M-F, to help with connectivity and other PC related issues.
2. City of El Paso committed:
    a. $100,000 to purchase equipment.
    b. Access to vertical assets and the fiber backbone.
3. El Paso Independent School District and the Housing Authority of the City of El Paso each committed:
    a. $25,000 to purchase equipment.
    b. Access to vertical assets and the fiber backbone.
4. El Paso Electric Company contributed a project manager to add the electric taps to the city owned light poles.
5. Intel Corporation contributed $10,000 initial seed funding for the site survey, design, and installation support.

6. Cisco Systems, Inc. extended significant discounts on the price of purchased wireless equipment. It also provided engineering design and troubleshooting assistance.

IMPLEMENTATION CHALLENGES

When DEP was launched, the team watched anxiously to see how much the community would use the free wireless Internet and began monitoring bandwidth utilization. Utilization was very low. We were getting some tourists, but the residents of the area did not have the disposable income to purchase PCs, nor did they have basic computer literacy training. Once we realized the implication of these issues, we began to work on overcoming these barriers to entry. It took almost a year to fully develop programs to meet these needs and launch critical social inclusion programs to enable the community to start using the wireless network.

Three critical support programs have evolved:

Computer Recycling Program

Local businesses contributed used PCs. This gave these businesses a "green" recycling option. The community college channeled the PCs through their student technology program. The PCs were refurbished, upgraded to include a wireless card, and loaded with the Microsoft Office suite and a Web browser. The PCs were then given to La Fe, a community based Non-Governmental Organization (NGO), to be contributed to families in the POC coverage area. If the PCs failed after they were distributed, users could return them to the NGO where it would be swapped out for another one. The returned PC would then re-enter the refurbishing program.

In the first year of operations, student interns from the local community college and participants from the Student Technologies Program worked over 2,600 hours to refurbish and distribute 80 refurbished PCs at no cost to the families in the POC. In addition, they maintained 40 PCs in their inventory ready to be distributed to needy families. La Fe was in line to receive about 2,000 PCs donated by local businesses as part of their refresh cycles.

Technology Literacy Program

Given the tight-knit community, many with limited access to transportation, the team developed a community-based and family-centered approach. A NGO located in the area of the POC offered their technology and cultural center as a training center for the community. Together with the community college they collaborated to develop basic computer literacy training in both Spanish and English. Classes included basic computer literacy, Internet browsing, e-mail, and printing skills. To encourage families, classes were offered on the PC they would ultimately take home upon class completion.

The first cohort of 20 families completed the required 40 hours of free training to earn a free refurbished PC that they could take home with them. In the first year of operations, 100 families were trained in cohorts of 20 families per class and over 3,000 hours of training was delivered to the community. Bandwidth utilization in the POC began to trend upward.

In addition, the local Cultural and Technology Center is situated so that as parents took their children to health clinic they could also encourage their children to use the computers available in the clinic lab. We hope that when these children grew older, they could participate in the Cisco Network Academy, create video in the TV studio, or use Center's business micro-incubator center to make flyers, posters, and brochures for small business.

Affordable PC Purchase Program

The price of desktops fell significantly and the price of laptops began to be more affordable. Mobility was beginning to influence PC trends.

Another NGO stepped in and added an offer to their portfolio for the purchase of new PCs, using Community Development Funding Institution (CDFI) backed loans. Applicants were required to qualify under federal poverty guidelines and meet loan eligibility. This was a new way to engage the community in understanding how to work with legitimate financial institutions. It also gave them a conduit to deliver financial literacy. Intel provided an additional $5,000 in grant funding for TV advertising and print media, including brochures, flyers, and posters. This was leveraged by the NGO in a grant request to a fund Volunteers in Service to America (VISTA). They were awarded the grant and volunteers began to operate in key community sites to market and support the POC's social inclusion programs. TV advertising time was donated by another NGO.

In the first year of the affordable PC purchase/ loan program, the NGO leveraged their *existing* $635K revolving loan fund to extend 32 PC loans and drive program awareness through television advertising, posters, and brochures that were distributed by VISTA volunteers housed in El Paso schools.

Up to this point, the focus of the team had been on social inclusion. Programs were in place to resolve the barriers to entry, but the challenge of scaling the network and growing the programs remained. In 2009, the team turned their focus to identifying and testing a productivity application over the wireless network. This could potentially fund the build out of the network. They identified a wireless crosswalk flasher and video surveillance application that used the same model that had been successfully deployed in Midland, Texas in 2008. The intent was to test an application that could potentially share the DEP bandwidth and help cost justify and fund expansion of the coverage area. This was the only potential anchor municipality application that was explored as part of the POC (Walker, 2009). The test was successful but the ability to work through the complexities of working through multiple agencies was overwhelming, so the work was tabled.

Also in 2009, The American Recovery and Reinvestment Act was signed into law by the newly elected President Barack Obama. The goal was to direct funds to technology infrastructure programs through a competitive grant called the Broadband Technology Opportunity Program (BTOP) to expand broadband access to un-served and underserved communities across the U.S. and to stimulate investments in wired and wireless technology and infrastructure to provide long-term economic benefits. BTOP would provide grants to fund broadband infrastructure (wired and wireless) public computing centers and sustainable broadband adoption projects to support education and healthcare delivery, public safety, and job creation.

DEP was aligned to these same goals, with public sector entities, private companies, and NOGs were eligible to apply. PPPs were strongly encouraged and projects had to be substantially complete within two years of the award. The DEP team seized the opportunity to submit grant proposals. If the grant request was awarded, it would provide funding to expand the wireless network countywide and scale the social inclusion programs to support a larger population.

Because it was difficult to meet the tight application submission deadline, the team suspended work on the DEP social inclusion programs, but left the free wireless Internet access in place. All of the work to date on DEP was used as the foundation for the grant. The core team grew to include more community partners to collaborate and submit the BTOP grants. In Round 1 the team submitted grant requests for approximately $40 million to fund four major initiatives: 1) critical citywide fiber back haul and connections to all city sites, 2) countywide microwave and wireless infrastructure, 3) public computing and training centers, and 4) and a sustainability program that would provide refurbished PC loaners tied to completion of computer literacy training, affordable PC loan/purchases tied to financial literacy training, and workforce training through the community college.

The DEP collaboration, cost sharing, and lessons learned were used as the cornerstone for all four requests. The grant requirements mapped almost identically to the social inclusion model developed by the DEP team. The Infrastructure structure request made the short list, but funding was not awarded. No feedback was provided by the BTOP to the DEP applicants.

In 2010, the team regrouped and submitted to the Round 2 BTOP grants, requesting grants to fund the same projects requested in Round 1. The request for public computing centers was awarded approximately $8.4 million to fund the "Virtual Village: Digital El Paso's Pathway to Success" project. The announcement cited El Paso's need for such a project:

The City of El Paso, Texas and the surrounding area face significant challenges because of its poverty level, sparse geographic distribution, and limited access to tools necessary for economic development. Among those tools are computer access, broadband connectivity, and other advances in technology necessary to improve education, job training, and health. A 2009 Digital El Paso survey showed that only 35 percent of El Paso-area households have high-speed Internet access, compared to over 60 percent of households nationwide. The city's Virtual Village project addresses these concerns through an extensive overhaul and expansion of its public computing capacity. The project will engage hundreds of agencies, community institutions, and local organizations to target vulnerable populations, particularly at-risk youth, the elderly, the unemployed, and minorities, with training in and access to computer technology (Broadband Technologies, 2010).

The grant funded wireless Internet access and technology literacy training for 96 public computing sites in El Paso County. Twelve public library branches and 18 partner sites, including NGOs and healthcare organizations were upgraded to provide training. This included 326 PCs, 211 of which were refurbished. Usage in the 13 library sites was 9,923 sessions per week and 2,931 community-based training hours were delivered. In essence, the POC social inclusion programs received funding and were being expanded. The infrastructure request was not funded, but the microwave network and fiber backhaul plan developed for the grant is being constructed and implemented by the City and County as funding becomes available.

The Digital El Paso POC has a proven track record and excellent collaborative community support. The project was executed as planned, but the challenges we faced are universal. We have limited resources of people, time, and, money to get the work done.

The largest remaining challenge is how to sustain the network and fund/direct the organization. The BTOP award has been such a boost to the project and the economy, but it has taken much of the DEP team's resources to bear on the task of implementing it within the two year grant requirement. The team who worked on the POC infrastructure is continuing to work within each of their respective organizations to build out infrastructure. Their participation in DEP and the BTOP efforts gave them a keen awareness of the importance of spotting opportunities to cost share fiber and wireless network as they are built out. The successful results of team has earned them credibility and support in the eyes of their leadership as we work together to ensure that all of our community has access and skills to become citizens in the global digital community.

PROPOSED BUSINESS MODEL

The covered area included a density of un-served and underserved populations. Similar to other community wireless projects hence PPPs were a key component of our plan (Wireless Minneapolis, 2006; Forlano, 2008; Karisny, 2010; Settles, 2011). The private sector brought PC donations and potential future funding opportunities for the programs. The NGOs brought access and understanding of the communities they served; without their participation, we could not have succeeded. In particular, Centro Salud Familiar La Fe (La Fe) was key to the success of the programs. La Fe has served the community for many years and had been very innovative in adding technology to their community based healthcare delivery services.

The initial success of DEP attracted local ISPs. At first, they were not supportive because, despite the fact that the initial stage covered only under-served areas, they thought that we were taking away their business. Eventually, they came to realize that if we were successful, we would educate a whole new market for them. In particular, as this market outgrew the basic free bandwidth, end users would see the value and be willing to pay for upgraded services. The local ISPs and competitive local exchange carriers embraced this quickly; the incumbent local exchange carrier did not.

To address the key challenges (e.g. sustainability and funding) facing community wireless networks, we propose the business model shown in Table 1.

Suggested Programs and Committees

We identified the following needed programs during the POC:

1. **Hotspot Central Exchange:** This program would create a catalog of all public hotspots willing to participate in DEP. In addition, the team had planned to change the broadcast Service Set Identifier (SSID), a 32-character unique identifier (SSID) sent over the wireless network.

2. **A Common "Digital El Paso" SSID:** This SSID would be used for all city, county, university, and college public hotspots. This would give a single interface to all DEP members and increase community awareness.

3. **Video Content for Education and Healthcare:** El Paso exists in extremely isolated and remote area, we preferred a family-centric approach for community wireless networks. The only world that many of our children know is their immediate neighborhood. There is no opportunity for them to travel and experience other places, traditions, or cultures. In addition, we live in a desert climate. Many people have never seen the ocean or large aquarium except on TV. Therefore, high quality videoconferencing would add a window to their world. If we engage children at a young age, this experience could serve to reinforce the importance of staying in school and not just getting a high school diploma. We also believe that integrating ICTs in education would prepare, and encourage students, to go to college and obtain a higher degree. This would reinforce education as the key to a better and more productive life. The same delivery method could be used for the delivery of community-based healthcare.

4. **Web Presence and Social Media:** Our population includes many first time Mexican/American residents. They use e-mail and social media heavily and this will enable them to communicate with friends and family members in Mexico. This is critical to maintain family ties. As the new economy continues to evolve, Internet access is needed to support entrepreneurial small businesses and NGOs or innovate new ways to tap into new markets and provide new service offerings.

5. **Accessibility Technologies:** El Paso is a home to many soldiers returning home with disabilities. Mobil technology and special IT-based programs are important them to overcome their disabilities.

6. **Smart Grid and Green Technologies:** Companies who are developing smart technologies could use our POC as a test-bed for sensor based technologies and help

Table 1. A proposed public-private business model for CWNs

Who Owns the Network?	Who Manages the Network?	Who Uses the Network?
Public Sector City Led Public Sector Consortium (Consortium)	Private Sector Master Service Provider (MSP)	Community, local businesses, and public sector

• Share construction and ongoing maintenance costs
• Contract with MSP to construct network and provide public sector Internet access
    -30% reserved for consortium use,
    -provisions to add 5% annually, up to a max of 50%
• Make contract provisions to provide electricity on illumination poles owned by city
• Grant MSP access to all publically owned vertical assets and rights of way
	• Manage/maintain network
	• Perform all billing and report to consortium
	• Option to co-locate on network
	• Guarantee consortium 50% bandwidth reservation for public use at $5 USD per month (this provides a base revenue stream for the MSP)
	• Sell bandwidth not reserved by consortium
	• Agree to dynamically provision <u>all</u> bandwidth to consortium Tier 1 (State & Local Government) in case of natural/manmade disasters
		Tier 1
		Free Internet Access for State & Local Government

		• Provide network for traffic management systems, parking meter systems
		• Support physical security -video surveillance
		• Enable Public sector workforce mobility
		    -Code Enforcement/
		    -Permitting
		    -Healthcare Delivery
		    -Work orders
		• Provide redundant network for emergency responders

		Tier 2 Free/Reduced Internet Access for the Community
		Tier 3 Free/Reduced Internet Access for: • Un served communities in pockets of poverty • Public spaces, parks, recreation centers, libraries, etc. • Tourism • Community centers, NGOs

underwrite the expansion and support of the network. They could also engage local university students to intern/participate in these projects.

However, we were not able to implement these programs because of lack of resources:

We also conclude that a community wireless network would benefit from a formal non-profit. We recommend the following organizational structure for a non-profit status for community wireless networks:

Executive Director

The Executive Director (ED) would oversee all aspects of the organization and report to the Board of Directors. The following managers would report to the ED.

1. A technical Operations Director (OD) who would oversee the network;
2. A Programs Director (PD) who manages all of social inclusion programs;
3. A Communications DIRECTOR (CD); and
4. A Financial Director (FD).

Executive Board of Directors

This board is comprised of local government, civic, and business leaders would give the executive director access to key state and local government, community, and private sector decision- makers and funding sources. This would give us access to a large pool of potential grants and funding sources.

The organizational strategy would be to leverage existing programs and resources wherever possible and should include at a minimum the following committees:

Research and Development (R&D) Committee

Reporting to the technical Operations Director (OD), R&D would be comprised of three teams working together to identify, test, evaluate, and recommend technology adoption. Membership should include technology leaders and partnerships with the private sector and ISP technical staff.

1.   The Network team would focus on emerging next generation wireless technologies and oversee ongoing operations of the network including maintenance and upgrades. They would also identify potential opportunities to cost share construction of infrastructure and identify and secure access to vertical assets and electricity required for access points and bridges.
2.   The Business Process team would operate as a cross-functional team to meet with public sector users to identify, test, and evaluate mobility software applications that could significantly improve the productivity and delivery of municipal services. Emphasis would be placed on applications that were multi-jurisdictional and could make the flow of information more integrated such as Geographic Information Systems (GIS) that are critical to understanding and managing many shared critical, business processes.
3.   The Application team would engage with participating entities to identify and prioritize software/productivity applications specific to their unique requirements that require mobility. For example, city specific applications may include automated traffic management and meter reading, permitting, emergency responder systems, public safety, pavement and street management, health and human services, code enforcement and inspections, and GIS. The team could help identify, evaluate, and test the applications in the POC to understand how they would function in our unique environment.

The R&D team would also provide input into the development of regional broadband policy and a plan for a fiber backbone and wireless "last mile." At a minimum, this should align to any federal or state's broadband policies and include recommendations for policy changes in the city permitting and subdivision ordinances to protect horizontal and vertical assets to be used by the city and county for fiber and wireless countywide build-out. Implementation of this policy would need to be done through the legislative process.

The Program Director (PD) would oversee all support programs including:

1.   **Small Business Technology Service:** Local NGOs would lead the efforts to identify and coordinate Public/Private Partnerships (PPPs) to provide small business incubator services and resources including understanding the best fit technology purchase for their business, including access, WAN/LAN PC and software purchases, access to capital, technology training, printing, website and brochure development, and social media marketing. Over 95% of El Paso's businesses have 20 employees or less. Most are minority and women owned businesses and most are novice technology buyers/users. Adding technology to their businesses has the potential to enable them to grow significantly. Internet access can give them new ways to do routine purchases of goods and supplies to run their business, as well as add new ways for them to market their business and sell their goods, such as social media marketing.
2.   **Help Desk:** This could be outsourced to the community college and other educational partners as jobs for student interns. It is a critical part of creating an industry ready workforce. All support calls would to be documented from intake to resolution; trending the calls could point out opportunities to tweak training to meet reality.

The Communications Director (CD) would be responsible for fostering and coordinating advertising and communications with media and public information officers from member organizations. The committee would oversee the following committees:

1. **Web Presence and Social Media:** This group would develop and maintain the website and social networking activities. It would be the place where people, information, and community come together.
2. **Content Creation:** Socially relevant content for each neighborhood could be added to the website.

The Financial Director (FD) would be responsible for all finances and business operations and oversee the following committees:

1. **Grants and Funding Committee:** This group would identify, fit, and prioritize potential grants and donations to further the community goals.
2. **Review and Sustainability Committee:** The committee would gather data from and report the outcomes from each of the programs and services as replicable models to support, grow, and sustain programs for large-scale deployment.
3. **Vendor Relationships and Contract Management Committee:** This committee would engage and negotiate contracts and Volume Purchase Agreements (VPAs) for hardware, software, by leveraging the combined community purchasing power and expertise. This will ensure the best price for packages defined by the affordable PC Purchase Program and could include ISP, local and wide area network (LAN/WAN) and affordable loan products. All products could also be extended small business in the community.

The POC of El Paso network was successful in large part because wherever possible we leveraged *existing* assets and relationships, and we aligned to key community leadership initiatives. All costs, hard and soft, were absorbed by PPPs. This is an adequate starter model, but in order to scale the model, a funded organization to manage some critical components of a comprehensive plan must be in place.

CONCLUSION

Community Wireless Networks are developed by individuals with common goals and visions. The creation, evolution, and success of these networks depend highly on the relationships among participants and the synergy of the community. The key to their success lies in awareness and community involvement. The implications of public/private partnerships and collaboration to achieve this common goal are significant. Shared costs, community engagement, and the synergy of exchanging ideas are among the true gems we discovered in our journey. Many of the idealistic projects launched across the U.S. are no longer functioning. Given the incredible growth of mobile devices and cloud-based services, we are poised for a new surge of growth using new technologies to build out wireless networks. It is the drivers for adoption and the community focus and support that make success. We also conclude that there is no single model for success

DEP received the prestigious Texas Association of Government IT Manager's (TAGITM, 2008) First President's award for the "Best Collaborative Project in the State of Texas." TAGITM leadership traveled to El Paso to present the award to the team at the annual El Paso Hispanic Chamber of Commerce BizTech Conference. El Paso County, the City of El Paso, The Housing Authority of the City of El Paso, El Paso Independent School

District, Intel, and Cisco received the award on behalf of the team.

In 2009, DEP received the prestigious award of El Paso Hispanic Chamber BizTech Technology Hall of Fame recognizing the DEP team as technology visionaries who made a significant impact on the community. We hope that our case would spark ideas and inspire this next wave of community wireless networks.

REFERENCES

Abdelaal, A., & Ali, H. (2008). A graph theoretic approach for analysis and design of community wireless networks. In *Proceedings of Americas Conference on Information Systems (AMCIS)*. Toronto, Canada: AMCIS. Retrieved October 15, 2011 from http://aisel.aisnet.org/amcis2008/310/

Abdelaal, A., Ali, H., & Khazanchi, D. (2009). The role of social capital in the creation of community wireless networks. In *Proceedings of the 42nd Hawaii International Conference on Systems Sciences*. Waikoloa, HI: IEEE Press.

Breitbart, J. (2008). The Philadelphia story: Learning from a municipal wireless pioneer. *New America Foundation*. Retrieved October 11, 2011, from http://www.newamerica.net/publications/policy/philadelphia_story

Broadband Technology Opportunities Program. (2011). The virtual village: Digital El Paso's pathway to success. *Broadband USA*. Retrieved October 15, 2011, from http://www2.ntia.doc.gov/grantee/city-of-el-paso

Carroll, J., & Rosson, M. (2007). Participatory design in community informatics. *Design Studies*, *28*, 243–261. doi:10.1016/j.destud.2007.02.007

Cisco Systems, Inc. (2006a). Municipalities adopt successful business models for outdoor wireless networks. White Paper (C11-325079-00). *Cisco*. Retrieved October 20, 2011, from http://www.cisco.com/en/US/prod/collateral/wireless/ps5679/ps6548/prod_white_paper0900aecd80564fa3.pdf

Cisco Systems, Inc. (2006b). Evolution of municipal wireless networks. White Paper (C11-378713-00). *IDG Connect*. Retrieved on October 26, 2011, from http://www.idgconnect.com/view_abstract/2254/evolution-municipal-wireless-networks

City of El Paso. (2006). *The digital El Paso project*. Retrieved November 10, 2011, from http://www.elpasotexas.gov/it/_documents/3T%20Quality%20of%20Place%20Proposal_Final.pdf

City of El Paso. (2011). *El Paso Lyceum*. Retrieved November 1, 2011, from http://www.elpasotexas.gov/mayor/elpaso_lyceum.asp

El Paso Region Creative Cities Leadership Project. (2011). *City of El Paso*. Retrieved October 31, 2011, from http://www.elpasotexas.gov/mcad/cclp.asp

Florida, R. (2002). *The rise of the creative class: And how it's transforming work, leisure, community and everyday life*. New York, NY: Basic Books.

Forlano, L. (2008). Anytime? Anywhere? Reframing debates around community and municipal wireless networking. *Community Informatics*. Retrieved November 3, 2011 from http://ci-journal.net/index.php/ciej/article/view/438

Gorney, D. (2010). Minneapolis unplugged. *The Atlantic*. Retrieved March 27, 2012 from http://www.theatlantic.com/personal/archive/2010/06/minneapolis-unplugged/57676/

Karisny, L. (2010). Year-end review: Economic recovery through municipal wireless networks. *MuniWireless*. Retrieved November 3, 2011 from http://www.muniwireless.com/2010/01/01/year-end-review-economic-recovery-through-municipal-wireless-networks/

Lavallee, A. (2008). A second look at citywide wi-fi. *The Wall Street Journal*. Retrieved March 27, 2012 from http://online.wsj.com/article/SB122840941903779747.html

Lawson, S. (2012). San Jose tries again with free downtown wi-fi. *NetworkWorld*. Retrieved March 27, 2012 from http://www.networkworld.com/news/2012/031212-san-jose-tries-again-with-257199.html?page=1

Quinn, P. (2006). Community wireless networks: Cutting edge technology for Internet access. *Center for Neighborhood Technology*. Retrieved November 1, 2011 from http://www.cnt.org/repository/WCN-AllReports.pdf

Rideout, V., & Reddick, A. (2005). Sustaining community access to technology: Who should pay and why! *The Journal of Community Informatics, 1*(2), 45–62.

Rubio, F. (2009). Public computer centers program – Sustainable adoption program. *Broadband USA*. Retrieved November 1, 2011 from http://digitalelpaso.com/documents/Sustainability%20(Non%20Infrastructure)%20Grant%20App%20-%2020090819.pdf

Settles, C. (2011). Should public-private partnerships be in your broadband future? *Government Technology*. Retrieved October 20, 2011 from http://www.govtech.com/wireless/Public-Private-Partnerships-Broadband-Future.html

Simpson, L. (2005). Community informatics and sustainability: Why social capital matters. *The Journal of Community Informatics, 1*(2), 102–119.

Simpson, L., Wood, L., & Daws, L. (2003). Community capacity building: Starting with people not projects. *Community Development Journal, 38*(4), 277–286. doi:10.1093/cdj/38.4.277

TAGITM. (2008). *2008 award recipients*. Retrieved November 10, 2011, from http://www.tagitm.org/?page=2008Awards

Vos, E. (2010). Philadelphia city council committee approves purchase of citywide wi-fi network. *MuniWireless*. Retrieved March 27, 2012 from http://www.muniwireless.com/2010/06/10/philly-approves-purchase-of-citywide-wifi-network/

Walker, B. (2009). Wireless traffic management network for downtown renewal. *Industrial Ethernet Book, 52*, 31–32.

Wireless Minneapolis. (2006). *Municipal broadband business case*. Retrieved October 9, 2011 from http://www.ci.minneapolis.mn.us/wirelessminneapolis/MplsWireless_Business-Case_V3.pdf

ADDITIONAL READING

Abdelaal, A., & Ali, H. (2007). A typology for community wireless networks business models. In *Proceedings of the Thirteenth Americas Conference on Information Systems (AMCIS)*. Retrieved July 25, 2012 from http://aisel.aisnet.org/amcis2007/308

Abdelaal, A., & Ali, H. (2012). Human capital in the domain of community wireless networks. In *Proceedings of HICSS*, (pp. 3338-3346). IEEE Press.

Ballon, P. (2007). Business modelling revisited: The configuration of control and value. *The Journal of Policy, Regulation and Strategy for Telecommunications. Information and Media, 9*(5), 6–19.

Caragliu, A., Del Bo, C., & Nijkamp, P. (2009). Smart cities in Europe. In *Proceedings of the 3rd Central European Conference on Regional Science (CERS)*. Retrieved July 25, 2012 from http://www.cers.tuke.sk/cers2009/PDF/01_03_Nijkamp.pdf

Chambers, J. (2011). *2011 commencement address*. Paper presented at the Duke University Commencement. Durham, NC. Retrieved July 25, from today.duke.edu/2011/05/commence2011

Charny, B. (2007). San Francisco formally ends citywide wifi effort. *Market Watch*. Retrieved July 15 2012, from http://www.marketwatch.com/story/san-francisco-formally-ends-citywide-wi-fi-effort

Coleman, S. (2007). E-democracy: The history and future of an idea. In Robin Mansell, C. A., Quah, D., & Silverstone, R. (Eds.), *The Oxford Handbook of Information and Communication Technologies* (pp. 362–382). Oxford, UK: Oxford University Press. doi:10.1093/oxfordhb/9780199548798.003.0015

Eisenmann, T. (2007). *Managing proprietary and shared platforms: A life-cycle view*. Cambridge, MA: Harvard Business School.

Evans, D., Hagiu, A., & Schmalensee, R. (2005). A survey of the economic role of software platforms in computer-based industries. *CESifo Economic Studies, 51*(2-3), 189–224. doi:10.1093/cesifo/51.2-3.189

Gardner, D. (2008). *EarthLink to shut down New Orleans' municipal wifi*. Retrieved July 18, 2012, from http://www.informationweek.com/news/mobility/muni/207402189

Gawer, A., & Cusumano, M. (2002). *Platform leadership: How Intel, Microsoft, and Cisco drive industry innovation*. Boston, MA: Harvard Business School Press.

Gillet, S. (2005). *Municipal wireless broadband: Hype or harbinger?* Paper presented at the Symposium on Wireless Broadband: Is the US Lagging? Washington, DC.

Graham, S. (2002). Bridging urban digital divides: Urban polarization and information and communication technologies. *Urban Studies (Edinburgh, Scotland), 39*(1), 33–56. doi:10.1080/00420980220099050

Hollands, R. (2008). Will the real smart city please stand up? *City, 12*(3), 303–320. doi:10.1080/13604810802479126

Hoorens, H., Elixmann, D., Cave, J., Li, M., & Cattaneo, G. (2012). *Towards a competitive European internet industry: A socio-economic analysis of the European internet industry and the future internet public-private partnership*. Cambridge, UK: Rand Europe.

Komninos, N. (2008). *Intelligent cities: Building 3rd generation systems of innovation*. (MSc Thesis). University of Sheffield. Sheffield, UK.

Porter, M. (2001). Strategy and the internet. *Harvard Business Review, 79*, 62–78.

Rizk, N., & Kamel, S. (2012). ICT strategy 4 development: Public-private partnerships—The case of Egypt. *International Journal of Strategic Information Technology and Applications, 3*(2), 72–90. doi:10.4018/jsita.2012040105

Rogoway, M. (2010). *Portland set to dismantle, donate abandoned wi-fi antennas*. Retrieved on July 22, 2012, from http://blog.oregonlive.com/siliconforest/2010/01/portland_set_to_dismantle_dona.html

Schiff, A. (2003). Open and closed systems of two-sided networks. *Information Economics and Policy, 15*, 425–442. doi:10.1016/S0167-6245(03)00006-4

Zimmerman, J. N., & Meyer, A. (2005). Building knowledge, building community: Integrating internet access to secondary data as part of the community development process. *Community Development, 36*(1), 93–102. doi:10.1080/15575330509489874

Chapter 7
Motivations and Barriers of Participation in Community Wireless Networks:
The Case of Fon

Giovanni Camponovo
University of Applied Sciences of Southern Switzerland, Switzerland

Anna Picco-Schwendener
Università della Svizzera Italiana, Switzerland

Lorenzo Cantoni
Università della Svizzera Italiana, Switzerland

ABSTRACT

Wireless communities are an interesting alternative to 3G networks to provide mobile Internet access. However, the key success factor for their sustainability is whether they are able to attract and retain a critical mass of contributing members. It is thus important to understand what motivates and dissuades people to join and participate. This chapter analyzes motivations, concerns, usage, and satisfaction of members of Fon. Fon is the largest wireless community in the world. This study employs a mixed research method, combining qualitative exploratory interviews with a quantitative survey. Members are mainly motivated by a mix of utilitarian (getting free connectivity) and idealistic motivations (reciprocity and altruism), whereas intrinsic and social motivations are less relevant.

INTRODUCTION

We live in an increasingly mobile and connected society. People traditionally accessed the Internet via fixed-line services. Currently, with the diffusion of a new generation of mobile devices like, laptops, smartphones and tablets, the need for having affordable Internet access anytime and anywhere becomes stronger. This fostered a massive adoption of wireless technologies for connecting to the Internet, to the point that they overtook fixed broadband subscribers in 2008 (International Telecommunications Union, 2009).

DOI: 10.4018/978-1-4666-2997-4.ch007

The 3G networks, offered by Mobile Network Operators, are by far the most widely adopted solution. They are ubiquitous and reliable, but are slow and expensive. For its distinctive advantages, Wi-Fi is an interesting alternative despite its limitations. It has limited range, but is faster and cheaper. Moreover, it operates on unlicensed spectrum and hence allows many alternative business models (Bharat, Rao, & Parikh, 2003). Network operators use it to complement 3G by offering paid fast connections in crowded venues like airports and hotels. Individuals can integrate their private Wi-Fi access points into wireless communities providing free wireless connectivity to each other and the public at large. Other for-profit companies may try to blend commercial and community aspects into hybrid communities where the company supports members who share their own access points in exchange of being able to operate and cash in on the community network.

For these communities to be viable it is fundamental to attract a critical mass of members. As a result, it is important to understand:

why people join and actively contribute to wireless communities?

Understanding why community members contribute and participate is very important for offering suitable incentives. Even though researchers have recognized this to be the most critical research issue on wireless communities (Bina & Giaglis, 2005), existing research mostly focused on pure non-commercial communities.

The purpose of this chapter is to understand what drives people to join and actively participate in a hybrid wireless community. The distinction between pure and hybrid communities is important because the presence of a supporting firm can influence members' motivations and participation, ultimately determining the success of the community. Moreover, while most pure communities struggle achieving a critical mass (the largest one, NYC Wireless, only has 40,000 participants),

hybrid communities appear to be more successful (the largest one, Fon, claims to have more than 4 million users). A possible reason is that the latter are better at motivating people by offering an attractive mix of incentives and support.

To attain this research purpose, a mixed method approach was employed. The Fon community has been chosen because it is the largest and most successful case of hybrid wireless community. In a first phase, a qualitative content analysis of Fon community forums and 40 exploratory semi-structured interviews with Fon members were conducted. This chapter complements these qualitative insights with a quantitative survey of 292 members about their participation, motivations, and concerns with the Fon community.

THE FON COMMUNITY

Fon (Fon Wireless Ltd.) is a for-profit company founded in 2005 by Martin Varsavsky. Its mission is to create "a Wi-Fi network built by the people" where "you share a little bandwidth with others and millions more share with you."

Fon initially provided a free software solution that could be used to convert Linksys routers into Fon hotspots, but then quickly started selling its own custom "Fonera" router to provide an easier way to create community hotspots. The idea was to generate revenues through the sale of routers and antennas and access fees from non-sharing members wanting to use the Fon network and advertising.

Fon received funding from important firms including Google and Skype. This allowed it to heavily promote its activity by distributing their routers at a low cost or even for free, thus seeding the community network and enabling its growth. However, with that course of action, "Fon has been losing large amounts of money since its inception" (Middleton, et al., 2008).

Over time, Fon regularly adjusts its business model to adapt to evolving market conditions.

In particular, it recently started to focus more on selling its routers as a source of revenue, relying less on promotions and more on building partnerships with telecom firms to further expand its network. The first collaboration started in 2007 with BT Group (UK), followed by SFR (France), ZON (Portugal), Comstar (Russia), Belgacom (Belgium), and others. These companies typically integrate Fon's software in their routers to allow customers to participate in the Fon community without having to buy an additional Fonera Router.

Fon has a broad target. Everybody who has a broadband Internet connection at home can join the community and become a Fonero. There are three different types of membership. "Linus" members share their home connection through a Fonera (or a compatible model) for free and in turn can access other Fon Spots for free. "Bill" members are like Linus but also get 50% of the net revenue (i.e. after subtracting fees and taxes) generated by passes bought at their spots. "Alien" members do not share their Internet connection and have to purchase passes to access Fon Spots. By segmenting members in this way, Fon may appeal to a variety of users: "Linus" for those who value community principles like free sharing, "Bill" for those who want to earn money or find fair that Fon shares revenues generated by their hotspot, "Alien" for those who do not want to share but want to occasionally use Fon Spots.

The core value proposition proposed by Fon is to get free connectivity to its community network and can be summarized by their slogan "share a little bandwidth and roam the world for free." In marketing its offering, Fon advertises above all its utilitarian aspects by promising "free access to over four millions Fon Spots worldwide," "speedy connection to all your devices" and the possibility to "make some money by selling access to non members." At the same time, Fon tries to address potential user concerns by promoting that it is "easy to join" (with plug-and-play hardware), "secure" (by providing one encrypted private signal just for its owner and another public signal

for registered Fon members), and allowing users to limit the shared bandwidth. Whether these claims are actually maintained is a debatable issue that goes beyond the goal of this chapter (see Middleton & Potter, 2008, for a critical discussion of those aspects).

Fon also marginally promotes itself as "a community network built by the people," even though in reality it provides members very limited control on the community network. Essentially, they can only limit the bandwidth shared with other members, visualize who connects to their spots and exchange messages with other members through the community forum.

Otherwise, with the notable exception that the network infrastructure is provided by individual members, Fon operates like a regular ISP in that it controls the development of the technical solution and operates central network elements like the authentication and billing system. Fon also maintains a central database of hotspots that is used to provide an interactive map that members can use to find Fon Spots and download their locations to GPS or other mobile devices.

BACKGROUND

Wireless communities emerged at the turn of the 21st century: while wireless carriers were struggling with deploying 3G cellular networks. Wireless communities is a grassroots movement quietly began to deploy open hotspots and organize itself in wireless communities. They aggregate individuals offering free Wi-Fi Internet access to each other and the neighboring population. Fueled by cheap equipment and flat-rate Internet connections, wireless communities started to grow and become an interesting alternative to operator-centric networks for providing wireless broadband, especially in densely populated areas (Schmidt & Townsend, 2003).

Some traditional operators soon realized that Wi-Fi could complement their slower but ubiquitous 3G networks and a few start-ups tried to exploit the low entry barriers of Wi-Fi to enter the mobile industry (Camponovo, Heitmann, Stanoevska, & Pigneur, 2003; Bharat, Rao, & Parikh, 2002).

This resulted in various approaches for deploying Wi-Fi: 1) an *operator-centric* approach where a firm builds the network and charges access to its users, 2) a *pure community* approach where individuals organically share their own access points with each other and the public for free, and 3) an *hybrid* approach blending both commercial and community aspects. In contrast to pure communities that are exclusively built and operated by its members, hybrid communities are built by members but operated by a firm. Members add their own access points to the community network in exchange of incentives like revenue sharing, subsidized equipment, or free network access. In return, the firm is allowed to commercially operate the network, e.g. by selling equipment, connectivity or advertising.

Quickly researchers began to investigate this phenomenon. A literature review on wireless communities (Bina & Giaglis, 2005) examined 40 peer-reviewed papers published before 2004, drawing a research agenda with critical technological, economic, and individual research challenges. Among the latter, they state that "the main research question refers to the assessment of the role of individuals [...] in the formation, growth and survivability of wireless communities" and more specifically "what are the motivational incentives that drive participation and contribution to a wireless community."

The literature review found eight papers addressing this question. The earliest four papers use theoretical arguments (range limitations, self-organized nature, and network externalities) to support that reaching a critical mass of active members is vital for wireless communities' growth and sustainability. It is thus crucial to understand their motivations and design proper incentives to

attract them and sustain their participation (Camponovo, et al., 2003; McDonald, 2002; Bharat, Rao, & Parikh, 2003; Readhead & Trill, 2003). Two papers (Auray, Charbit, Charbit, & Fernandez, 2003; Schmidt & Townsend, 2003) describe a set of motives such as to create a spirit of cooperation, gain prestige in the community, break free from telecom firms and promote free communication. The two last papers study two potential conflicts of interests between individuals and the community: inducing members to contribute to the community instead of free riding (Sandvig, 2004) and limiting them to fair usage practices (Damsgaard, Rao, & Parikh, 2006). However, most of these studies are conceptual and provide little empirical evidence.

To address this limitation, Bina and Giaglis (2006a) developed a model proposing that members are driven by a mix of intrinsic motivations (enjoyment, competence, autonomy, or relatedness), obligation-based motivations (reciprocity or other community values), extrinsic motivations (get free connectivity, develop skills, get appreciation by others, feel altruist or pursue ideological goals). On the other side, members are discouraged by the perceived cost and effort to join and participate in the community. This model has been tested with two online surveys submitted to members of wireless communities in Greece (Bina & Giaglis, 2006b) and Australia (Lawrence, et al., 2007). They found that although different groups of members participate for different reasons, members generally tend to participate to communities more for intrinsic than extrinsic reasons.

Two studies on Wireless Toronto also analyze motivation and barriers. The first (Wong & Clement, 2007) suggests that people have "positive feelings about the benefits of sharing, especially when using others' signals, but serious reservations about making their own signals open" because they consider it difficult, lack trust in strangers, worry for security or their available bandwidth. However, sharing becomes more viable if these concerns are addressed and people get tangible benefits like cost reduction or increased reliability. The second study (Cho, 2008) suggests that personal motivations (having fun, learning technical skills, social networking, getting free Wi-Fi access) are complemented by public interest motivations like promoting inclusion in the information society, media democracy and civic activism.

Abdelaal et al. (2009) focuses on the various types of contributions from members (time, money, expertise, sharing, hardware, software) and shows the importance of social capital besides technical and economic benefits, proposing that communities "were built by technology developers to obtain expertise [but] have been redirected to achieve social objectives."

Recently, a few authors tried to expand research on members' motivations and hindrance factors of hybrid communities. Biczók et al. (2009) proposes a theoretic game-theory model to illustrate motivations of the various stakeholders of a hybrid community: members, community operator and ISPs. Shaffer (2010) surveys members from both pure and hybrid Wi-Fi communities finding various motivations (expand broadband access, use technical skills and get connectivity, but not to save money or challenge ISPs) and concerns (signal reliability, speed, security, and privacy). She also suggests differences between motivations of members of each type of community.

Finally, Camponovo and Picco-Schwendener (2010, 2011) conducted a qualitative study on motivations of members of the Fon community. Participation appears to be driven by tangible rewards (especially free nomadic connectivity, but also revenue sharing or cost-effective equipment), idealistic reasons (the appeal of altruistic and reciprocity values embedded in the concept of sharing and to promote free Internet) and technical interest, whereas social and intrinsic motivations tend to be weak. On the other hand, members are generally aware of potential risks like security, abuse, and legality, but are only mildly concerned as the presence of a firm supporting the community plays a key role in reassuring them.

METHODOLOGY

The literature review above shows that existing research covers several aspects of motivations of wireless community members, but has some relevant shortcomings. Most notably, motivations in hybrid wireless communities have insofar been explored to a lesser extent and only through qualitative methods. As a result, this chapter intends to address this issue by presenting the results of a quantitative study conducted on members of the Fon community.

Research Model and Hypotheses

This study is the third part of a research project aiming at understanding motivations and barriers in hybrid wireless communities. Firstly, a theoretical model of motivations and barriers for participating in these communities was developed based on previous research. The model was then refined through a content analysis of 1100 threads of Fon community forums (Camponovo & Picco-Schwendener, 2010) and 30 semi-structured exploratory interviews with members of the Fon community (Camponovo & Picco-Schwendener, 2011). The resulting model is depicted in Figure 1 and briefly justified thereafter.

This model is theoretically grounded on previous studies on wireless communities (especially Bina & Giaglis, 2006a, 2006b) and a set of motivation theories including Self Determination Theory (SDT) (Deci & Ryan, 1985), Volunteer Functions Inventory (VFI) (Clary, et al., 1998), and the Unified Theory of Acceptance and Use of Technology (UTAUT) (Venkatesh, et al., 2003). As a result, the following six hypotheses are proposed:

1. **Utilitarian Motivation Positively Affects Participation:** As explained by SDT (extrinsic motivation), UTAUT (performance expectancy), and VFI (instrumental function), a behavior can be motivated by the

Figure 1. Motivation and barriers affecting participation in hybrid wireless communities

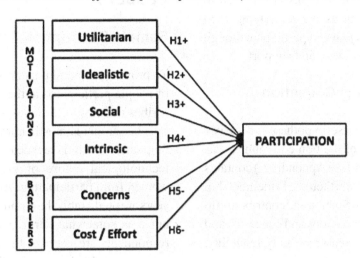

expectation to get something of value in return. In pure communities, this motivation appears to be weak. However, in a hybrid community utilitarian motivation is expected to be important, especially free network access, and maybe also revenue sharing, as their network is bigger and thus more valuable.

2. **Idealistic Motivation Positively Affects Participation:** SDT (identification) and VFI (enhancement and values) suggest that people are also motivated by psychological rewards like enhancing self-esteem or attaining fulfillment by pursuing idealistic goals. In pure wireless communities these motivations appear to be quite important; in hybrid communities idealistic motivations like endorsing values of sharing, reciprocity or promoting free wireless connectivity are expected to be present, even though they may be limited if the sponsoring firm is perceived more as a business.

3. **Social Motivation Positively Affects Participation:** As suggested by SDT (relatedness), VFI (social function), and UTAUT (social influence), people can also be driven by social motives like feeling part of a group or gaining approval by others. In pure communities, these motivations are

important. In hybrid communities, we expect them to be weaker due to the firm-supported resource-oriented nature of the community (Camponovo, 2011) and their larger size (Olson, 1971).

4. **Intrinsic Motivation Positively Affects Participation:** SDT (intrinsic motivation) explains that people can be motivated by the enjoyment obtained by performing an interesting task for itself. While this is one of the most important motivations in pure wireless communities, in hybrid communities it may be mitigated by the use of extrinsic rewards, which can have a negative effect on intrinsic motivation (Gagné & Deci, 2005).

5. **Concerns Negatively Affect Participation:** As explained in the literature review, people may be reluctant to participate in a wireless community due to a variety of concerns like security, bandwidth consumption and legal concerns. In a hybrid community, these concerns are expected to be mitigated by the presence of a supporting firm.

6. **Effort Negatively Affects Participation:** As suggested by UTAUT (effort expectancy), people are keener to do an activity if they think it requires low effort. While in pure communities, effort is a significant barrier,

in a hybrid community this is expected to be less important as the underlying firm makes it easy to join and participate though standardized hardware and support.

Instrument of Data Collection

To empirically test these hypotheses, a survey directed at members of the Fon community was developed. The survey (see Appendix A) contains questions about four main themes: 1) membership and experience, 2) participation and contribution to the community, 3) motivations and concerns, and 4) demographic data. To ensure validity, reliability, and comparability of results, the questions have been developed based on earlier surveys (Bina, 2007; Shaffer, 2010) and tested measurement scales as shown in Appendix B.

The survey was principally addressed to Foneros in Switzerland. To contact them, Fon has agreed to send an invitation to fill in the Web-survey as part of its April 2011 newsletter to all Swiss Foneros. In addition, the survey was advertised through the Fon Twitter channel and posted on the official Fon forum, where it stayed on top of all posts for two months. In that way, it was also possible to tap into Fon users from other European countries. This was useful to check for particularities in the sample and extend the generality of the results.

The survey was published on the project website (www.wi-com.org) from April to October 2011. It was available in English and in the three Swiss official languages (German, French, and Italian).

It obtained 292 complete and usable responses. A descriptive statistical analysis of these responses is presented in the following section. To enrich their interpretation, they will be complemented with the findings from the 40 semi-structured interviews conducted in the previous phase of the project.

RESULTS

Sample Description

To provide a general understanding of what kind of people participated at the survey, Table 1 describes the most relevant aspects of the sample.

The sample is mostly composed of male participants, which is expected as the perceived technological nature of Fon may discourage women from participating. Adults from 25 to 50 years are predictably the dominant age class (75%), but it is somewhat surprising that only 9% of respondents are younger than 25, whereas 16% are over 50. With regard to the country, almost half of respondents are Swiss (as expected, given that the survey was advertised mainly to them) and half from other countries. Since we checked that responses are not significantly different among the countries, the whole sample is used for the analyses below. It is also interesting that 45% respondents are Linus (without revenue sharing), whereas 34% are Bills and only 3% are Alien (passive member). Finally, by looking at the year of entry, it emerges that those joining in the early years are more numerous than in the following years. This is surprising considering that the growth of the number of members has accelerated through the years, but may partly be explained by the fact that Fon reduced promotional activities, especially in Switzerland, since 2007.

Motivations

From the survey emerged the groups of motivations shown in Figure 2.

These results fit well into the motivational model described above. Coherently with Fon's business model, which combines business and community aspects, members are motivated by a mix of utilitarian and idealistic motivations like getting free connectivity and respecting community ideals of reciprocity and sharing. Members are also motivated, albeit to a lesser extent, by

Table 1. Sample description

Question	Answer	Count	%	Question	Answer	Count	%
Gender	Men	258	88%	Education	Primary/secondary	25	9%
	Women	34	12%		Upper secondary	106	36%
					Tertiary	161	55%
Age	<18	2	1%	Membership type	Linus	130	45%
	18-24	22	8%		Bill	99	34%
	25-34	100	34%		Alien	9	3%
	35-50	121	41%		Ex Member	19	7%
	>50	47	16%		Not yet member	35	11%
Country	Switzerland	131	45%	Membership year	2006	91	31%
	Italy	60	21%		2007	58	20%
	France	32	11%		2008	41	14%
	Germany	23	8%		2009	35	12%
	UK	15	5%		2010	32	11%
	Other countries	31	10%		Not yet member	35	12%

intrinsic motivations such as enjoyment and technical interest. However, social motivations, getting a cheap router and revenue sharing play a marginal role. In the following sections these results will be interpreted with the help of interviews with Fon members.

Utilitarian Motivations

Getting free Internet *connectivity* through the Fon network stands out as the strongest motivation, with 81% of respondents. This is in line with our previous interviews and consistent with Fon's marketing message emphasizing the benefit of

Figure 2. Motivations of Fon members

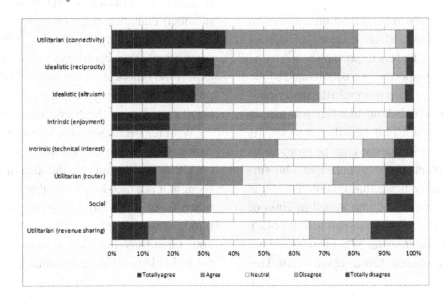

getting "free access to over 4 million Fon Spots worldwide."

In the interviews, this emerged even more clearly as the main motive. Many members explained that "what really made me participate is the fact that it allows me to use other Fon Spots worldwide and for free. Still today this is the key point." Members value both the extension and the internationality of the community network. However, many contend that the network is not widespread enough and that it is difficult to find working Fon Spots when travelling. They would like Fon to support network expansion through partnerships (60% of respondents find them positive, only 8% find them negative), promotional activities or other means. Members want coverage in areas that are useful to them. For many this implies having access in places like city centers, hotels or transportation stations. Others see Fon's value in offering connectivity in residential, industrial, and rural areas that are neglected by commercial providers.

With the development of 3G and flat rates subscriptions, a lot of members now are less interested in Fon for national use and perceive it as being useful only abroad, where the expensive roaming fees of mobile network operators still make it worthwhile to make the effort to search for access points.

On the other hand, other utilitarian motivations like acquiring a cost-effective *router or revenue sharing* appear to be less important.

Only 43% of respondents are motivated by the possibility to get a cheap *router*. In the interviews, some members pointed out that this was true in the beginning when Fon heavily promoted its routers, but is no longer the case as promotions decrease and other routers got cheaper. Our survey does indeed indicate that this motivation is weaker among members who joined in 2008 or after (39%) than those who joined before (44%)

As for *revenue sharing,* 32% of members are motivated by it, whereas 35% disagree with it. This

is further supported by the fact that more members choose to be Linus (45%) than Bill (34%), even though the latter has the same advantages and also get a share of revenues. Two possible explanations emerged from the interviews. Some members like the idea of revenue sharing, but do not think that they can earn much in their location: only 27% of respondents think that their Fonera reaches areas that are attractive for other Foneros. For others, getting a financial pay-off is in contrast with ideological motivations: for instance "what I like less is the commercial aspect. I am a Linus type, like Linux, who offers it for free, but the mean thing is that most people still have to pay, because they are not members of Fon.*"

Ideological Motivations

Ideological motivations also play a key role for Fon members: 76% of respondents are motivated by reciprocity and 68% consider altruism an important aspect of the community.

In this community, *reciprocity* (that is the mutual exchange of connectivity between members) is a key value. It is not surprising that it emerges as one of the most important motivations. For some members reciprocity has mainly a fairness connotation: they consider it right to share given that other Foneros do so. On the other side, other members consider reciprocity simply a means for getting connectivity, reinforcing their utilitarian motivations: they contribute to Fon because they expect the others to do the same *("I don't have a problem with sharing my connection, since in this way I can also use the connections of others").*

With regard to *altruistic motivations*, various nuances emerged from the interviews and were confirmed by the survey: the idea of sharing *("I thought it is a nice idea to be able to share it with others"),* of providing universal Internet access ("today you cannot live without Internet, how can you? I don't say that it has to be a universal right, but it should be easily accessible"), of supporting an alternative to commercial operators ("it is not really rebellious but it is a sort of declaration of war to the big mobile hot spots, so I wanted to participate and operate a free hotspot myself, so that people can connect") and of better exploiting existing infrastructures ("I like it because anyway during the day I don't use my bandwidth, so why shouldn't other people use it, too?").

Intrinsic Motivations

Participants are also moderately motivated by intrinsic reasons. 61% of respondents perceive participating in the community as *enjoyable and interesting*. A member expressed it nicely: "for me Fon is cool, they offer me something that is interesting." Furthermore, 55% of participants are attracted because of *technical interest*. This can be explained by the high percentage of IT specialists among respondents: 56% work in the IT field and 25% are open source contributors. They are naturally curious to see how the community works technically and want to learn and apply their technical skills: "I try to follow and try out, to a certain extent, all IT and social trends, so that as an IT manager I have a clue and feeling on what is going on."

Social Motivations

Finally, it is worth noticing that members are not really attracted by social aspects (32%). This is surprising as it contrasts with the concept of a community. Many members did indeed express a "lack of community feel," especially when compared to other communities like open source. In the case of Fon, this might be explained by the fact that the community aspect mainly lies in sharing connectivity and not in interacting: "a community without interacting, where you simply share something with others." Only 20% state that they interact with other community members through one of the channels offered by the community (forum, messages between members, meetings etc.). In fact, Fon does not put a lot of effort in building social ties among members: forums are strongly moderated and the messaging system is only internal, meaning that members are only notified about messages when they log into their Fon account, resulting in low usage: "I wrote to others, but I never received an answer."

Barriers: Concerns, Cost, and Effort

In addition to motivations, we also investigated several potential concerns for participating in Fon. Their importance is illustrated by Figure 3.

Figure 3. Concerns of Fon members

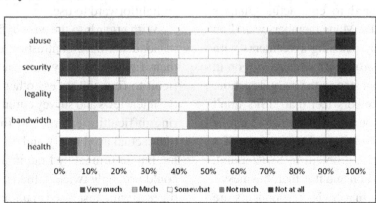

Members are quite aware of these potential risks, but are somewhat reassured by noticing that Fon actively tries to address these concerns by proposing some solutions.

Participants are mainly concerned by the possible *abuse* of their shared connection to make illegal or immoral activities (44% being much concerned). The fact that Fon is only open to registered members and claims on its website that "if anyone tries to do anything illegal with your Internet connection, we block them" comforts many users, as confirmed by several members during the interviews (e.g. "yes, clearly, it is an open network where you log in but you have to identify yourself. And then, I don't believe that someone is doing some bad things using this hotspot").

Security and privacy issues are also quite salient (39%). Fon strongly claims to be "safe and secure," emphasizing on his website that "la Fonera protects your connection with two secured WiFi signals" (an encrypted signal for the owner and a public one accessible to other Foneros). Although its actual security is a debatable topic (Middleton and Potter, 2008), this is enough for many Foneros (e.g. "I have no doubt, I have read what they described on the Internet and I saw that one can be quite safe, knowing that the access is separate from the one which I have on the home PC"). However, some doubt the security and prefer to implement additional security measures like "always operating the Fonera in front of a firewall."

Members are also concerned by the *legality* (34%) of sharing their Internet connection, especially with regard to contractual clauses imposed by their ISPs. Many members want Fon to solve this issue by reaching agreements with ISPs. The current partnership strategy of Fon in various countries is a step in the right direction. However, some members feel that "ISPs don't have any interest in pursuing their customers" or that the responsibility falls on Fon anyway: "if it should ever become a problem in Switzerland, they should contact Fon and it is their business." It is worth noticing that Swiss respondents are

less concerned by legality (27%) than respondents in others countries like France (69%) and Italy (34%) with stricter laws affecting Wi-Fi sharing.

On the other hand, *bandwidth* consumption seems to be of no significant concern to respondents (13%) as they can restrict the bandwidth dedicated to sharing and have large broadband connections that they only partially use. *Health* concerns are similarly unimportant (14%) even though they emerged a few times during the interviews showing a certain sensibility towards "energy efficiency" and "radiations."

Beyond these concerns, *effort and cost* are traditionally considered hindrance factors in the adoption of a technology solution. With regard to cost, during the interviews it became apparent that this is not a problem for most members. Fonera routers are quite cheap and connection costs would be paid regardless of Fon. With regard to effort, we measured several aspects tied to perceived *ease of use* as shown in Figure 4.

A large majority of respondents (80%) agrees that the initial *setup* of the Fonera is easy. The interviews confirmed that this can be achieved without any difficulties ("the solution is well implemented, it works, you don't have to be a specialist, you plug it in and log yourself on the webpage and then it works"). A few interviewees, however, expressed some concerns with the reliability of the Fonera router.

Fon Spots are similarly considered *easy to use* by 70% of the respondents. The landing page and login procedure are generally found to be straightforward to use.

With regards to the *ease of finding Fon Spots*, respondents have contrasting opinions: 44% of them find it relatively easy and 36% find it difficult to find and access other Fon Spots. From the interviews and survey comments, some recurring difficulties emerged in finding Fon Spots: their actual availability and diffusion ("I travel a lot, but unfortunately I ran into a Fon Spot that I could use, only twice"), the lack of reliability of the map with misplaced or unreachable Fon Spots

Figure 4. Ease of use of the Fon network solution

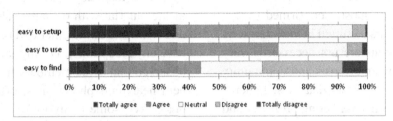

("some disappointments, when you are sure you will find a connection and then when you arrive it does not work"), and signal strength ("Fon Spots are not very well placed. Most people place it so that they have good reception inside the flat but then it is mostly just enough to see it on the street but not for the other to use it properly"). A frequent comment is that "there should really be easier ways to find out about Fon Spots around the world" otherwise members would "never use Fon Spots while travelling cause it is too troublesome to actually find them."

Usage, Satisfaction, and Future Intentions

The difficulty of finding Fon Spots also exerts an influence on their *actual usage*. The reported number of Fon Spots accessed by respondents is quite low: in average respondents accessed 2 Fon Spots in the last 12 months, with 43% them not accessing any and only 6% accessing more than ten.

In our survey, we also asked respondents what they do when connected to a Fon Spot. The vast majority of members mainly check their *email* (77%) or simply browse the *Web* (75%). Many look for specific *local information* (65%) such as "getting tourist information on the region, organizing something for the evening or [finding] something to visit." This suggests that free Wi-Fi networks may have valuable applications in the tourism sector. Fon is also used for phone calls via VoIP *communication* services like Skype (48%), using "Wi-Fi in order to do phone calls at reasonable prices." Yet, some members find the quality of

such calls not always satisfactory. On the other hand, heavy bandwidth-consuming applications like audio/video streaming (24%) and file sharing (10%) are used less. Many members do not want to consume too much bandwidth keeping in mind that the community is based on respectful sharing and that Fon Spots are "the wrong place for doing downloads." In addition, Fon Spots seldom reach places that are comfortable for connecting, resulting in "mostly short connections [...] when I stay online for longer time then I usually look for a more comfortable place."

The most common *devices* used to access Fon Spots are notebooks (81%) and smartphones (74%), followed by far by tablet PCs (20%) and gaming consoles (9%).

It is interesting to note that, inside their country, respondents use mainly 3G (57% use it at least some times per week), much more often than WiFi (27%) to connect to the Internet. In contrast, abroad they tend to prefer WiFi. This is explained by the diffusion of 3G subscriptions that allow flat-fee consumption of data nationally but charge expensive roaming costs abroad. This is in line with the above suggested interpretation that wireless communities may play an important role in the field of tourism and travelling by supplying people with cheap Internet connectivity abroad, while 3G will probably dominate the domestic use of mobile Internet access.

The survey also tried to measure the satisfaction of respondents. A large majority of members declared to be satisfied with their experience with Fon (70%) and even more would recommend it to their friends (77%). Such a high satisfaction likely results from the fact that Foneros expectations about using Fon are mostly confirmed (57%). While many members admitted in the interviews that their expectations are not necessarily very high, given the low entry costs and community nature of Fon, this suggests that most Foneros are not fooled by Fon's marketing promises and mostly get what they expect.

Finally, respondents were asked about their future intentions. 81% of respondents state that they intend to remain an active member, at least for the next year, while 78% expect to actually use Fon in that period.

Types of Contributions

Figure 5 shows how members participate in various ways in the Fon community.

The main contribution of members is predictably sharing their Internet connection with other members. Approximately 70% of respondents claim to keep their Fonera *active* most of the time and believe that they installed it in a way that it is *easily accessible* by other Foneros. This is somewhat surprising as it contrasts with the reported difficulties of finding working Fon Spots. A possible explanation is that Fonera have limited range and their signal is strongly attenuated

by walls ("It already had some reception problem inside our flat with 6 rooms. The router was in one room and at the other end of the flat I didn't have any reception"). During the interviews, we observed that some members simply put the Fonera near their phone lines, often not an ideal place. Yet, some really make an effort to place their Fonera well, for instance "a bit outside [so that] it really reaches far," or reinforce the signal with external antennas. It is also worth noticing that about 30% of respondents claim to have more than one active hotspot.

Only 27% of respondents consider that their Fonera *reaches attractive places* for other Fon members. Most members live in residential areas that are not close enough to points of interests like bars, restaurants, stations or tourist attractions where "it really could have a chance to be used."

This is also supported by data on the number of accesses to participants' Fon Spots, with an average of only 4 visitors in the last 12 months. Nearly 50% of them had no visitors at all, while only 6% had more than one visitor per month. The number of visits is coherent with the perceived attractiveness of their position: those declaring to live in unattractive places report less visits than those in more attractive ones.

Figure 5. Contributions of Fon members to the community

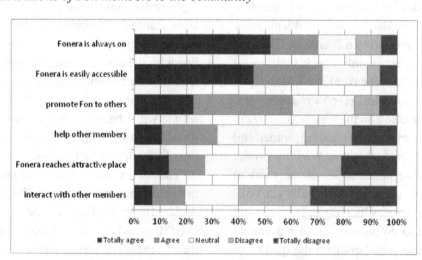

Another way in which members contribute to the Fon community is by *promoting or recommending* Fon to friends and people they know. 60% of respondents state to promote Fon to potential new members and even more to recommend to their friends (77%). The interviews showed that they do it in various ways. Many members simply talk to their friends about it and try to convince them to join. Others use signs to advertise their Fon Spots like "a shield with Login and Password in front of my house" or "a Fon sticker on my mailbox." Some members even give friends extra Foneras and "regularly ask and check whether they actually use it or at least forward it to someone else when they don't use it anymore." This is interesting as it helps the community grow, resulting in benefits for both Fon and members. The former can sell additional routers; the latter can have access to more Fon Spots.

On the other hand, it may be quite surprising for a community that members do not engage much in "social" activities like *helping other members* (32%) and *interacting* with each other (20%). However, this is coherent with the observation that social motivations are weak and that most respondents do not exchange messages at all with other members (86% have received no messages from other members in the last 12 months).

DISCUSSION AND RECOMMENDATIONS

Given that connectivity is the most important motivation of Fon members, extending the community must be considered as the key success factor. Fon should, therefore, continuously put effort in attracting new members and thus enlarging the community at a worldwide level. To do so, it can employ various means.

Their current strategy of extending their network of *partnerships with telecom* operators in various countries seems to be an effective means to quickly enlarge their community. This strategy

is positively regarded by a large majority of our respondents (60%), while only a few find it negative (8%). However, many of our interviewees indicated that these deals are welcome only as long as they respect reciprocity. It is important that all Foneros can benefit from them by accessing the spots of the partner's customers and not only vice versa. Members would be upset if "you participate and try to make everything work on your spot and then [partner's clients] can use your spot but you cannot use theirs. This I don't like."

Members also suggest extending *collaborations to commercial partners* who own places where an Internet connection would be prized. These include companies in the travel sector like transportation companies, airports, bus or train stations ("It's a pity that you cannot find Fon Spots in train stations, you always find Swisscom but never Fon, it's a pity because there you really need it"). Restaurants, hotels and commercial centers are also seen as interesting venues. Members find that "the concept of Fon would be better if it was more deployed in commercial centers, cafés and so on." Another suggestion is to look for partnerships with *municipal / public institutions* to "go much more into the public space." In the last years, several cities created their own wireless networks for their citizen and tourists. Usually they are confined to the central areas of the city and to a few points of interests. As Fon's business model is currently based on private people sharing their Internet access, most access points are in residential areas, leaving attractive places in the center uncovered. In this sense, looking for collaborations with cities could increase Fon's attractiveness and make people more aware of the community.

Foneros also believe that elaborating commercial deals with *Local distributors* would be a good idea. This would make the Fonera routers more accessible to the large public and increase its visibility. In this way you don't need to know Fon beforehand, but you can find it as a possible alternative in the local shop when buying a router.

Furthermore, not everybody likes to buy equipment online and paying expensive delivery costs. It seems that in some countries this is already possible and appreciated: "In Japan I saw Fon Spots for sale in various shops. You don't have to order it in an online shop in Spain with 20 Euro delivery cost to Switzerland. This is certainly a big obstacle."

Even partnerships with other *companies developing routers* might be interesting in order to "be able to activate Fon on a standard router without necessarily having to buy a Fonera."

Concerning the *Fonera router*, most members are actually quite satisfied with it. They mostly consider it secure and easy to use. However they would like Fon to include a more powerful antenna or at least "it should not be necessary to buy the external antenna extra." They would like Fon to invest more in increasing the signal strength of their routers: this is fundamental to make the signal reach the streets and make it actually available to others. Some members also notice that an improvement in this sense should be possible, as other routers seem to have stronger signals and reach farther. Furthermore a better education of members in how and where to best place the router inside the house and in how to install them may be useful to increase availability and reliability of Fonera WiFi signals to other members.

Members also strongly expressed in both the survey and interviews that it is crucial to be able to *easily find active Fon Spots*. Many of them indicated that there should really be easier ways to find Fon Spots around the world, otherwise members would never use them, as it may be too troublesome to actually find them. They need effective means to locate them and want Fon to focus on "improvement of search tools." They consider that Fon maps are not reliable enough and expect that Fon engages more in "checking whether the Spots really are where they are marked, whether they are usable and possibly providing a solution for evaluating each spot." They would also like to

be able to download such maps on mobile phones or GPS devices.

Several participants also suggested that Fon should *facilitate interaction* among community members and with the company. Members generally appreciate the availability of an official Forum that can be used to discuss with other members and get some support from them, but many members find that Fon should provide a more effective and timely support when they encounter problems. Another aspect that should be improved is Fon's internal *messaging system*. As a community, it should be easy for members to get in touch with each other without having to log into their Fon-account regularly. As it is currently implemented, it seems to be nearly impossible to get in contact with other members through the messaging system; it is a "communication system that does not really work." It would be appreciated by community members to receive an email whenever a Fon Message arrives and be able to easily answer it.

Finally, many members are interested in how Fon works from a technological point of view and would like to be able to play and experiment with the technology. Unfortunately, the community does not leave much room for these kinds of experiments as most technical aspects are controlled by Fon itself. However it might be beneficial for the community to allow a higher involvement of the community (many members are active open source software contributors) in the evolution of the service and welcome technological contributions and inputs.

RESEARCH LIMITATIONS AND FUTURE DIRECTIONS

This research focuses on one specific wireless community: Fon. Even though it is the largest one, with more than 4 million members worldwide, it might be interesting to look for other hybrid communities and repeat the study with their members in order to better understand if the results of this study can be generalized to all types of hybrid wireless communities.

It would also be interesting to compare motivations between pure and hybrid wireless communities. Existing literature provides some interesting studies on pure communities, suggesting that motivations are partly similar but differ in their importance. In particular, intrinsic and social motivations appear to be more important in pure communities, whereas utilitarian motivations seem to be stronger in hybrid ones. However, a direct comparison of these results is difficult as they refer to communities in different times, diverse cultural contexts and employ various methodologies. This makes it impossible to tell whether these differences are caused by the presence of a supporting firm or by other differences. It would therefore be interesting to conduct a new study, similar to this one, on pure communities.

Researchers can even go a step further and analyze if similar motivations and barriers apply to other types of online communities such as peer-to-peer networks, open-source projects, communities of practice, social networks, user-generated content communities and other forms of Web 2.0 collaborations.

CONCLUSION

This chapter analyzed motivations and barriers influencing participation in a hybrid wireless community like Fon, based on the data collected from a quantitative analysis of 292 members and interpreted with the help of the qualitative insight collected through semi-structured interviews with 40 Foneros.

It appears that members of hybrid wireless communities are essentially motivated by a mix of utilitarian, idealistic, and—to a lesser extent—intrinsic motivations.

The first and most important motivation is the *utilitarian* value of the community. In particular,

members value the possibility to get Internet *connectivity* from other members at a worldwide level. Especially when being abroad, where 3G roaming prices are generally found excessive, wireless communities offer a much cheaper solution to get Internet connectivity. However, other utilitarian motivations like getting cheap equipment and revenue sharing are quite weak.

A second important group of motivations is *idealistic*. It includes *reciprocity* as well as *altruism*. The first is a fundamental aspect of a resource-based community where members are supposed to contribute to and not only benefit from the community network. The latter is also important as members value the spirit of helping and sharing with others, allowing them to contribute to make Internet accessible to everyone. They also like the idea of supporting an alternative to traditional commercial operators. Finally, members appreciate the environmental gains (e.g. lower radiation and electricity savings) by better exploiting existing infrastructure through sharing.

Intrinsic motivations like technical interest and enjoyment are also present, albeit to a lesser degree, but are still stronger than *social motivations*. While in other communities, social aspects play a fundamental role and represent the community spirit, here it is not the case: members do not really interact with each other. The community is consequently reduced to a reciprocal exchange of Internet connectivity.

Regarding *concerns*, members are generally aware of the risk that someone may use their hotspot for illegal activities (abuse), that someone might be able to access and use their data (security) and that participating in the community might not comply with contractual clauses of their ISP (legal). However, most members feel sufficiently reassured by the various efforts done by Fon to deal with these issues.

This chapter also analyzed member *contributions* to and *usage* of Fon. Members principally contribute by sharing their own connections, trying to keep their router on, and making it easily accessible to others. They also try to help by promoting the community, but engage less in socializing, helping, and interacting with other members. On the other hand, actual *usage* of the community is quite low. This is influenced by the difficulty of finding active Fon Spots and the increasing adoption of 3G with flat rate contracts. In spite of this, members are generally satisfied with their experience of participating in Fon and find that their expectations are mostly met.

ACKNOWLEDGMENT

This research is supported by the Swiss National Science Foundation, under grant number 100014-127006. We would like to particularly thank Fon for promoting our survey to all Swiss Foneros through its April 2011 newsletter, Forum, and Twitter channel.

REFERENCES

Abdelaal, A., Ali, H., & Khazanchi, D. (2009). The role of social capital in the creation of community wireless networks. In *Proceedings of the Hawaii International Conference on System Sciences (HICSS)*. IEEE Press.

Auray, N., Charbit, C., Charbit, C., & Fernandez, V. (2003). *Wifi: An emerging information society infrastructure*. Retrieved from http://ses.telecom-paristech.fr/auray/Auray%20Beauvallet%20Charbit%20Fernandez.pdf

Biczók, G., Toka, L., Vidacs, A., & Trin, T. A. (2009). *On incentives in global wireless communities*. Paper presented at the International Conference on Emerging Networking Experiments and Technologies Archive. New York, NY.

Bina, M. (2007). *Wireless community networks: A case of modern collective action*. (PhD Thesis). Athens University of Economics and Business. Athens, Greece.

Bina, M., & Giaglis, G. M. (2005). Emerging issues in researching community-based WLANS. *Journal of Computer Information Systems, 46*(1), 9–16.

Bina, M., & Giaglis, G. M. (2006a). *A motivation and effort model for members of wireless communities*. Paper presented at the European Conference on Information Systems. Göteborg, Sweden.

Bina, M., & Giaglis, G. M. (2006b). *Unwired collective action: Motivations of wireless community participants*. Paper presented at the International Conference on Mobile Business. Copenhagen, Denmark.

Camponovo, G. (2011). *A review of motivations in virtual communities*. Paper presented at the European Conference on Information Management and Evaluation (ECIME). Geneva, Switzerland.

Camponovo, G., Heitmann, M., Stanoevska, K., & Pigneur, Y. (2003). *Exploring the WISP industry: Swiss case study*. Paper presented at the 16th Bled eCommerce Conference. Bled, Slovenia.

Camponovo, G., & Picco-Schwendener, A. (2010). *A model for investigating motivations of hybrid wireless community participants*. Paper presented at the 2010 Ninth International Conference on Mobile Business / 2010 Ninth Global Mobility Roundtable. New York, NY.

Camponovo, G., & Picco-Schwendener, A. (2011). *Motivations of hybrid wireless community participants: A qualitative analysis of Swiss Fon members*. Paper presented at Tenth International Conference on Mobile Business 2011. Athens, Greece.

Cho, H. H.-N. (2008). Towards place-peer community and civic bandwidth: A case study in community wireless networking. *The Journal of Community Informatics, 4*(1).

Clary, E. G., Snyder, M., Ridge, R., Copeland, J., Stukas, A., & Haugen, J. (1998). Understanding and assessing the motivations of volunteers: A functional approach. *Journal of Personality and Social Psychology*, *74*(6), 1516–1530. doi:10.1037/0022-3514.74.6.1516

Damsgaard, J., Rao, B., & Parikh, M. (2006). Wireless commons perils in the common good. *Communications of the ACM*, *49*(2), 104–109. doi:10.1145/1113034.1113037

Deci, E. L., & Ryan, R. M. (1985). *Intrinsic motivation and self-determination in human behavior*. New York, NY: Plenum.

Esmond, J., & Dunlop, P. (2004). *Developing the volunteer motivation inventory to assess the underlying motivational drives of volunteers in Western Australia*. Perth, Australia: CLAN WA.

Gagné, M., & Deci, E. L. (2005). Self-determination theory and work motivation. *Journal of Organizational Behavior*, *26*(4), 331–362. doi:10.1002/job.322

International Telecommunications Union. (2009). [*ICT facts and figures*. Retrieved from http://www.itu.int]. *WORLD (Oakland, Calif.)*, *2009*

Lawrence, E., Bina, M., Culjak, G., & El-Kiki, T. (2007). *Wireless community networks: Public assets for 21st century society*. Paper presented at the International Conference on Information Technology. Las Vegas, NV.

McDonald, D. W. (2002). *Social issues in self-provisioned metropolitan area networks*. Paper presented at the ACM Conference on Human Factors in Computing Systems. Minneapolis, MN.

Middleton, C., & Potter, A. B. (2008). *Is it good to share? A case study of Fon and Meraki approaches to broadband provision*. Paper presented at the 17th Biennial International Telecommunications Society Conference. New York, NY.

Olson, M. (1971). *The logic of collective action public goods and the theory of groups*. Cambridge, MA: Harvard University Press.

Rao, B., & Parikh, M. (2002). *Wireless broadband experience: The U.S. experience*. Paper presented at the First International Conference on Mobile Business. Athens, Greece.

Rao, B., & Parikh, M. (2003). Wireless broadband networks: The U.S. experience. *International Journal of Electronic Commerce*, *8*(1), 37–53.

Readhead, A., & Trill, S. (2003). The role of ad hoc networks in mobility. *BT Technology Journal*, *21*(3), 74–80. doi:10.1023/A:1025159115207

Ryan, R. M. (1982). Control and information in the intrapersonal sphere: An extension of cognitive evaluation theory. *Journal of Personality and Social Psychology*, *43*, 450–461. doi:10.1037/0022-3514.43.3.450

Sandvig, C. (2004). An initial assessment of co-operative action in wifi networking. *Telecommunications Policy*, *28*(7-8), 579–602. doi:10.1016/j.telpol.2004.05.006

Schmidt, T., & Townsend, A. (2003). Why wi-fi wants to be free. *Communications of the ACM*, *46*(5), 47–52. doi:10.1145/769800.769825

Shaffer, G. L. (2010). *Peering into the future: How wifi signal sharing is impacting digital inclusion efforts*. Philadelphia, PA: Temple University.

Venkatesh, V., Morris, M. G., Davis, G. B., & Davis, F. D. (2003). User acceptance of information technology: Toward a unified view. *Management Information Systems Quarterly*, *27*(3), 425–478.

Wong, M. (2007). *Wireless broadband from backhaul to community service: Cooperative provision and related models of local signal access*. Paper presented at the 35th Research Conference on Communication, Information and Internet Policy. Arlington, VA.

Wong, M., & Clement, A. (2007). *Sharing wireless internet in urban neighbourhoods.* Paper presented at the Third Communities and Technologies Conference. East Lansing, MI.

ADDITIONAL READING

Abdelaal, A., & Ali, H. (2012). Human capital in the domain of community wireless networks. In *Proceedings of HICSS*, (pp. 3338-3346). IEEE Press.

Bar, F., & Galperin, H. (2004). Building the wireless internet infrastructure: From cordless ethernet archipelagos to wireless grids. *Communications and Strategies, 54*(2), 45–68.

Battiti, R., Lo Cigno, R., Orava, F., & Pehrson, B. (2003). *Global growth of open access networks: From warchalking and connection sharing to sustainable business.* Paper presented at the 1st ACM International Workshop on Wireless Mobile Applications and Services on WLAN Hotspots. San Diego, CA.

Camponovo, G., & Cerutti, D. (2005). *WLAN communities and internet access sharing: A regulatory overview.* Paper presented at the 4th International Conference on Mobile Business. Sydney, Australia.

Camponovo, G., & Pigneur, Y. (2006). *From hype to reality: A case study on the evolution of the Swiss WISP industry.* Paper presented at the 5th International Conference on Mobile Business. Copenhagen, Denmark.

Clark, B. (2002). *The wireless napsterization of broadband access.* New York, NY: Gartner Group.

Giussani, B. (2001). *Roam: Making sense of the wireless internet.* New York, NY: Random House.

Kalakota, R., & Robinson, M. (2002). *M-business: The race to mobility.* New York, NY: McGraw-Hill.

Lehr, W., & McKnight, L. (2002). Wireless internet access: 3G vs. wifi? *Telecommunications Policy, 27,* 351–370. doi:10.1016/S0308-5961(03)00004-1

Lyytinen, K., & Yoo, Y. (2001). *The next wave of nomadic computing: A research agenda for information systems research.* Retrieved from http://sprouts.aisnet.org/221/1/010301.pdf

Powell, A., & Shade, R. L. (2006). Going wi-fi in Canada: Municipal, and community initiatives. *Government Information Quarterly, 23*(3-4), 381–403. doi:10.1016/j.giq.2006.09.001

Rao, B., & Parikh, M. (2003). Wireless broadband drives and their social implications. *Technology in Society, 25,* 477–489. doi:10.1016/j.techsoc.2003.09.025

Verma, S., Beckman, P., & Nickerson, R. (2002). *Identification of issues and business models for wireless internet service providers and neighborhood area networks.* Paper presented at the Workshop on Wireless Strategy in the Enterprise. Berkeley, CA.

Wireless Commons. (2002). *The wireless commons manifesto.* Retrieved October 30, 2011, from http://www.wirelesscommns.org/manifesto

APPENDIX A: SURVEY

Table 2.

	Membership
1	Are you currently a member of Fon ?
2	When did you join the Fon community ?
3	What kind of Fonero are you ?
4	What are your future intentions about Fon ?
	I intend to remain an active Fon member in the next 12 months
	I expect to use Fon in the next 12 months
5	Do you know other wireless communities ?
6	Which ones?
7	Do you participate in (are a member of) some of them ?
8	Do you participate in Open Source communities?
	Contribution and usage
9	Please indicate how much you agree with the following statements regarding your contribution to the Fon community:
	My Fonera is always on and connected to the Internet
	My Fonera is installed in a way that it is easily accessible by other members
	My Fonera reaches attractive or frequented public places (parks, cafés etc.)
	I interact with other community members (Fon messages, forums, meetings)
	I promote or recommend Fon to potential new members
	I volunteer my skills to help members or improve the Fon offering
10	Please answer the following questions, if possible by looking at your Fon statistics on your account:
	How many active Fon Spots do you have ?
	How many Fon Spots did you access in the last 12 months?
	How many users did use your Fon Spot in the last 12 months ?
	How many messages did you receive from other members in the last 12 months?
11	Which bandwidth limits have you set for your Fon Spots ?
12	Please, indicate how often you use the following applications when you are connected to a Fon Spot of another member:
	Email or chatting
	Searching local or touristic information
	Web browsing
	Voice communication services (VoIP, skype calls, ...)
	Bandwidth-consuming applications (audio/video streaming, gaming, ...)
	File sharing (Peer to Peer)
13	Which devices do you use to connect to a Fon Spot ?
	Notebook / netbook Notebook / netbook
	Tablets / iPad Tablets / iPad
	Mobile phone / smartphone / PDA Mobile phone / smartphone / PDA
	Other (es. gaming devices) Other (es. gaming devices)

continued on following page

Table 2. Continued

14	Please, indicate how often you perform each of the following activities
	Using Fon or other WiFi spots (in my country)
	Using GPRS or 3G data networks (in my country)
	Using Fon or other WiFi spots (abroad)
	Using GPRS or 3G data networks (abroad)
Motivations and experience	
15	Please indicate how much you agree with the following statements:
	1. Participating in Fon is fun
	2. I would describe participating in Fon as interesting
	3. Participating in Fon is quite enjoyable
	4. Participating in Fon allows me to learn or apply technical skills
	5. I am interested in Fon from a technical viewpoint (to see how it works)
	6. I would like a chance to interact with other Foneros more often
	7. I feel close to the other members involved in the Fon community
	8. I feel like I can trust other people in the Fon community
	9. Being appreciated by other Foneros is important to me
	10. I feel that it is important to receive recognition for my contribution to the community
	11. Participating in Fon is useful to get free Internet access when not at home
	12. Participating in Fon enables me to get free Wi-Fi access worldwide
	13. Participating in Fon allows me to get a cheap router
	14. I would like to earn some money in exchange for the connectivity I offer
	15. It is important to get some compensation for sharing with the community
	16. Participating in Fon allows me to do something for a cause that is important to me
	17. I participate in Fon because I feel it's important to give connectivity to others who need it
	18. I like the idea of sharing and helping others through my involvement in Fon
	19. Participating in Fon makes me feel like a good person
	20. Participating in Fon makes me feel useful
	21. Participating in Fon is a way to support an alternative to mobile operators
	22. I like sharing in order to help better exploiting existing infrastructure
	23. I can use other people's access points, so I desire to give back
	24. I know other Foneros share their access with me, so it's fair to share my connection too
	25. When I contribute to the Fon community, I expect others to do the same
	26. Since other Foneros can use my access point, I expect to be able to use theirs
16	Please indicate how much you agree with the following statements concerning your involvement in Fon
	The Fonera is easy to setup
	It is easy to find other Fon Spots
	Fon Spots are easy to use

continued on following page

Table 2. Continued

17	Please indicate how much you are concerned with the following topics regarding sharing your Fonera with other members
	Security or privacy (viruses, hackers, access to personal data etc.)
	Abuse (illegal or immoral activities)
	Legality (of sharing my Internet connection)
	Bandwidth consumption from other members
	Health concerns (from wireless waves)
18	Please indicate how much you agree with the following statements concerning your experience with Fon:
	I am satisfied of the experience with Fon
	I would recommend Fon to my friends
	Overall most of my expectations from using Fon were confirmed
19	Please indicate how much you agree with the following statement concerning Fon
	I perceive Fon as a profit-oriented company
	Commercial agreements with telecom operators (Neuf, BT etc.) are positive
	As there is a company (Fon) behind the community, I feel reassured about sharing my Internet connection
Demographic data	
20	Gender
21	Age
22	Education (please indicate the highest level completed)
23	Which field best describes your studies ?
24	Do/did you work in/for the IT or telecom sector ?
25	Country of residence
26	email (optional)
27	Would you like to add some comments ? (optional)

APPENDIX B: SURVEY SCALES

Table 3.

#	Questions	Scales
1-3	Membership	
4	behavioral intention	UTAUT behavioral intention (Venkatesh, et al., 2003)
5-8	Other communities	
9	Contribution	Social capital contributions (Abdelaal, et al., 2009)
10	Community network use	Community activities (Bina, 2007)
11-14	Fon/Wi-Fi behavior	Developed from our interviews
15.01-03	Intrinsic motivation	IMI enjoyment/interest (Ryan, 1982)
15.04-05	Competence	BPN competence (Bina, 2007)
15.06-08	Relatedness	IMI relatedness (Ryan, 1982)

continued on following page

Table 3. Continued

#	Questions	Scales
15.09-10	Recognition	VFI recognition (Esmond & Dunlop, 2004)
15.11-13	Usefulness	IMI value/usefulness (Ryan, 1982)
15.14-15	Revenue sharing	Rewards (Bina, 2007)
15.16-18	Values	VFI values (Esmond & Dunlop, 2004)
15.19-20	Self esteem	VFI self-esteem (Esmond & Dunlop, 2004)
15.21-22	Idealistic motivation	Based on interviews (Camponovo & Picco-Schwendener 2001)
15-23-26	Reciprocity	Reciprocity (Bina, 2007)
16	Effort	UTAUT effort expectancy (Venkatesh, et al., 2003)
17	Concerns	Concerns (Matthew Wong, 2007)
18	Satisfaction	
19	Fon perceptions	Based on interviews (Camponovo & Picco-Schwendener 2001)
20-27	Member data	

Chapter 8
Social Cohesion and Free Home Internet in New Zealand

Jocelyn Williams
Unitec Institute of Technology, New Zealand

ABSTRACT

This chapter discusses community outcomes of free home Internet access. It draws on case study research on Computers in Homes (CIH), a scheme established in New Zealand in 2000 for the purpose of bridging the digital divide, particularly for low-income families who have school-aged children. The government-funded CIH scheme aims to strengthen relationships between families and schools, improve educational outcomes for children, and provide greater opportunities for their parents. CIH achieves this by working with many primary (elementary) schools, each of which selects 25 families who will benefit from the program. Each family receives a refurbished computer, software, and six months free Internet, as well as twenty hours of free IT training and technical support so that all adults are equipped to make effective use of the Internet. The scheme has evolved to deliver much more than technology. It has become a contributor to social capital in the communities where it has been established. This chapter uses a case study research approach to demonstrate and theorize this process of community building using a construct of social cohesion, which appears to be strengthened by the CIH intervention. Where stronger social networks, volunteerism, and civic engagement were documented in the research, leader figures also mobilized to act on shared goals. These findings highlight the value of existing social resources within communities for achieving community goals while also maximizing community Internet longevity.

DOI: 10.4018/978-1-4666-2997-4.ch008

INTRODUCTION

The information revolution is changing the way many of us live and work; yet digital inclusion remains a pressing issue at the heart of a socially inclusive society based in the information age. It is imperative to ensure that no one gets left behind, but despite being a more economically developed nation New Zealand (NZ) does have poor and disadvantaged communities, and digital inequality. The image of egalitarianism and inclusive opportunity in NZ is less robust when examined closely; in fact the most recent household economic survey shows "inequality …rose from 2010 to 2011 to its highest level ever" (Perry, 2012, p. 1). With respect to global ranking of technological readiness - the capacity to fully benefit from information and communication technologies that enhance the nation's competitiveness and the daily lives of citizens - NZ appears comparatively well off, being listed at number 18 out of 138 countries in 2011 (Dutta & Mia, 2011, p. 230). Yet, acute disparities in digital opportunity continue to exist. In a nation of slightly more than 4 million people in 2006, 116,000 households with school-aged children remained without Internet access at home (Statistics New Zealand, 2012). In the wake of the more recent global economic recession, income and other gaps such as health and education are widening sharply (Collins, 2012), making the need for targeted efforts to promote digital opportunity even more acute.

This chapter discusses community outcomes of free home Internet access in the Computers In Homes (CIH) scheme. This scheme was created for the purpose of overcoming the social and economic consequences of an emerging digital divide, an issue that has been a focus of attention for NZ politicians, policy makers and practitioners, just as it has been a global issue from the late 1990s. CIH was founded in early trials in 1996 in Wellington, NZ's capital city, which already had an established history of successful community-based ICT projects (Newman, 2008,

p. 3; Zwimpfer, 2010), and it thrives today with a mission "to provide all NZ families, who are socially and economically disadvantaged, with a computer, an Internet connection, relevant training and technical support" (Computers in Homes, 2011, p. 2). By 2000, the CIH scheme was being developed further and piloted by the 2020 Communications Trust, a charitable organization that wanted to develop a successful community Internet model aimed at raising the literacy level of children from low decile[1] schools. Its purpose was to "promote dialogue and understanding through local action" (2020 Communications Trust, 2009), to provide leadership in ICT and deliver programs that address issues of digital literacy, skills and inclusion.

The non-profit 2020 Communications Trust has achieved this through partnerships with national and local government agencies and businesses to obtain funding for its activities. The Trust identifies gaps in digital inclusion in NZ communities, devises possible approaches and then seeks to partner with other agents to take action. CIH is managed within this context and has become a key intervention program in the 2020 Trust's line-up of initiatives. In 1999, the 2020 Communications Trust began to draw together sponsors and partners who could collaborate to provide the resources to begin work in the pilot CIH communities. These partners included Computer Access NZ (CANZ), an agency set up by the 2020 Trust to access and refurbish used computers to an as-new standard. This was necessary because of the cost of new computers for low decile schools, which

*…can be beyond the resources of cash-strapped schools and not-for-profit community organisations. To help solve the problem, the Computer Access NZ Trust (CANZ) was set up in 1999. It was an initiative of the **2020 Communications Trust,** supported by the Ministry of Education. CANZ accredits **computer-refurbishing companies,** which use the CANZ quality brand…Accredited refurbishers sell used equipment donated by com-*

mercial and government organisations, usually about three years from brand new. This equipment, all quality 'name' brands, is refurbished, upgraded as necessary and sold with a warranty and after-sales service (Computer Access New Zealand, 2009).

With refurbished computers made available in this way, and with the support of the Ministry of Education, pilot CIH projects were set up in two low-income communities: Cannon's Creek in Wellington and Panmure Bridge in Auckland, NZ's largest city. These first CIH projects were concerned with improving educational outcomes, in particular through building parents' confidence to be more involved in school life and providing them with skills to help their children with their schoolwork. In these first projects, numbers of parent helpers at school events and attendance at parent-teacher interviews were monitored by CIH and school staff to evaluate progress toward the desired outcomes of the 2020 / CIH pilots, and in this way to make improvements. For example, in Flaxmere, a suburb of a smaller provincial city in the North Island of NZ, CIH was the most visible and successful initiative of "a series of innovations related to improving home-school relations within and between...Flaxmere schools" (Clinton, Hattie, & Dixon, 2007). Fully funded by the Ministry of Education, the Flaxmere project experimented with a range of strategies such as homework centers and home-school liaison persons (Perry, 2004) alongside CIH to improve the relations of the five local schools with their communities, and to engage parents more fully in the learning activities of their children (Clinton, et al., 2007). Thus, CIH has always been committed to developing stronger communities, and seeking improvements to CIH practice that work within the different community contexts.

By 2003, the government of the time had consulted widely on and produced a draft Digital Strategy document (Ministry of Economic Development, et al., 2004) which was formally adopted in 2007 (New Zealand Government, 2007), committing to a variety of initiatives and funding support for community-based schemes such as CIH. A feature of the draft Digital Strategy (2004) as well as the upgrade referred to as the Digital Strategy 2.0 (Ministry of Economic Development, 2008) was an implicit assumption that communication technologies are important for "social connectedness" (Department of Internal Affairs, et al., 2002, p. 3) on the basis that:

A modern cohesive society is an essential building block for a growing and innovative economy and society...people who feel socially connected also contribute towards building communities and society (Department of Internal Affairs, et al., 2002, p. 3).

The goal of stronger community often appears to underlie community interventions programs like CIH. Yet, the way in which that goal is to be achieved is generally not at all explicit. Assumptions about a relationship between Internet access and cohesive communities is seen in statements such as "measures of social connectedness [include] access to telephones and to the Internet" (Department of Internal Affairs, et al., 2002, p. 3), set in the same context as "people who feel socially connected also contribute towards building communities and society" (Department of Internal Affairs, et al., 2002). Communication technologies are generally viewed in a policy context as tools for empowering individuals and communities and integral to the functioning of a progressive society (Ministry of Social Development, 2010). Thus, the first indicator of social connectedness listed by the Ministry of Social Development is "telephone and Internet access in the home" (Ministry of Social Development, 2010). The Digital Strategy also clearly relates Internet access closely with social connectedness.

This chapter examines what relationship may exist between Internet access and social connectedness, using data gathered in CIH communities during 2003 – 2005 in Auckland in case study research. It provides valuable insights into the role of the social context in enhancing the longevity of Internet initiatives. The following section provides an overview of how these insights were derived through longitudinal community research using an eight-dimension schema of social cohesion. Later sections of the chapter summarize the results that emerged through use of the schema. The chapter ends with a discussion of practical implications for community Internet practitioners.

LITERATURE REVIEW

Literature relevant to the relationship between Internet access and social outcomes draws on a number of related fields. Hampton (2007) acknowledged that "existing research on how information and communication technologies influence neighbourhood relationships has been explored in three complimentary [sic] …research traditions: community informatics, sociology, and communications" (p. 717). Reflecting on this necessary inter-disciplinarity, this chapter now reviews material relevant to the research goal (examining how Internet access and social cohesion are related) within several broadly inter-connected fields of study.

The use of a range of similar terms in the literature can create ambiguity about how community Internet interventions are intended to work, and meanings can be nebulous (Kearns & Forrest, 2000). While in a broad sense there is some agreement that a cohesive society is, essentially, one that "hangs together" (Kearns & Forrest, 2000, p. 996), care must be taken to distinguish how each term is understood, and how they inter-relate. Some of the similar terminology includes *community, social capital* (Forrest & Kearns, 2001; Pigg & Crank, 2004), *social cohesion* (New Zealand Government, 2007), *community renewal* (Housing New Zealand Corporation, 2006), *community capital* (Williams, 2006), *community capacity* (Casswell, 2001; McKnight & Kretzmann, 1996), and *com-*

munity building (Bimber, 1998; Toyama, 2007). An overview of the terms and their relationships is provided in the following section.

Community

Assumptions about social benefits that may be expected from improved Internet access are premised on an ideal of community that is rarely defined. In 2004, the New Zealand government's Digital Strategy document claimed that "communities will be strengthened by being connected to fast global communications networks" (Ministry of Economic Development, et al., 2004, p. 6) and "our businesses and communities will possess the skills and confidence to utilize national and local information resources" (Ministry of Economic Development, et al., 2004, p. 6). The CIH scheme, cited in this policy context as a success story (Ministry of Economic Development, et al., 2004, p. 37), aims to build stronger community in a broad sense. One of its key goals is to "empower low socio-economic communities to become active participants in the online world" (Computers in Homes, 2011). Yet some questions are implied here, such as: what does community in this context mean? How is the term community used in the literature? How can we know that a community has become strong?

At the most basic level, community can be thought of as a "group or collectivity" (Vergunst, 2006, p. 1) in society. However Loader and Keeble (2004) note that the term tends to be "ambiguous" (p. 36) and:

It frequently appears to mean different things to different people despite the fact that the term is often used as if in common agreement. In one sense, it is imbued with the aura of companionship and human warmth which derives from its linguistically related concept of 'communication.' Consequently any technologies which foster more and perhaps better communications between people contribute to a greater sense of community (Loader & Keeble, 2004, p. 36).

Postill (2008) argues that too much emphasis has been placed on the concept of community because of the global process of Internet localization, concluding that it is "a polymorphous folk notion widely used both online and offline, but as an analytical concept with an identifiable empirical referent it is of little use" (Postill, 2008, p. 416). Although community often tends to be associated with the ties of support that exist between people within a geographical area, increasingly this connection is contested. Keith Hampton, ethnographic researcher in "Netville," wrote that:

When one defines communities as sets of informal ties of sociability, support and identity, they are rarely neighbourhood solidarities... Communities consist of far-flung kinship, workplace, interest group and neighbourhood ties that together form a social network that provides aid, support, social control and links to multiple milieus (Hampton, 2002, p. 228).

While community is generally understood to be fostered by interpersonal contact, by reciprocity, by meeting together (Williams, 2006), certain qualities of community can, of course, be facilitated and sustained across distances using a variety of communications media, so that people can be constantly in touch if they wish to be (Castells, Fernandez-Ardevol, Qiu, & Sey, 2006). Wellman (1999, p. xiv) asserts that individuals *are* their relationships, no matter how these are enacted, and furthermore "the trick," he states, is "to conceive of community as an egocentric network, a 'personal community,' rather than as a neighbourhood" (Wellman, 1999). In this sense, while one might at first think of community as necessarily being about place, it may be better understood as a feeling an individual experiences in relation to their unique set of personal networks which includes distant ones as well.

Loader and Keeble (2004), while acknowledging that the concept of community remains contested, highlight the important role of community

as a space in which to negotiate one's place in the larger world: "an 'intermediate space' between the individual/family and larger social structures, such as government" (p. 4) and thus it is "important for fostering many life opportunities" (Loader & Keeble, 2004). Meegan and Mitchell describe neighbourhood in a very similar way as "a key living space through which people get access to material and social resources, across which they pass to reach other opportunities" (p. 2172). Thus the idea that both community and neighbourhood are a form of conduit to social opportunity underscores the potential value of interventions like CIH at this level, as they may point people towards engagement with life beyond the immediate vicinity or their nearby social networks.

The neighbourhood itself can have "heightened importance" (Meegan & Mitchell, 2001, p. 2174) in the context of social exclusion for certain groups such as the unemployed, and can thus be a "place-based community" (p. 2179). Wellman also points out "neighbourly relations are especially important when poverty or disability leads people to invest heavily in local relationships" (p. 11). In this regard, Gaved and Anderson argue that "place still matters," citing "an increasing trend towards considering ICT initiatives as part of existing social interactions rather than separate, purely online virtual communities of interest" (p. 5).

Social Capital

Robert Putnam defined social capital as "features of social life—networks, norms, and trust—that enable participants to act together more effectively to pursue shared objectives" (Putnam, 1996). He became known for his view that social capital had collapsed in the US (2000, 2002) based on statistics showing a dramatic decline in the numbers of people involved in clubs, churches, sports groups and the like. The idea that the glue holding society together was coming undone became attributed, in some circles, to the rise of media consumption, especially television (Moy,

Scheufele, & Holber, 1999), which was thought to be eroding community at that time.

An apparent deficit of social capital, called a "crisis in social cohesion" (Forrest & Kearns, 2001, p. 2126) has been a strong theme in the literature on community. Building on the ideas of Coleman (1988) who identifies three types of capital—physical, human, and social—and views social capital as a resource that can be "mobilized for collective action" (Pigg & Crank, 2004, p. 60), Onyx and Bullen (2000) consider social capital has five dimensions: networks, reciprocity, trust, shared norms, and social agency. Williams (2006) has referred to "confusion in the literature about whether social capital is a cause or an effect" although some support is found for the idea that social capital is generated when there is some already existing.

Social capital has been described as a "contentious and slippery" (Williams, 2006) term. Even Putnam tends to use a range of conceptually similar expressions including "community engagement," "civic trust," and "social trust and reciprocity" (Saguaro Seminar, 2007). Social capital and civic engagement are used almost interchangeably by Putnam, as in the title of a Web page about social capital to which he is a key contributor, called Civic Engagement in America (Saguaro Seminar, 2007). In these contexts, "the usual premise is that [it] is a good thing, so it is conveniently assumed that further elaboration is unnecessary" (Kearns & Forrest, 2000, p. 996).

Social capital is also understood as the value derived from social ties: out of our social relationships comes the impetus to do things for one another (Putnam, 2007). This impetus to reciprocate is a resource, generally understood to be like financial capital, in that a community needs to use it in order to grow more of it (Williams, 2006). Thus the idea that social capital is a necessary building block of social action (Pigg & Crank, 2004) has become orthodox, and yet it is often viewed as being in short supply (Putnam, 1995). This may be because of a social policy focus on

disadvantaged neighborhoods in research and literature on community, contributing to "deficit theory syndrome" (Forrest & Kearns, 2001, p. 2141) or a "deficiency-oriented social service model" (McKnight & Kretzmann, 1996, p. 1). Within this paradigm, communities are "noted for their deficiencies and needs" (McKnight & Kretzmann, 1996) and therefore are seen to lack certain resources or exhibit less robust processes rather than actually having assets. Tuck (2009) calls this a "damage-centered" approach that must change, because it perpetuates a view of communities as "depleted" (p. 409), "defeated and broken" (p. 412), a perspective that is inherently disempowering.

Putnam focuses on the importance of associational activity for participation and democracy, conceiving of social capital as both the social networks themselves, and the positive outcomes of them (Williams, 2006). Other commentators understand it to be either the networks *or* the outcomes (ibid., p. 2). In this sense, social capital operates at individual level and at community level: individuals can leverage networks for their own advantage such as in deriving social support for themselves (Wellman & Berkowitz, 1988), while social capital in action is a collective asset that improves social outcomes at a community level (Ferlander, 2003).

Social Cohesion

The term *community cohesion* (Vergunst, 2006) alludes to a notion of strong community by "address[ing] the characteristics (and the strength in particular) of the bonds between the individuals who constitute that collectivity or group" (Vergunst, 2006, p. 1). However, the term *social cohesion* (Das, 2005; Forrest & Kearns, 2001; Friedkin, 2004) is more frequently used. The literature on cohesion features an emphasis on the ability of a cohesive group to mobilize toward a collective goal. A focus on collective action "historically… enabled citizens to efficiently pursue common goals, often creating community wide gains" (Shah & Scheufele, 2006, p. 2) in a socially cohesive setting. Friedkin (2004) highlights collective action as a characteristic of cohesion as follows:

The members of a highly cohesive group, in contrast to one with a low level of cohesiveness, are more concerned with their membership and are therefore more strongly motivated to contribute to the group's welfare, to advance its objectives, and to participate in its activities (Cartwright, 1968; cited in Friedkin, 2004, p. 412).

Spoonley et al. (2005) cite a Canadian definition of a socially cohesive society as "one where all groups have a sense of belonging, participation, inclusion, recognition and legitimacy" (Jenson, 1998; in Spoonley, et al., 2005, p. 88) and suggest that social cohesion is "interactive" (Spoonley, et al., 2005, p. 88). Additionally, the degree of cohesiveness in a group contributes to social influence: "in cohesive groups, conformity pressures are greater because individuals value the opinion of other group members" (Vishwanath, 2006, p. 327) and hence "in such groups, individual internal attitudes and beliefs converge with that of the group" (Vishwanath, 2006). Thus, interpersonal influence plays a key role in social cohesion, which may also be linked to the presence of influential individuals or opinion leaders, and is centrally concerned with dialogue. For example, Burt (1999) describes cohesion as "the strength of the relationship between [the receiver] and [the sender]" (p. 3).

NZ government policy discourse through the 2000s has been increasingly specific about social cohesion, listing the ways in which it may be recognised. A 2006 Statistics NZ document on social cohesion stated that it is evident where people feel a part of society, relationships are strong, differences are respected, people feel safe and supported by others, and they feel a sense of belonging, identity, and willingness to commit to shared tasks. This list was made measurable

for government research purposes in 2008, with key indicators of social cohesion stated as formal unpaid work outside the home, rate of death from assault, impact of fear of crime on quality of life, voter turnout at general and local elections, representation of women in Parliament and local government, and trust in government institutions (Statistics New Zealand, 2009, p. 117). Forrest and Kearns characterise social cohesion in a very straightforward way as "getting by and getting on at the more mundane level of everyday life" (p. 2127), a definition that values the domestic continuities of everyday life as much as the more actively altruistic expressions of engagement in community life. In their view, people's ability to cope with day to day life is a feature of cohesion that helps maintain order and stability, and we "may underestimate the importance of the lived experience of the dull routine of everyday life" (p. 2127) for its role in "ongoing 'repair work' and 'normalisation'" (Forrest & Kearns, 2001). These elements are included in the design of the Computers In Homes (CIH) case study research explained further on in the chapter.

However Forrest and Kearns (2001; Kearns & Forrest, 2000) have proposed a more structured model of social cohesion incorporating five elements, one of which is social capital. In this sense, social capital is a specific outcome of social cohesion, rather than being more or less the same thing. Following this reasoning, where "a cohesive society is one in which dilemmas and problems can be easily solved by collective action" (Kearns & Forrest, 2000, p. 1000), social cohesion is more likely to be developed if social capital exists. Civic engagement is also an element of social cohesion, expressed through associational activity in neighbourhood and community organizations (Kearns & Forrest, 2000). This Kearns and Forrest (2000) model, highlighting the importance of *existing relationships and networks* to "sustain the expectations, norms and trust which facilitate…solutions" (p. 1000), underlies the research design in the CIH study. Finally, civic participation, defined by Shah and Scheufele as "public involvement in efforts to address collective problems" (p. 2), is another term used in the literature that touches on the theme of collective action. It appears to be synonymous with civic engagement as understood by Putnam, and is understood to be one of the key behavioural outcomes of a cohesive group.

Connectedness and Cohesion

Elsewhere the phrase *social connectedness* is strongly associated with social cohesion (Department of Internal Affairs, et al., 2002). Indeed even Putnam uses this term, apparently as a surrogate for social capital, exploring "why education has such a massive effect on social connectedness" (Putnam, 1996). Social connectedness may be a construct inspired by the UK approach to social policy, where a cohesive community

... is one where: there is a common vision and a sense of belonging for all communities; the diversity of people's different backgrounds and circumstances is appreciated and positively valued; those from different backgrounds have similar life opportunities; and strong and positive relationships are being developed between people from different backgrounds and circumstances in the workplace, in schools and within neighbourhoods (Local Government Association, 2002, 2004; cited in Institute of Community Cohesion, 2009).

In the NZ setting, social connectedness—"the relationships people have with others" (Ministry of Social Development, 2010)—is valued also because "people who feel socially connected also contribute towards building communities and society" (Ministry of Social Development, 2010). Furthermore, "several studies have demonstrated links between social connectedness and the performance of the economy" (Ministry of Social Development, 2010). The Ministry of Social Development (MSD) now identifies six indica-

tors of social connectedness[2], most of which were incorporated in the Computers in Homes (CIH) research design[3]. The first two of those listed are telephone and Internet access in the home, and contact with family and friends (Ministry of Social Development, 2010). The fact that Internet access is prioritized as an indicator of social connectedness is based on the view that "The Internet in particular is becoming an increasingly important means of accessing information and applying for services" (Ministry of Social Development, 2010). This explicitly assumed role for Internet access in cohesion is a key reason for measures of Internet "connectedness" (Kim, Jung, Cohen, & Ball-Rokeach, 2004) being included in the CIH research. Participation in unpaid work outside the home, another social connectedness indicator listed in the 2006 version of the MSD Social Report, no longer available online, was also built into the design because it indicates a willingness to volunteer in the community.

The range of definitions and dimensions of cohesion reviewed so far highlights the *individual* experience. That is, what feelings and behaviours does a cohesive community generate for individuals? Moving to a more subtle interpretation, Friedkin (2004) notes that social cohesion can be defined either through individual level behaviours and attitudes (such as volunteerism and participation), or through group level "conditions" (Friedkin, 2004, p. 410) and outcomes. If cohesion exists on two levels in this way, do the individual level behaviours / attitudes arise *because* of the group level conditions? That is, do people feel better and behave in altruistic ways as individuals because there is an overall sense of support and care within the group? Alternatively, do the cohesive group level conditions arise *because* of the individual level behaviours and attitudes? Friedkin suggests, "groups are cohesive when group-level conditions are producing positive membership attitudes and behaviours, and when group members' interpersonal interactions are operating to maintain these group level conditions" (p. 410). Thus in Friedkin's view, *group level conditions* have primary importance and, in turn, they generate *individual responses* (Friedkin, 2004). He argues "we should discard the idea that group-level conditions indicate social cohesion and instead treat these conditions as *antecedents*[4] of particular individual membership attitudes and behaviours" (p. 416). In a sense, therefore, social cohesion accumulates recursively, with the group conditions laying the foundation for individual behaviours that build upon it, as responses to the right conditions.

The logic that 'cohesion,' or overall wellbeing, at community level drives the development of the 'capital' or resource located in social networks, is reinforced also in NZ's policy discourse. For example:

Relationships and connections can be a source of enjoyment and support. They help people to feel they belong and have a part to play in society. People who feel socially connected also contribute towards building communities and society. They help to create what is sometimes called "social capital," the networks that help society to function effectively (Ministry of Social Development, 2010).

Thus it seems agreed that social cohesion is required to generate social capital, a perspective echoing the priority placed on group-level conditions by Friedkin (2004) above, in which for example the number of interpersonal ties and the pattern of social networks (p. 416) are the antecedents for individual behaviours, such as joining a group or volunteering at school.

Community Capital and Capacity

Community capital has a specific local or contextual focus, but it is conceptually related to social capital, loosely understood to mean informal social ties (Hays & Kogl, 2007) and shares denotations of trust, mutuality, tolerance, and other regard (Coleman, 1988; Fukuyama, 1995; McClenaghan, 2000). In a group or community setting, such behaviours and attitudes may be directed towards the achievement of shared goals or the enhancement of shared values. Civic engagement, used in the social capital literature, is "people's connections with the life of their communities, not only with politics" (Putnam, 1996), or more specifically "associational activity in neighbourhood and community organisations" (Kearns & Forrest, 2000, p. 1000). In general then, civic engagement becomes manifest in specific actions of individuals that are oriented toward community and civic life, such as volunteering one's time for school events or committees.

Community capacity is used to describe the relative ability of a group to mobilize resources, to plan, and reach toward collaboratively derived goals. Although the phrase tends to be used without clear definition (Casswell, 2001), the nearest may be a community's "capacity to identify and address social and health issues at the community level" (p. 23), implying a consideration of civic infrastructure and community resources such as skilled, available people, and time. Community capacity is used as a means of describing the capability latent within a community to frame its own solutions in a societal context where "the hard truth is that development must start from within the community and, in most of our urban neighbourhoods, there is no other choice" (McKnight & Kretzmann, 1996, p. 2). This type of view reverses the assumptions of a deficit, or "damage-centred" (Tuck, 2009, p. 409) approach to community building, which tends to focus on what is lacking. Moreover, community capacity could be understood as an embodiment of all the other dimensions of a cohesive society reviewed so far, as shown in Figure 1. Social cohesion or connectedness at the source is a dynamic interplay of resources and relationships operating on individual and group levels. Social capital and civic engagement—a willingness to volunteer and to act collectively—are outcomes of social cohesion.

Figure 1. Dimensions of community capacity

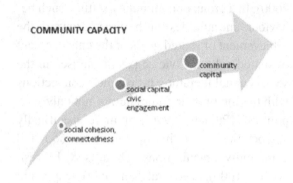

Together these processes constitute both latent and manifest community capital. Finally, all of these components together comprise *community capacity*, the relative ability of a community to mobilise resources, to plan, and reach toward collaboratively derived goals.

The Social Cohesion Schema

Thus, a comprehensive review of the literature of community informatics, sociology, and communication studies shaped the design of a framework for investigation of social cohesion in relation to Internet use in the CIH study. The essence of what was common across the literature was included in a framework of social cohesion (see below) that would guide the design, research methods and data analysis for the CIH case studies, and underlies the interpretation of results presented later in the chapter.

1. **Social connectedness** (Ministry of Social Development, 2010) is characterized especially by levels of Internet access, levels of interaction with family and friends, and the number of community members carrying out unpaid work;
2. **Routine day-to-day life** (Forrest & Kearns, 2001) encompasses activities relating to getting on at the everyday level, such as walking the children to school, and ensuring the needs of a household are met;

3. **Inclusiveness** (Ministry of Social Development, 2010) is created when diversity is valued and community members feel part of society as a result;
4. **Support** (Ministry of Social Development, 2010) is a feature of communities in which people feel safe and know they have people to call on when in need;
5. **Place attachment and identity** (Spoonley, et al., 2005) is evident where people express feelings of belonging, identity, and willingness to commit to shared tasks;
6. **Networks of mutual support** (Vergunst, 2006) are created as outcomes of the bonds holding people together in kinship, friendship or other groups;
7. **Social capital** (Kearns & Forrest, 2000), sometimes understood as civic engagement (Putnam, 2002), is generated by interaction and reciprocity, associational activity, and volunteerism within the community;
8. **Social solidarity** (Friedkin, 2004; Kearns & Forrest, 2000) is present in a community where dilemmas and problems seem to be easily solved by collective action: people show the ability to mobilize as a group.

Data relevant to these dimensions were collected through a combination of methods, the most important of which was in depth interviews with adult participants and staff in the CIH scheme. The interviews were conducted in two waves of data collection about one year apart[5] with adult caregivers (usually parents, occasionally grandparent caregivers). Both open-ended reflective questions and survey items on Internet use were included. Each interview required about 90 minutes to complete. Out of 30 interviews with participants from two CIH communities in Auckland, 23 were complete and, therefore, used for analysis. Researcher observation also generated notes of meetings, training sessions and discussions over the course of about two years of research within the two CIH communities. Iterative processes of review and reflection, thematic analysis, summary and aggregation of data for each of the two cases, followed by comparison of results for each, produced the findings on social cohesion reported in the following section.

CASE DESCRIPTION

The research was conducted in two communities where CIH was being launched in 2003 in Manukau[6], and continued until 2005. In writing about "Auckland, a city divided by income," Collins (2012) declares a growing recognition that:

Auckland has changed from an equal city to an unequal one in less than a generation with the income gap between rich and poor widening dramatically over the past 25 years...Many economists ... see reducing inequality as a prime economic goal—both to harness our full human potential and to dampen boom/bust cycles caused by excessive lending by people who have more than they need to people who need the loans but can't afford them (Collins, 2012).

Social inequality has continued to deepen in some low-income urban areas within the larger Auckland metropolitan area such as Manukau (Statistics New Zealand, 2012). Government agency Statistics NZ prepared data analysis for the 2020 Communications Trust showing Manukau had 15,000 families of school aged children with no Internet at home in 2006, shortly after the case study research was completed. Some of the issues creating this digital disparity include affordability of landline telephone connections and broadband subscriptions for low-income households, and family experiences of low educational achievement or failure. CIH aims to improve these educational outcomes by working with primary school[7] communities where it sets up a structure of action in which school staff, guided by a CIH Regional Coordinator, work alongside families to provide digital opportunity (technology, training, and

support). The overarching goal is to enable these communities to become technologically autonomous and be able to use ICTs for self-determined objectives such as further training, setting up a business or finding employment.

Although the earlier discussion shows the term "community" can have a variety of meanings, it is used in the CIH research to mean a group of people who live in the same geographic locality, because "most of the resourced and evaluated community initiatives in NZ have worked with geographical communities" (p. 25). Although the complete research project (Williams, 2009) included two detailed case studies referred to as Case A and Case B, key findings suggesting a relationship between Internet scheme sustainability and social cohesion were particularly apparent in the results for Case A.

Case A is a community in a relatively new suburb in an outlying industrial locality between Auckland's southern motorway and a tidal mudflat coastline in Manukau. It grew from large subdivisions of cheap group housing built in the early 1980s. Case B is a community located in the southernmost semi-rural fringes of the greater Auckland metropolitan area. However, while the housing is older and the suburb is much more established, the Case A and B populations are comparable. They both feature low levels of educational achievement (Statistics New Zealand, 2006a, 2006b), considerable ethnic diversity including a high proportions of Maori[8] (Statistics New Zealand, 2006b), and Pacific peoples (Statistics New Zealand, 2006a), and comparatively high unemployment (Statistics New Zealand, 2006a) around double that of the greater Auckland region.

During the 2003 launch of CIH for 20 selected families from the school at Case A, the community was dealing with an unusual school restructure that was, for a time, a distraction from CIH scheme implementation. Until 2004, the Case A community was served by one large primary school with two co-Principals. However, at the beginning of 2004, the school was divided into two separate schools a few streets apart. This meant that at the time CIH was being launched in October 2003, two school leaders shared the task of ensuring the CIH scheme was well managed and delivered. The school's co-Principals were offered the standard CIH package of 25 computers for 25 families for their community. However, they decided to limit the scheme to 20 families. A few months later, the school restructure occurred so that at the start of the school year in 2004, half of the school's children and their families had been relocated to a new school a few streets away. Thus, the twenty families involved in CIH in this community were dispersed across two separate schools.

Data Collection and Analysis

Extended interviews including open-ended and survey questions were conducted with 30 volunteers from across both the Case A and Case B communities. All interviews were transcribed to capture the open-ended comments and anecdotes, and quantitative data such as attitudinal responses using Likert scales were input to Excel (refer to Williams, 2009). Although all participants' contributions were recorded and retained some quantitative data were not included in analysis because they were incomplete. For example, an "Internet Connectedness Index" (Williams, 2009, pp. 111-113) using survey questions assessed each participant's "relationship with the technology" (Jung, Qiu, & Kim, 2001, p. 513) and comprised several quantitative items. Where data were absent for a participant, it was not possible to determine an Internet Connectedness Index for them. Therefore, after missing data were taken into consideration, the analysis focused on ten respondents from Case A, and thirteen from Case B.

Participation in CIH by families at Case A and Case B, as well as their involvement in the research, showed a high attrition rate from the beginning of the study to the end. Fourteen of the original 23 research participants for whom data were complete had dropped out of CIH by the end of the study in 2005. Additionally, a slight decrease in Internet engagement was found across the whole

sample over the period of the research. Yet two interesting trends stand out in respect of Case A: first, a greater proportion of the parents who were still active Internet users at the conclusion of the research were from Case A (Williams, 2009, p. 175). The second finding that sets Case A apart is that social cohesion was not only more evident there at the beginning of the study than at Case B, but it was even more marked at the conclusion of the research. A later section in the chapter, 'Findings relating to social cohesion,' reflects on what may account for this outcome.

Technology Components

The model of technology provision that has evolved in CIH is as follows. CIH undertakes an agreement with the school to provide the hardware, software, Internet connection, and training to use a computer and the Internet, for a cost of $50 per family. The school principal and staff consider which families do not have a computer or Internet at home and where it is felt this is an issue for the child's learning, select them to be involved in the scheme. If the family is unable to pay $50 straight away, arrangements are made, for example, for the parent to bring $5 each week to the ten training sessions. Involvement in the scheme includes the following items for each family:

1. A refurbished desktop computer;
2. Free broadband Internet for six months;
3. Weekly two hour training sessions for ten weeks;
4. Technical support provided by the school with paid technicians as needed;
5. Software including Windows XP, Office 2007, Antivirus and Spyware, Adobe Reader, Flashplayer, Shockwave and Java, and desktop shortcuts to Internet safety and learning sites.

Training in all CIH communities follows a standard pattern that has evolved over recent years to equip families effectively to begin making educational use of the Internet at home. The standard approach is that one parent from each family must attend a two hour training session once a week for ten weeks at the school, generally in a computer classroom, in order to "graduate" and receive a computer to take home. Initially, the basics of computer use such as powering on and off, opening Word and saving documents, accessing the Internet, setting up an email account, and opening and saving attachments are covered. Topics discussed at the training sessions include Internet risks such as viruses and scams, and Internet safety for children. The aim is to ensure that when the family takes their computer home after graduating from the 20 hours training, they will be able to set up the computer, power on and use the Internet independently and safely. Parents may not graduate and take their computer home if they have not attended all the sessions.

For the parents at Case A, training was provided by the local Housing NZ (HNZ) Community Renewal center, as well as by a school staff member. In this case, the sessions were neither well attended nor closely managed, and the involvement of HNZ was problematic. For example, subsequent feedback from parents recalled their feelings at the time, that the training "…was just too rushed… the manual is a waste of time. I had trouble reading it" (Williams, 2009, p, 315). Another mother agreed, saying the training was

…inadequate. And I'm not the only person that thinks that - T who was the school source behind it also feels that way. She wants the families to be able to [see]—'OK, this is what we are going to do'—like a lesson plan—this is what we are going to learn today and we are all going to learn it. Yeah just a bit more support [is needed]… several weeks into this program it was discovered that lots of people had trouble actually getting initial Internet access (Williams, 2009, p. 315).

Another participant at Case A suggested that "a lot more communication, better communication, like - calling frequently, you know, like really

communicating with them" (Williams, 2009, p. 279) would be a helpful tactic on the part of school staff responsible for the scheme. This kind of follow-up action has, in subsequent years, become a norm in CIH practice. Recent enthusiastic feedback from parents in another CIH community suggests the teacher responsible for delivering CIH in the school is crucial to the scheme's success:

The upside for me doing Computers in Homes was having the right person with computer smarts at the helm. We had that person in D---. His positive vibes gave me the encouragement to do and try things with a computer, which I never thought I had. Tena koe[9] D---.

A particularly important point about Case A is that despite the complex issues the school faced at the time CIH was being implemented in 2003 (including the school restructure, the involvement of Housing NZ, and the less than optimal CIH training) it went on to experience very positive outcomes from the scheme. As will be shown in the next section, social cohesion was readily apparent in this community as the research began. CIH parents knew one another, some were active in volunteering both within the school and in other community activities, and some actively took roles in mentoring others who were less confident with computers. These features may have had a relationship with successful longer-term outcomes for family Internet use in Case A, because those who have a drive to operate collectively can address problems. A problem-solving attitude is seen, for example, in the thoughts of a Case A mother who was motivated to want to change things for the next group of CIH families:

I don't feel that this present intake has quite got enough out of it, and if they're talking about a new intake I want to be involved in making it smoother running – and the process and the support behind it is there (Williams, 2009, p. 313).

This mother went on to become a leader among the group of CIH families, volunteering to take on administrative duties so that the scheme could run smoothly and take on new families. Social solidarity among motivated parents who feel part of a group, share a sense of purpose, and mobilize toward a shared objective are the elements of a socially cohesive group.

FINDINGS RELATED TO SOCIAL COHESION

Research design for the CIH study was guided by a framework for social cohesion including eight dimensions, derived from a comprehensive literature review outlined earlier in the chapter. The overarching research goal for the entire study focused on the relationship between Internet access and social cohesion. Thus, data collection methods also addressed specific aspects of Internet use so that trends for this could be examined in relation to trends for social cohesion.

Data collection and analysis in respect of Internet use are touched on in less detail in this chapter, as its primary focus is on the community outcomes in terms of social cohesion. Briefly and for the purposes of context, the participants' use of the Internet, or "Internet connectedness" (Jung, et al., 2001) was also assessed at Time 1 and Time 2 using an aggregation of data from eight survey items, to determine how intensely people engaged with the Internet. In summary, there was a large attrition rate in the number of parents involved in CIH at both case study sites by Time 2. For the whole group of nine remaining participants, Internet connectedness declined slightly over the two phases of research. However many more of the dropouts were from Case B rather than Case A, and *proportionately more of those still committed to continuing to use their home Internet were from Case A* (see Williams, 2009, section 4.1, p. 134). This apparently more successful outcome for Internet use at Case A is paralleled by the findings for social cohesion.

Social Cohesion Framework

To review, the eight dimensions of social cohesion condensed from a literature review were: social connectedness, routine day-to-day life, inclusiveness, support, place attachment and identity, networks of mutual support, social capital, and social solidarity. They are divided into behaviors that can be demonstrated in a community context at *individual* and *group* levels, because as described earlier in the chapter, Friedkin (2004) offers a useful assertion that group level behaviors relating to social cohesion are the antecedents for individual behaviors.

Data relevant to these eight dimensions were gathered in the in depth interviews, using both survey and open-ended questions. While the interview predominantly used closed questions, such as "Do you feel you have many people to turn to when you really need help?" (Williams, 2009, p. 307), answers often led to open-ended responses and a great deal of narrative from participants who talked about their lives within the neighbourhood and community.

As an insight into the comprehensive nature of data collection and inductive analysis, *social connectedness* was assessed by inviting respondents to talk about the unpaid work they contributed such as helping at the school, and using survey items such as frequency of contact with family members. *Routine day-to-day life* was documented in a variety of ways using survey items (such as frequency of visiting and calling people) and interview techniques, which opened the opportunity for participants to share anecdotes about their lives in the neighbourhood and their feelings about it. *Inclusiveness* was touched on in survey and open-ended questions relating to people's feeling of belonging in the neighbourhood. *Support* was assessed using, for example, data on the number of people known by name in the neighbourhood, and interview data about support networks. Survey questions on the length of time in the neighbourhood, home ownership, and intention to stay, and others, created a picture of *place identity and attachment*. Data relevant to *social capital* were gathered from questions on membership of churches and clubs, involvement in community projects, and others. Finally, *social solidarity* was assessed in a number of ways from researcher observation and discussion with key figures in the CIH setting. The entire set of methods matched with the social cohesion framework is detailed in Williams (2009, p. 117) as well as in the interview schedule (Williams, 2009, p. 269).

Social Cohesion Results

Thus, evidence relevant to social cohesion was gathered using a variety of methods at Time 1, when the families first took their computers home, and at Time 2, about one year later (Williams, 2009, p. 229-232). Data were analyzed and aggregated over time into trends at individual and group levels for both cases. Comparison between the cases was thus readily achieved. For example, research participants in each case study were asked questions relating to place attachment, a sense of belonging in the neighbourhood. One of these was "How likely is it that you will leave this area in the near future (i.e. next 1-2 years)?" using a 5-point Likert scale for rating, and "How many people do you know by name and say hello to in this immediate neighbourhood?" (Williams, 2009, p. 307). A trend was found (see Table 1) that Case A participants knew many more neighbors by name than participants in Case B, implying that an important group level condition relating to social cohesion (Friedkin, 2004) was satisfied at Case A but much less so at Case B. In Case B, families were more likely to rent rather than own their home (Friedkin, 2004, p. 206), an interesting result in view of the fact that physical rootedness "has been found to positively affect community attachment" (Ball-Rokeach, et al., 2012, p. 5). In these ways, comparisons were drawn between the cases.

Table 1 summarizes the trends for social cohesion across the two case studies from Time 1 to Time 2, in each of the eight social cohesion dimensions shown at individual and group levels.

In summary, readily identifiable features of social cohesion characterized Case A at both group and individual levels. Case A also featured leader figures who mobilized to ensure the CIH scheme remained active and effective. One mother in this community cited earlier in the chapter, and her husband, had a number of ideas that were subsequently put into practice, including better planning—"the next time is *not* going to be rushed. That we start the planning process in June, for the next intake in November..." (Williams, 2009, p. 315)—and, they thought, "A home visit for each person where they can get set up at home" (Williams, 2009). These suggested improvements are now norms of CIH practice. It appears possible then that social cohesion may have a relationship with the presence of influential individuals who have been described as technology opinion leaders (Vishwanath, 2006) in the diffusion of innovations in which "interpersonal contacts are especially important for new communication technologies" (p. 3). By this definition, cohesion was not so evident in Case B where social networks were much less strong, while in Case A certain group members were influential, and mobilized to direct the fortunes of the group. Social influence as embodied in the relationship between opinion leaders and opinion seekers plays a key role in creating a cohesive setting and is more effective within it, a point made by Vishwanath (2006) as mentioned in the literature review earlier in the chapter. This hypothesis appears to be borne out in the results of the CIH study where the strong presence of influential leaders and greater social cohesion coincided in Case A, where comparatively better success with Internet use was also an outcome.

Table 1. Summary of case study results for social cohesion at individual and group levels

INDIVIDUAL LEVEL BEHAVIORS associated with social cohesion in a group	CASE A	CASE B
Social connectedness	• Strong evidence was present of unpaid work outside home, and voluntary activity; • Internet uptake was more successful here initially, and retained by more families; • One half felt more connected with family & friends after Internet provided.	• Most parents were not engaged in unpaid work outside the home; • There were more Internet "low-connectors" here; • One third felt more connected with family and friends after the Internet was provided.
•unpaid work outside the home		
• household access to telecommunications *(NB: all households in this study had Internet access provided)*		
• frequency of interaction with family and friends		
Routine day to day life	Presence of proactive individuals who exerted agency in their lives, such as by starting up a business, and in the CIH context for example through helping other parents with the Internet.	Presence of more passive individuals.
Inclusion	More evidence of positive neighbourly attitudes were apparent.	Evidence of participants being more private, showing disinterest in or suspicion toward neighbours.
Support	Stronger neighbourhood networks existed, and these were comprised of greater numbers, such as neighbours known by name. Trust and life satisfaction were higher.	Fewer neighbours were known; people were more insular, managing by themselves; there was a sense of distrust of others, or a lack of interest.
Place attachment and identity	More permanency (home ownership), attachment (pride in neighbourhood), and willingness to commit to shared tasks were apparent.	More renters; similar levels of pride and interest in neighbours. For no apparent reason, there was little evident interest in being part of a group – people were more self-contained.
GROUP LEVEL CONDITIONS & OUTCOMES associated with social cohesion in a group	CASE A	CASE B
Networks of mutual support	Stronger evidence over time of social networks: closer relationships, and trust between individuals who knew one another well.	Network ties were present, but there was less of an observed sense of familiar and close relationships with one another.
Social capital	Much stronger evidence of networked individuals making active efforts to volunteer in a range of ways.	Less involvement in community action; less 'networked' as a group.
Social solidarity	Evidence of collective action through parent/school/neighbourhood networks, mobilizing to ensure the continuation of Computers in Homes for a new group of families.	The Computers in Homes initiative was carried and managed by one person (the school principal) rather than a group sharing the task.

Success Factors

CIH is a community Internet scheme built on principles of social constructivism, an approach that suggests people learn best when they do so in the context of relationships with others. The CIH community Internet model takes every opportunity to contextualize learning within social relationships, such as at the group training sessions, family meetings at the school where stories can be shared and friendships and support networks built over hot drinks and supper, and the highly social and celebratory graduation from the program at the end of ten weeks training. The school setting at Case

A along with the CIH emphasis on socialization arguably fostered positive Internet attitudes and behaviors among the researched group, such as those seen among families attending CIH events, and in evidence of volunteering, peer mentoring and the group mobilization to keep the scheme going. Thus, the success of the Internet intervention in Case A appears to be related to the fact that it was 'bedded in' to a fertile social context that helped it flourish and in turn strengthen social ties.

The neighbourhood setting of Case A was characterized by several features of social cohesion such as supportive networks that, according to Hampton (2002), "encourage[s] place-based community" (p. 230). In this neighbourhood, the CIH scheme was initially launched as part of the work of the Housing NZ (HNZ) Community Renewal project, which was housed in an ordinary domestic dwelling. During the research it was clear that meetings, such as one held to plan CIH, another for the local community group, and other community events made this house a hub of community activity and participation. This HNZ facility could be described as an "institutional opportunity for social contact" (Hampton, 2007) where groups met over tea and shared food for various meetings, that was one important element among several at Case A that fostered the formation of local social networks. Hampton (2007) argues that research on neighbourhood has shown:

The provision of neighbourhood common spaces increases local tie formation, stronger local ties, and higher levels of community involvement … [and] planning advocates the use of neighbourhood common spaces, front porches and other design factors to encourage surveillance, community participation and a sense of territoriality (Hampton, 2002, p. 230).

Thus the free home Internet service offered to some families through CIH, which includes meetings and training at the school, was set within a "neighbourhood where context [already] favours local tie formation" (Hampton, 2007, p. 739) and a relationship appears to exist between the Internet scheme and increased social cohesion. Hampton's point is that where neighbourhoods already have resources, and feature "an interest in building community, with the neighbourhood context to back it up, [they] are most likely to profit" (Hampton, 2007, p. 740) from Internet services. This appears to be so in the Case A neighbourhood.

Opinion Leaders

Case A showed positive outcomes for a relatively cohesive group of opinion leaders operating in a situation of ambiguity and uncertainty caused by the unfamiliarity of the technology itself, the school restructure, and domestic transience. While the nature of this case study research limits its significance in terms of a relationship between opinion leadership and civic engagement, the possibility of such a relationship is supported by other recent research (Shah & Scheufele, 2006). Shah and Scheufele (2006) suggest that civic participation leads to opinion leadership, and that opinion leaders "tend to seek out informational content on television, newspapers, and the Internet, likely as a way to maintain their environmental surveillance and structural influence" (Shah & Scheufele, 2006, p. 1). An example of this in Case A is Participant A3[10] whose interviews demonstrate her strategic management of their household's media environment, supported by her husband, in a way that exemplifies the point made by Shah and colleagues, above. At least four Case A research participants exhibited characteristics of opinion leaders and took up responsible roles in the running of CIH in mid-2005 when the research concluded. Their willingness to do so meant that the scheme continued to flourish long after this time. What made them stand out as key figures?

First, they were high-connectors in terms of Internet use; they were individuals who inspired others and acted as mentors where CIH parents were anxious about the technology. They were socially confident, even garrulous, and already socially networked. They were conscious of taking a valuable opportunity and making deliberate use of it, to see where it could take them. This is a characteristic echoing Bourdieu (1999) who allows that, despite the limitations of learned social structures he calls *habitus* and the larger forces of social reproduction, certain individuals are able to exert more agency in their lives because they somehow have what it takes to seize opportunities. In this sense, they seemed determined to leverage a better future out of the potential they could see in the technology, which in Bourdieu's terms is available to them as cultural capital from the education field.

Opinion leaders are key figures in an equivocal technological context. They have confidence, they are more competent communicators, they have influence, and they turn to the media readily as an information source. They also have a readiness to publicly "individuate" (Misra, 1990, p. 3) or stand out from the crowd, a characteristic of all the leader figures at Case A. Additionally, we can discern that their contribution to this particular purpose or cause (Computers in Homes) is likely to be one of a variety of ways that they mobilize. Postill (2008), as noted in the literature review, asks us to consider 'community' to be a limited theoretical construct because it is not a static single entity: we move within many inter-linked communities. The 'leader' figures at Case A mobilize in support of many causes: Participant A3 was highly involved in school, kindergarten and other educational groups, A4 was a Board of Trustees member and volunteer for Habitat for Humanity, and A5 worked at the school, while all of them were leaders within their CIH community too.

The classic two-step flow studies within the communication literature such as Robinson's (1976) investigation of voting in the 1968 US election afford insights that remain relevant to

opinion leadership, such as that "those who engage in conversations are more politically active" (p. 316). In civic settings such as those found in the CIH case study research, the process of collective community engagement was facilitated by conversation that was a natural by-product of CIH practice. An individual such as Participant A3 exemplifies the catalysing effect of networking and conversation both within her own life and in her effect on others, as seen in her interview (Williams, 2009, pp. 269-276). Parents who conversed at CIH meetings and among themselves became exposed to the reassuring experience and knowledge of those who were more confident. An emphasis on events and gatherings in CIH is therefore a distinct strength of the program.

Thus, there are indications that stronger social cohesion at Case A may have had a relationship with the CIH free home Internet intervention. What confidence can we have that the observed social cohesion effects were in any way related to Internet use? What was the evidence assembled?

Volunteerism and Mobilization

Evidence of a possible relationship between social cohesion and Internet use at Case A exists in four principal dimensions. The first dimension is volunteerism (Participants A3 A4, A5, A25, A9[11], and others) beyond the ordinary expectations of belonging to the CIH parent community. Volunteering is a form of behaviour associated with civic engagement, which in turn is a feature of social cohesion. These volunteers and leader figures were not only busy being visible, influential and helpful towards others, but also tended to be "high connector" Internet users. Second, there was clear commitment to a group goal established by the Case A participants listed above who expressed a fervent desire to see that the CIH free home Internet scheme would continue for future families. They worked out a plan for how to achieve that goal themselves, showing engagement with a shared vision of the community's future. This type of collective mobilization is a feature of a cohesive group. The third dimension of evidence is that the participants operationalized their shared goal by planning specific objectives, such as two members of the group (A5 and A25) agreeing to divide up the list of families and visit each address to collect unused computer hard drives for upgrading. Finally, the CIH national coordinator later reported (D. Das, personal communication, 23 November 2008) that these parents later took on responsible roles such as training other parents (Participant A4), and the school coordination roles (Participants A5 and A3, one at each of the Case A school sites). Implications of these four types of evidence include a possible fifth aspect, further social cohesion, which may accrue to the larger CIH group because of the activities of these individuals, such as stronger networks of mutual support.

While Internet access and use cannot be said to have a direct causal relationship with social cohesion, it may be that Internet access offered to a relatively cohesive community like Case A will be more successfully used because of the presence of strong networks and support. Case A already exhibited strong neighbourly networks, volunteerism, pride and a sense of belonging irrespective of the CIH intervention. Various kinds of 'capital' are already embedded within communities, as well as deficits. The outcomes of this study suggest that attention should be paid to leveraging the community resources, or community capital, that are already present. A community that is more cohesive in the first place is likely to provide the supportive social setting where new Internet users lacking in experience, knowledge and confidence can become confident in a social setting with others. They are likely to find that the neighbourhood or school networks augmented by the socially engaging nature of CIH will provide encouragement, advice, and informal mentoring volunteered by other families.

CHALLENGES AND THE WAY AHEAD

Findings from the CIH case study research in NZ (Williams, 2009) outlined in this chapter suggest that existing resources and resilience in a community can be harnessed by stakeholders working within a framework of needs defined by the community itself, rather than seeking to compensate for deficits defined by external agencies. The study was not designed to assess the broader impact of CIH on the wider community, but as shown in the previous section, the social networking and mobilization aspects of CIH over time appeared to have a positive ongoing effect for the Case A school community.

Communities in NZ that decide to implement CIH often struggle with issues of family engagement in the life of the school, which is critical to school children's educational achievement. CIH communities are those that have the least advantage in NZ society in terms of socioeconomic resources, and where educational achievement is lower than the norm. Numerous school Principals connected with the decade of research on CIH underpinning this chapter share a deep commitment to breaking down barriers to belonging experienced by parents who have 'failed' in the educational system. CIH has proven its worth in this respect in hundreds of communities, as the scheme provides an important opportunity for parents to become engaged in their children's learning. Therefore, the social dimension of feeling part of a group, feeling supported, feeling included and valued, is central to both the immediate and long-term effectiveness of the CIH community Internet scheme. It is a means of bringing family and school together in activities that create, reinforce, recognize, and celebrate the achievement of learning goals.

CIH community Internet strategies spring from a social constructivist approach, facilitating social support and learning in a group. Case A outcomes suggest careful attention should be given to the unique characteristics of each targeted community in a free home Internet scheme, beginning with an assessment of the social resources that already exist there. A key advantage of CIH in this respect is that its formal and informal social structures provide social connections for people learning to use the technology. In turn, these social connections help to develop motivation, engagement, belonging, and commitment to something bigger than an individual's own use of the Internet. This may be an answer to the question of how Internet access and social cohesion may be related in a community setting. Internet access, made available through the framework of CIH or a similar scheme, can facilitate social cohesion through structures and practices that exist to foster a socially supportive learning environment. The CIH model provides an operational structure, a highly social culture, and an environment of practice in which understanding of group cohesion processes and the role of individuals who may serve as opinion leaders, mentors, and champions can all play a part in enhancing sustainability of Internet use in communities.

Those responsible for establishing a free home Internet scheme like CIH should first identify the resources of 'place'—informal networks of support, leader figures, and manifestations and patterns of volunteerism—in order to understand how these resources can be recruited to the task of establishing Internet use. An understanding of existing social connectedness (Ministry of Social Development, 2010) in the community could create the conditions for a sensitive and collaborative approach to building social and cultural capital through digital literacy, in turn ensuring that the community Internet scheme becomes self-determining. This is more of a strengths-based approach than a deficit or damage-centered (Tuck, 2009) approach to community building, which tends to operate from a view that some communities are deficient in ways that external agencies can repair by simply delivering ready-made solutions. CIH has the structures in place to effect a shift in ownership from external bodies such as Housing

NZ, which focuses on providing the resources of technology and training, to the community itself, with its emphasis on building social networks through peer mentoring and socialization practices including family meetings and group training sessions. These are valuable tactics in managing ambivalence about technology among novice Internet users who may self-exclude because of low confidence or lack of awareness of what the Internet makes possible. Parents in the Case A study considered home visits and more technical support would be valuable ways to support parents using the Internet at home, echoing the view that "a sustainable ICT initiative requires resources dedicated to providing ongoing support" (Gaved & Anderson, 2006, p. 17).

Encouragingly, government funding for CIH through the Department of Internal Affairs has been considerably expanded, with CIH receiving a boost of "$6.6 million for three years from the Government's 2010 budget" (Department of Internal Affairs, 2010). This has occurred in large part because of the evidence generated by a number of research projects over the years, sufficient to convince the NZ Government that ongoing investment in this community Internet scheme is worthwhile:

Funding for the scheme has also been tied to research evidence. The small-scale early pilots were heavily reliant on community contributions and volunteers with minimal short-term funding from a number of interested government agencies such as education and economic development. Instruments developed over these ten years to measure economic and social benefits at the community level have in the last year (2010) produced sufficient evidence to guarantee ongoing annual funding for CIH at the Cabinet level involving multiple government agencies. Additionally the Ministry of Education funds a separate CIH program for refugee families as they settle into NZ communities (Craig & Williams, 2011).

The increase in funding and the prospect of being able to plan for the next three years on the basis of the budgeted $6.6 million has brought renewed vigor to the CIH scheme. With the backing of greater resources, there is also a determination to address the digital inclusion issue of "more than 100,000 NZ families with school-aged children in low income communities [who] still don't have a computer and Internet access at home" (Department of Internal Affairs, 2010). In the Auckland region, funding is being directed toward connecting and supporting several hundred families each year.

CONCLUSION

The resources of the community itself are highlighted in Case A as having considerable latent value. Rather than falling into the trap of "deficit theory syndrome," stakeholders should collaborate in order to evaluate and seek to leverage the community's existing capacity. The social "net"—the networks, the relational practices existing in social capital and the dynamics of social cohesion—is the true resource of high valence in the digital divide context that should be more highly valued, over and above the technological hardware and infrastructure. The CIH study (Williams, 2009) shows, above all, that resources exist in unexpected places, in a way that resonates with Eubanks (2007), who cites one of her community research participants saying "have-nots possess many different kinds of local knowledge: community knowledge, knowledge of 'the system,' ... more finely attuned social Geiger counters, as well as social networks, navigation skills, and an ethic of sharing" (Eubanks, 2007).

These are the relational resources existing in cohesive groups such that "technology ... can mediate across social structure by creating a network of equal exchange" (Eubanks, 2007). Case A outcomes highlight both the challenges and opportunities that inhere in community Internet interventions. On the one hand, outside-in solutions may override the needs of those who remain silent and unrepresented, while offering the very social structures that help harness dynamics of cohesion, frame locally-defined solutions and encourage the community voice to be heard.

REFERENCES

Ball-Rokeach, S. J., Gibbs, J., Hoyt, E. G., Jung, J. Y., Kim, Y. C., Matei, S., et al. (2002). *Metamorphosis project white paper #1 - The challenge of belonging in the 21st century: The case of Los Angeles*. Retrieved 2 September, 2002, from http://www-scf.usc.edu/~matei/stat/globalization.html

Bimber, B. (1998). The internet and political transformation: Populism, community and accelerated pluralism. *Polity*, *31*(1), 133–161. doi:10.2307/3235370

Bourdieu, P. (1999). Structures, habitus, practices . In Elliott, A. (Ed.), *The Blackwell Reader in Contemporary Social Theory* (pp. 108–118). Malden, MA: Blackwell.

Burt, R. S. (1999). *The social capital of opinion leaders*. New York, NY: American Academy of Political and Social Science.

Casswell, S. (2001). Community capacity building and social policy – What can be achieved? *Social Policy Journal of New Zealand*, *17*, 22–35.

Castells, M., Fernandez-Ardevol, M., Qiu, J. L., & Sey, A. (2006). *Mobile communication and society: A global perspective*. Cambridge, MA: MIT Press. doi:10.1111/j.1944-8287.2008.tb00398.x

Clinton, J., Hattie, J., & Dixon, R. (2007). *Evaluation of the flaxmere project: When families learn the language of school*. Auckland, New Zealand: Ministry of Education.

Coleman, J. (1988). Social capital in the creation of human capital. *American Journal of Sociology*, *94*, S95–S120. doi:10.1086/228943

Collins, S. (2012). Auckland: A city divided by income. *New Zealand Herald*. Retrieved from http://www.nzherald.co.nz/nz/news/article. cfm?c_id=1&objectid=10783692

2020Communications Trust. (2009). *2020 leader in communications ICT in New Zealand and the Pacific*. Retrieved March 12, 2009, from http://www.2020.org.nz/

Computer Access New Zealand. (2009). *Computer access trust NZ: Refurbishing office computers for schools and the community*. Retrieved January 31, 2012, from http://www.canz.org.nz/

Computers in Homes. (2011). *Website*. Retrieved April 16, 2011, from http://www.computersin-homes.org.nz/

Das, D. (2005). *How do we measure if closing the digital divide addresses barriers to social inclusion?* Wellington, New Zealand: Victoria University of Wellington.

Department of Internal Affairs. Department of Labour, Department of Prime Minister & Cabinet, Ministry of Agriculture & Forestry, Ministry of Economic Development, Ministry of Education, et al. (2002). *Connecting communities: A strategy for government support of community access to information and communications technology*. Retrieved from http://www.dol.govt.nz/PDFs/cegBooklet2000.pdf

Department of Internal Affairs. (2010). 3,000 New Zealand families to learn new digital skills. *Computers in Homes*. Retrieved from http://www.computersinhomes.org.nz/

Dutta, S., & Mia, I. (2011). *The global information technology report 2010–2011: Transformations 2.0*. Washington, DC: World Economic Forum-INSEAD.

Eubanks, V. E. (2007). Trapped in the digital divide: The distributive paradigm in community informatics . *Journal of Community Informatics*, *3*(2).

Ferlander, S. (2003). *The internet, social capital and local community*. Stirling, UK: University of Stirling.

Forrest, R., & Kearns, A. (2001). Social cohesion, social capital and the neighbourhood. *Urban Studies (Edinburgh, Scotland)*, *38*(12), 2125–2143. doi:10.1080/00420980120087081

Friedkin, N. E. (2004). Social cohesion. *Annual Review of Sociology*, *30*, 409–425. doi:10.1146/annurev.soc.30.012703.110625

Fukuyama, F. (1995). Social capital and the global economy. *Foreign Affairs*, *74*(5), 89–103. doi:10.2307/20047302

Hampton, K. (2002). Place-based and IT mediated "community". *Planning Theory & Practice*, *3*(2), 228–231. doi:10.1080/14649350220150099

Hampton, K. (2007). Neighborhoods in the network society: The e-neighbors study. *Information Communication and Society*, *10*(5), 714–748. doi:10.1080/13691180701658061

Hays, R. A., & Kogl, A. M. (2007). Neighbourhood attachment, social capital building, and political participation: A case study of low- and moderate-income residents of Waterloo, Iowa. *Journal of Urban Affairs*, *29*(2), 181–205. doi:10.1111/j.1467-9906.2007.00333.x

Housing New Zealand Corporation. (2006). *Community renewal*. Retrieved August 29, 2012, from http://www.hnzc.co.nz/about-us/research-and-policy/housing-research-and-evaluation/summaries-of-reports/community-renewal-march-2006/

Institute of Community Cohesion. (2009). *The nature of community cohesion*. Retrieved May 20, 2009, from http://www.cohesioninstitute.org.uk/Resources/Toolkits/Health/TheNatureOfCommunityCohesion

Jung, J., Qiu, J., & Kim, Y. (2001). Internet connectedness and inequality. *Communication Research*, *28*(4), 507–535. doi:10.1177/009365001028004006

Kearns, A., & Forrest, R. (2000). Social cohesion and multilevel urban governance. *Urban Studies (Edinburgh, Scotland)*, *37*(5-6), 995–1017. doi:10.1080/00420980050011208

Kim, Y.-C., Jung, J.-Y., Cohen, E. L., & Ball-Rokeach, S. J. (2004). Internet connectedness before and after September 11 2001. *New Media & Society*, *6*(5), 20. doi:10.1177/146144804047083

Loader, B. D., & Keeble, L. (2004). *Challenging the digital divide? A literature review of community informatics initiatives*. York, UK: The Joseph Rowntree Foundation/YPS, for the Community Informatics Research and Applications unit at the University of Teesside.

McClenaghan, P. (2000). Social capital: Exploring the theoretical foundations of community development education. *British Educational Research Journal*, *26*, 565–582. doi:10.1080/713651581

McKnight, J. L., & Kretzmann, J. P. (1996). *Mapping community capacity*. Chicago, IL: The Neighborhood Innovations Network.

Meegan, R., & Mitchell, A. (2001). It's not community round here, it's neighbourhood: Neighbourhood change and cohesion in urban regeneration policies. *Urban Studies (Edinburgh, Scotland)*, *38*(12), 2167–2194. doi:10.1080/00420980120087117

Ministry of Economic Development. Ministry of Health, Ministry of Research Science and Technology, Ministry of Education, Department of Labour, The NZ National Library, et al. (2004). *Digital strategy: A draft New Zealand digital strategy for consultation*. Retrieved from www.beehive.govt.nz/Documents/Files/Digital%20Strategy.pdf

Ministry of Economic Development. (2008). *Digital strategy 2.0: Smarter through digital*. Retrieved from http://www.beehive.govt.nz/release/digital-strategy-20-%E2%80%93-smarter-through-digital

Ministry of Social Development. (2010). *The social report: Te purongo tangata 2010 - Social connectedness*. Retrieved August 28, 2012, from http://www.socialreport.msd.govt.nz/social-connectedness/index.html

Moy, P., Scheufele, D., & Holber, P. (1999). Television use and social capital: Testing Putnam's time displacement hypothesis. *Mass Communication & Society*, *2*(1/2), 27–46.

New Zealand Government - Ministry of Communications and Information Technology. (2007). *Digital strategy: Creating our digital future*. Retrieved 2 November, 2007, from http://www.digitalstrategy.govt.nz/

Newman, K. (2008). *Connecting the clouds: The internet in New Zealand*. Auckland, New Zealand: Activity Press in association with InternetNZ.

Onyx, J., & Bullen, P. (2000). Sources of social capital. In Winter, I. (Ed.), *Social Capital and Public Policy in Australia* (pp. 105–135). Melbourne, Australia: Australian Institute of Family Studies.

Perry, B. (2012). *Household incomes in New Zealand: Trends in indicators of inequality and hardship 1982 to 2011*. Retrieved from http://www.msd.govt.nz/about-msd-and-our-work/publications-resources/monitoring/household-incomes/index.html

Pigg, K. E., & Crank, L. D. (2004). Building community social capital: The potential and promise of information and communications technologies. *The Journal of Community Informatics*, *1*(1).

Postill, J. (2008). Localizing the internet beyond communities and networks. *New Media & Society*, *10*(3), 413–431. doi:10.1177/1461444808089416

Putnam, R. D. (1995). Bowling alone: America's declining social capital. *Journal of Democracy*, *6*(1), 65–78. doi:10.1353/jod.1995.0002

Putnam, R. D. (1996). The strange disappearance of civic America. *The American Prospect*, *7*(24).

Putnam, R. D. (2000). *Bowling alone: The collapse and revival of American community.* New York, NY: Simon & Schuster.

Putnam, R. D. (2002). Bowling together. *The American Prospect, 13*(3).

Putnam, R. D. (2007). E pluribus unum: Diversity and community in the twenty-first century. *Scandinavian Political Studies, 30*(2), 137–174. doi:10.1111/j.1467-9477.2007.00176.x

Robinson, J. P. (1976). Interpersonal influence in election campaigns: Two step-flow hypotheses. *Public Opinion Quarterly, 40*(3), 304–319. doi:10.1086/268307

Saguaro Seminar. (2007). *Civic engagement in America.* Retrieved January 3, 2009, from http://www.hks.harvard.edu/saguaro/index.htm

Shah, D. V., & Scheufele, D. A. (2006). Explicating opinion leadership: Nonpolitical dispositions, information consumption, and civic participation. *Political Communication, 23*(1), 1–22. doi:10.1080/10584600500476932

Spoonley, P., Peace, R., Butcher, A., & O'Neill, D. (2005). Social cohesion: A policy and indicator framework for assessing immigrant and host outcomes. *Social Policy Journal of New Zealand, 24*, 85–110.

Statistics New Zealand. (2006a). *QuickStats about Clendon South.* Retrieved August 9, 2011, from http://www.stats.govt.nz/Census/2006CensusHomePage/QuickStats/AboutAPlace/SnapShot.aspx?id=3524822&type=au&ampParentID=1000002

Statistics New Zealand. (2006b). *QuickStats about Papakura East.* Retrieved August 9, 2011, from http://www.stats.govt.nz/Census/2006CensusHomePage/QuickStats/AboutAPlace/SnapShot.aspx?id=3525610&type=au&ampParentID=1000002

Statistics New Zealand. (2009). *Measuring New Zealand's progress using a sustainable development approach: 2008.* Retrieved from http://www.stats.govt.nz/searchresults.aspx?q=social%20cohesion&mp=20&sp=0&sort=r

Statistics New Zealand. (2012). *Census 2006 households with school-aged children.* Retrieved from http://www.stats.govt.nz

Toyama, S. (2007). *Local area SNS and community building in Japan.* Retrieved from http://www.ccnr.net/prato2007/archive/toyama%20135.pdf

Tuck, E. (2009). Suspending damage: A letter to communities. *Harvard Educational Review, 79*(3), 409–427.

Vergunst, P. (2006). *Community cohesion: Constructing boundaries between or within communities-of-place?* Paper presented at the Rural Citizen: Governance, Culture and Wellbeing in the 21st Century Conference. New York, NY.

Vishwanath, A. (2006). The effect of the number of opinion seekers and leaders on technology attitudes and choices. *Human Communication Research, 32*(3), 322–350. doi:10.1111/j.1468-2958.2006.00278.x

Wellman, B. (Ed.). (1999). *Networks in the global village: Life in contemporary communities.* Boulder, CO: Westview Press / Perseus Books Group.

Wellman, B. (2001). *The persistence and transformation of community: From neighbourhood groups to social networks.* Paper presented to the Law Commission of Canada. Ottawa, Canada.

Wellman, B., & Berkowitz, S. (Eds.). (1988). *Social structures: A network approach.* Cambridge, UK: Cambridge University Press.

Williams, D. (2006). On and off the net: Scales for social capital in an online era. *Journal of Computer-Mediated Communication, 11*(2). doi:10.1111/j.1083-6101.2006.00029.x

Williams, J. (2009). *Connecting people: Investigating a relationship between internet access and social cohesion in local community settings.* (Unpublished Doctoral Thesis). Massey University. Palmerston North, New Zealand.

Zwimpfer, L. (2010). Building digital communities: A history of the 2020 trust. In Toland, J. (Ed.), *Return to Tomorrow: Fifty Years of Computing in New Zealand.* Wellington, New Zealand: New Zealand Computer Society.

KEY TERMS AND DEFINITIONS

Community: In this chapter refers to a group of people living in a defined geographical area who define themselves as part of a community, a symbolic construct that is an outcome of what those people do together.

Community Informatics: Is a multidisciplinary field that combines community Internet practice and research, on the assumption that research can assist practitioners to evaluate and better position their community Internet interventions.

ENDNOTES

[1] Generally low decile is understood in NZ to mean areas where residents are on average earning low incomes, are more likely to be unemployed, have less educational attainment and less access to, or ability to access, the resources and services than those in other areas. The lower the decile rating, the more funding the school is entitled to, on the assumption that the families attending them are more likely to face barriers to educational achievement than those who attend higher decile schools (Williams, 2009, p. 10).

[2] Telephone and internet access in the home, contact with family and friends, contact between young people and their parents, trust in others, loneliness, and voluntary work (Ministry of Social Development, 2010).

[3] However at the time of planning the methodology an earlier MSD list of social connectedness indicators was used, which did not include "loneliness" or "contact between young people and their parents" (Ministry of Social Development, 2010). Indicators drawn from this source for the CIH study were Internet access, contact with family and friends, trust in others and participation in unpaid work outside the home.

[4] My emphasis added.

[5] Time 1 interviews were completed between November 2003 and April 2004; Time 2 interviews were completed with the original participants between October 2004 and March 2005 (Williams, 2009, p. 77).

[6] Manukau is a district within Auckland, NZ's largest city of 1.25 million people.

[7] In NZ, primary schools cater for children's learning from the age of five years old; elsewhere, this phase is called elementary school.

[8] Maori are the indigenous people of NZ.

[9] Tena koe is a phrase in Maori and means, here, thank you

[10] Each of the participants was coded with an 'A' or 'B' according to which case study community they came from, and a number that was unique to them, as a means of preserving anonymity. Thus, 'A3' was the third participant interviewed at Case A.

[11] Each of the participants was coded with an 'A' or 'B' according to which case study community they came from, and a number that was unique to them, as a means of preserving anonymity.

Chapter 9
Telecentre–Based Community Wireless Networks:
Empowering Rural Community in Uganda

Dorothy Okello
Makerere University, Uganda

Julius Butime
Makerere University, Uganda

ABSTRACT

This chapter shares the experiences of the Community Wireless Resource Centre (CWRC) as it embarked on the journey to address affordable connectivity for four telecentres in rural and underserved Uganda via telecentre-based community wireless networks. Telecentres have long played a key role in availing access to Information and Communication Technologies (ICTs) and in supporting the provision of universal access. With falling prices and new technologies increasing individual access to ICTs, the telecentre-based community wireless networks need to continually innovate in order to remain relevant to both the telecentres and the partners that together comprise the community wireless networks.

1. INTRODUCTION

The concept of community wireless networks describes a wireless communication infrastructure that is shared and managed by a community. It is based on the possibility for groups or communities to build self owned and operated networks (Wireless Networking in the Developing World, 2006). This chapter focuses on community wireless networks set up at four telecentres in Uganda, namely, Kachwekano and Kabale telecentres in Kabale district, Nabweru telecentre in Wakiso District, and the Lira CPAR telecentre in Lira District. Establishment of the networks involved engagement of the telecentres and of the Community Wireless Resource Centre (CWRC).

DOI: 10.4018/978-1-4666-2997-4.ch009

The CWRC is a research and training centre established in 2006 within the Department of Electrical and Computer Engineering, College of Engineering, Design, Art, and Technology (CEDAT), Makerere University with support from the International Development Research Centre (IDRC). The CWRC arose out of the need to reduce the high cost of connectivity in IDRC-supported telecentres in Uganda, and to explore optimal connectivity models such as sharing the existing bandwidth with neighboring institutions via outdoor wireless networks. It was anticipated that by managing collectively the costs of connectivity at each telecentre, more institutions could get access to Internet without heavy initial investments in satellite hardware and subscriptions. At the time, the predominant option for Internet access in rural communities was via expensive satellite-based connections. The aim of the CWRC was to make connectivity more affordable for telecentres by implementing community wireless networks.

1.1. Access to Communications in Uganda

While significant progress has been witnessed in Uganda's communications sector, attributed to increasing mobile cellular subscriptions and increasing access to broadband connectivity, access in rural and underserved areas remains a challenge. This is largely due to well-known challenges common to African countries including lack of access to communications infrastructure, lack of energy sources for users and for powering up the Information and Communications Technology (ICT) infrastructure, expensive communication infrastructure, and the high tariffs associated with ICT services and applications. Furthermore, Internet awareness is low—a 2006-2007 study found that less than 10 percent of the surveyed population in Uganda knew what the Internet is (Organisation for Economic Co-operation and Development, 2009). Figure 1 shows the slow but steady growth of Internet users in Uganda. It is also clear that mobile/wireless options for connecting to the Internet are the more practical option given the limited fixed line infrastructure in Uganda (UG) and across Sub-Saharan Africa (SSA) based on statistics from the International Telecommunication Union (ITU).

The low Internet penetration rates and high tariffs are primarily due to a lack of high-capacity international networks. Having multiple cables along the Eastern coastline as highlighted in Figure 2 should not only bring down prices, but should also serve to provide multiple options for countries and remove their reliance on any single network—satellite or cable. The cables are also significantly increasing the capacity available for Internet connectivity, as profiled in Table 1.

Figure 1. Trend of ICT usage in Uganda and in Sub-Saharan Africa (Source: ITU)

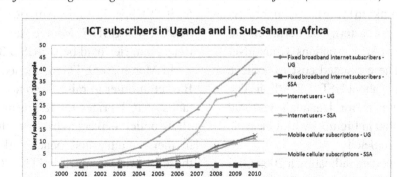

Figure 2. African undersea cables planned by 2012 (Source: manypossibilities.net)

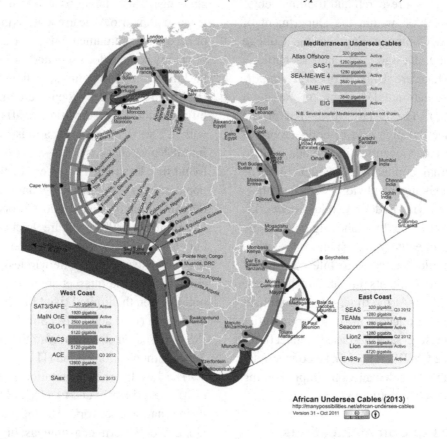

Even with the increasing number of high-speed undersea cables along the East Africa coastline, as a landlocked country, Uganda needs both transit access through neighboring countries to the undersea cables as well as a well distributed national backbone so that broadband connectivity is available nationwide. A 2007 study by the International Telecommunication Union (ITU) showed that there was a need for about 66,000 kilometers of fiber links to bring about connectivity among countries in eastern, southern, central, western, and northern African nations (UbuntuNet Alliance, 2009). The estimated cost to achieve this was, at that time, about USD 1.2 billion. The current status and plan for Uganda's national backbone is presented in Figure 3 with Phases 1, 2, and 3 already completed.

With World Bank data indicating a 2009 Gini index of the national distribution of income at 44.3, a poverty gap at $2 a day of 27 percent (PPP), and a 2010 GDP per capita of 509 current US$, Uganda is a low income country[1]. Yet, while Africa's ICT infrastructure penetration rates remain low, Sub-Saharan Africa has the highest Internet and telephony prices in the world (OECD, 2009). According to ITU and World Bank estimates, the average price of a broadband connection in Sub-Saharan Africa was about USD 110 for 100 Kbps. In Latin America and the Caribbean, it was USD 7. Middle East and North African countries paid below USD 30.

In March 2009, ITU launched the ICT Price Basket in order to raise awareness of the importance of ICT prices for ICT usage as well as to allow policy makers to evaluate the cost of ICTs in their country and benchmark them against those of other countries (ITU, 2009). The ICT Price Basket is made up of the average of three

Table 1. African undersea cables planned by 2012 along Eastern Africa coastline (Source: manypossibilities.net)

Cable	Capacity (Gigabits)	Status
SEAS	320	Q3 2012
TEAMs	1,280	Active
Seacom	1,280	Active
Lion2	1,280	Q2 2012
Lion	1,300	Active
EASSy	4,720	Active

sub-baskets, which measure the prices of fixed telephone, mobile cellular and fixed broadband Internet services. Table 2 presents the 2008 and 2010 ICT Price Baskets for East African countries for whom the prices were available (ITU, 2011). For comparison purposes, the prices for Egypt and South Africa are also included—the choice of these particular countries is guided by ICT statistics, which show that Northern Africa and South Africa generally perform much better in the ICT sector than East Africa countries.

To determine the ICT Price Basket, the cost of each sub-basket as a percentage of the monthly Gross National Income (GNI) per capita is

Figure 3. Uganda national backbone (Source: Connect Africa, 2010)

Table 2. 2008, 2010 ICT price baskets for East African countries (Source: ITU)

	ICT Price Basket		GNI per capita, USD, 2009 (or latest available year)	Sub-baskets					
				Fixed telephone (% GNI per capita)		Mobile cellular (% GNI per capita)		Fixed Broadband (% GNI per capita)	
	2010	2008		2010	2008	2010	2008	2010	2008
Egypt	3.5	4.4	2,070	1.7	2	4.1	5.6	4.6	5.5
South Africa	4.5	5.3	5,760	5.2	4.2	4.8	4.5	5.7	4.9
Uganda	30.2	61.8	460	22.8	34.9	31.8	50.4	35.9	374.9
Tanzania	31.4	57.0	500	21.1	28.1	23.2	43.1	50.0	174.4
Kenya	33.1	49.8	760	22.4	18.0	17.0	31.5	59.9	261.2
Rwanda	56.9	58.1	460	34.4	21.2	36.3	53.1	224.5	267.6

limited to a maximum value of 100. By providing the price baskets in relation to the GNI, one is able to access the affordability and hence the uptake of ICT services. Clearly, the ICT price basket being at more than 30% of the GNI means that ICT services remained unaffordable in East Africa even by 2010. Furthermore, at inception of the telecentre-based community wireless networks project in 2007, the ICT price basket in Uganda was over 60% of the GNI. It should be noted that in the case of Uganda, there has indeed been a significant drop of almost 50% for the ICT price basket between 2008 and 2010. Nevertheless, the costs in East Africa are significantly higher than those of the comparison countries, Egypt and South Africa—both of which are rated as middle-income countries.

Given the limited ICT infrastructure as well as the concerns of affordability, it becomes necessary to explore various options for availing public or shared access as a means of lowering the costs of access and use of communication services. These options include both infrastructure development as well as policy interventions. For example, policy makers could use public funds towards communication networks in support of universal access to communication targets for their countries. From an infrastructure perspective, the community wireless networks, which allow for a community to share the costs of Internet access into their communities are also an option. As observed from Figure 1, it

is also the more practical option given the wider spread of the mobile networks relative to fixed line networks.

2. THE CWRC AND TELECENTRE-BASED COMMUNITY WIRELESS NETWORKS

The general objective of the CWRC is to provide or enhance sustainable Internet connectivity infrastructure, particularly in rural or underserved areas in Uganda, by means of wireless technology. Affordable Internet access is widely recognized as essential for improving people's social and economic livelihoods. In line with this, community wireless networks emerged as connectivity solutions particularly for rural and underserved communities within 80 percent of the world's population that has limited access to the Internet (Abdelaal, et al., 2009a, 2009b; Ishmael, et al., 2008; Siochru, et al., 2005; Women of Uganda Network, 2006). Not only can wireless networks be established at much less cost than wired alternatives, but communities can also directly benefit from cheaper and easier access to information by sharing the costs of Internet access within the community (WNDW, 2006).

The telecentre-based community wireless networks were established with financial support from International Development Research Centre

(IDRC) and with technical support from IT+46, a Swedish based organisation with experience in establishing community wireless networks in developing countries. This initiative built upon an ongoing interest of IDRC in sustainability of the telecentres. In 2004, an independent consultant had been engaged by the IDRC to conduct surveys at six of the telecentres they supported in Uganda. In addition to the telecentres, neighbouring institutions were identified as potential partners for the community wireless networks. From inception in 2006, implementation of the telecentre-based wireless networks was done by CWRC staff together with third year undergraduate students pursuing degrees in Electrical Engineering at the Department of Electrical and Computer Engineering. The telecentre managers and partner representatives received basic training to enable them troubleshoot and maintain the network.

The review of CWRC and the telecentre-based community wireless networks in empowering rural communities in Uganda is presented around four areas, namely:

- **ICT Infrastructure:** The technology and infrastructure deployed to establish the telecentre-based community wireless networks.
- **Management:** Ownership and organisation of the telecentre-based community wireless networks.
- **Partners & Users:** A brief overview of the partners and users in the telecentre-based community wireless networks. The key objectives and anticipated outcomes of the telecentres engaged in this initiative
- **Research:** The CWRC is based in the Department of Electrical and Computer Engineering at Makerere University and, together with students, continues to interact with the telecentres in research on affordable Internet access.

2.1. ICT Infrastructure

While not a new technology, establishment of the community wireless networks at the four telecentres was an innovative way of increasing Internet access in the communities and at the same time enabling the service be sustainable through sharing of costs related to delivery of Internet access to the community. In 2006, the Very Small Aperture Terminal (VSAT) Internet connections were providing 64 Kbps at an average monthly cost of US$ 250. Previously such costs were being solely met by the telecentres who were having challenges meeting the monthly subscription charges after the IDRC project to establish the telecentres ended.

Not only were the telecentres challenged, but operating in areas with generally low incomes among the population meant that the telecentres could not recoup their costs through pricing of the services at high costs. At the time, the telecentre-based networks were being established, not only were telecentres experiencing low demand, as there was generally low awareness on the potential and benefits of using the Internet, but the easiest form of access to the Internet was via the expensive satellite-based connections as highlighted in Figure 4.

In general, the telecenter-based community wireless networks have been implemented in form of a star topology with the telecentre acting as the hub. The networks operate in the 2.4 GHz frequency band for both indoor and outdoor radio units. In this way, the CWRC networks are typical of community wireless networks that reduce their costs of setup and maintenance by taking advantage of the 2.4 GHz free/unlicensed spectrum. The radio units at the telecentres were set in the access point mode while the radio units at the partner sites were set up in client mode. The community wireless networks were established based on the 802.11b/g standard from the Institute of Electrical and Electronics Engineers (IEEE) 802.11 family of standards for wireless Local Area Network (LAN) technology. Most of the antennas

Figure 4. Topology for community wireless network

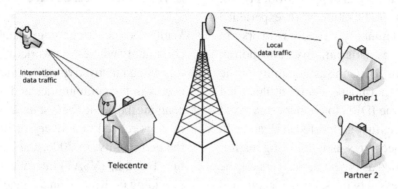

used to establish the wireless links were sectoral, with a beam width of 65 degrees and a gain of 9dBi. Having sectoral antennas allowed for coverage of the wireless networks to be targeted towards particular directions of interest. For example, as shown in Figure 4, the wireless network could be targeted in the direction of Partners 1 and 2.

Except for the wireless network established in Kabale, the community wireless networks were easy to deploy given the relatively flat terrain in Lira and Nabweru, and the good line of sight between these telecentres and their partners. The CWRC experience in deploying a network in Kabale, South Western Uganda, was peculiar owing to the hilly nature of the Kabale. In addition to the hilly terrain in Kabale, the population there was not evenly distributed. There was sparse population in some of the targeted areas and a dense population in others. The CWRC had the opportunity to set up a network covering both these diverse regions.

Due to close proximity of the Kabale and Kachwekano telecentres within the town of Kabale, it was planned to establish the community wireless network in Kabale town as a single network instead of as two separate networks at the individual telecentres. The hub of the community wireless network in Kabale town was at the Kachwekano telecentre, about 5 kilometers from Kabale town. This was chosen because of its raised nature and the presence of a tall mast of one of the telecom operators. This was done to

attain clear line of sight between the hub and the intended partners, and to facilitate future expansion to other partners. Two radio units with inbuilt sectoral antennas were mounted 30 meters up on the mast of about 50 meters. The radio units set up in access mode were facing opposite directions so as to cover the intended area of the Kabale community wireless network. Repeater stations were also used to reach clients who were not reachable by direct line of sight.

One of the issues that came up during implementation of the community wireless networks was the availability of reliable and stable power. This was not a challenge at the telecentres, as for example, the Kachwekano Community Multimedia Centre is connected to the national grid and also has a standby generator while the Lira and Nabweru telecentres are also connected to the national grid. However, not all the partners had ready access to the national grid and had to make do with alternate sources of energy. For example, one side of the Kabale network had partners with no access to the national grid. This meant that all of the implementations on this side were done using solar energy, which is unreliable as it is dependent on the availability of sunlight energy and the capacity of the solar battery packs. It should also be noted that most of the network equipment especially the radio units are delicate and not highly resistant to power fluctuations. This has resulted in a significant loss of equipment since the community wireless networks were set up. At the time of establishing the networks, most of the network equipment had to be sourced from abroad; however, some of the equipment is now locally available.

Prior to the community wireless networks, one needed to physically go to the telecentres in order to access the Internet. However, with the community wireless networks, partners could access the Internet in the comfort of their places via wireless links to the telecentre. Such convenience and ready access was expected to drive up the use of the Internet within the respective communities. In addition, the costs for the Internet access into the community would then be shared between the telecentres and the partners within the community wireless network. In this way, both the telecentres and the partners would gain access to the Internet at much reduced costs.

A key factor in addressing the issue of infrastructure is the role of government. Two key challenges faced by the telecentres are access to affordable telecommunications infrastructure and access to reliable energy supply—particularly renewable energy sources. Indeed government does have initiatives at national and local levels seeking to address this challenge. At national level, there is the Rural Communications Development Fund (RCDF) and the Energy for Rural Transformation Programme. For example, under RCDF, the government of Uganda is establishing Internet Points of Presence (POPs) at each of the district headquarters with the aim of availing Internet access to rural areas at similar quality and prices to Internet access in Kampala and other urban centres. In addition, rollout of the National Backbone Infrastructure is expected to provide broadband Internet access at district level nationwide. It should also reduce the cost of such access relative to the costs of Internet access via satellite. At local level remains the challenge of 'last mile' access which in part is already being addressed by the RCDF POPs—but this does not always translate into equitable access across a

district. With support of the International Telecommunication Union (ITU), the Ministry of Information and Communication Technology has also piloted a community wireless network with Nakaseke Telecentre as its hub. The Ministry also facilitated access to an E1 (2 Mbps) Internet link to the telecentre, which was already connected to the national grid and has a solar power system as backup. Nakaseke is also a CWRC partner, and indeed a number of experience-sharing meetings were held between the Ministry, telecentre and CWRC during setup of the Nakaseke community wireless network.

2.2. Management

The project to establish the community wireless networks was led by the CWRC. However, right from the beginning, the telecentre managers were involved in planning and implementation of the community wireless networks. The telecentre managers have remained responsible for handling and addressing matters at their telecentres. This includes collecting the monthly dues from the network partners and remitting the monthly Internet subscription charges.

The telecentres have experienced varying degrees of success in running the networks. For example, in one network, the telecentre manager and partners have collectively raised the necessary funds for Internet access and for technical maintenance of the networks. In another network, there have been challenges in developing local management of the community wireless network due to the multiple local government levels within the community. As a result, the collection of dues necessary to run and maintain the wireless networks has remained a challenge to-date. From this experience of the community wireless network that has had to deal with multiple local government levels, a key lesson learnt is the need for the CWRC to also engage with end users in promoting and explaining the case for community wireless networks.

2.3. Partners and Users

The primary partners in the community wireless networks initiative are the CWRC together with our students and the telecentres together with their network partners. The selection of partners at each telecentre built upon an earlier consultation undertaken in 2004, as previously noted. In 2006, site visits were undertaken to the six telecentres considered in 2004 namely, Buwama, Kabale, Kachwekano, Lira CPAR, Nabweru, and Nakaseke.

The selection of partners for each of the telecentres was based on assessment surveys undertaken at each of the six telecentres. These surveys considered both technical issues related to roll out of a wireless networks as well as sustainability and financial issues related to interest and ongoing operations of the wireless networks. Among the issues assessed during the 2006 site visits were:

- Monthly bandwidth (and its cost) available at the telecentre.
- Potential partners for the wireless network (including whether the partner had clear, partial or no line of sight access to the telecentre).
- Presence and height of any masts in the vicinity of the telecentre as well as ownership of the masts and possibility of the masts being used by the community wireless initiative.
- The kind of business already being undertaken at the telecentres.
- The availability and regularity of power at the telecentres and potential partners, for example, who had Uninterruptable Power Supplies (UPSs) to cater for power outages, who had alternative sources of energy such as solar power systems, etc.
- Perceived interest of the telecentres and potential partners in the community wireless network initiative ranging from whether there was demand for an immediate set-

up of the networks to whether telecentres and/or potential partners did not care much whether the networks were established or not.

From the site assessments, the telecentres and set of potential partners were selected based on who had the best chance to create a successful community wireless network. Given the generally low Internet demand, it was also important that the telecentres offer diverse services to their communities in addition to the Internet services. The telecentres selected offered a range of services including computer training, printing and photocopying services, binding, scanning, small public libraries, and typing services. Two of the telecentres also operated community radio stations – Box 1 highlights the services offered then at one of these telecentres.

While the telecentres offered a range of services, Box 1 reveals that the most beneficial services were photocopy and printing, ICT training and the community radio services. This further emphasized the need to ensure that telecentres selected were indeed offering multiple services because Internet access as a service was not the most popular service at any of the selected telecentres and hence the community wireless network would not, on its own, lead to sustainability of the telecentres. It is also important to note that the telecentre site visits also learnt that the telecentre customers varied from service to service (Mugume, 2006). While the youth primarily visited the telecentres to make use of the video, Internet, library and computer training services, for men and women the more popular services were radio, printing, and photocopying.

With respect to the partners, while site visits were made to the original partners list developed in 2004, some of these potential partners had either moved on to get their own Internet access or were no longer interested in participating in a community wireless network. So, the final list of partners for the telecentre-based community wire-

Box 1. Telecentre services and average income

- **Computer Training:** This was an activity carried out mainly as part of the telecentre's outreach program mainly in the schools located in rural areas. Computer training also took place at the telecentre premises. It fetched an average monthly income in the range US$81 (150,000 Uganda Shillings, Ug. Shs.) to US$108 (200,000 Uganda Shillings, Ug. Shs.).

- **Email/Internet:** This was a popular service among the relatively elite. It realized about 10-15 customers per day and the numbers shot up during school holidays when students in good secondary schools, universities and other institutions of higher learning returned home. It fetched an average daily income of Ug. Shs.3,500 – Ug. Shs. 5,000 per day.

- **Community Radio:** This generated income mainly from charging announcements and advertisements. It generated an average daily income of between Ug. Shs. 20,000 and Ug. Shs. Shs.30,000.

- **Telephone Services:** This was among the least income generating services offered by the telecentre. It fetched little income due to stiff competition from other calling booths or kiosks. It fetched an average daily income of between Ug. Shs.1,000 and Ug. Shs.2,000.

- **Library Services:** This was another low income generating service that generated less than a dollar per day.

- **Photocopying Services:** This was one of the highest income generating services and it attracted many customers. It fetched an average daily income of between Ug. Shs.25,000 and Ug. Shs. 30,000.

- **Printing Services:** This service also commanded a significant customer base and fetched an average daily income of between Shs.6,000 and Shs.10,000.

- **Distance Education:** Distance education was done via the World Space Program (WSP) and was therefore offered as a free service.

- **Box Services:** This is where schools are provided with boxes containing assorted reading materials like books, magazines, etc. Such schools will have subscribed as members to enjoy this service. The subscription fee is Ug. Shs.50,000 (about $27) per year.

- **Outreach Services:** These were accomplished by setting up outreach centres in every parish. These places were provided with radio receivers to be able to listen to news, announcements, debates, etc.

- **Indigenous Knowledge (IK):** Here, the telecentre imparted local skills to the residents free of charge. These skills were used mainly in agriculture, for example, how to use green manure instead of artificial fertilizers. Other areas in which training was offered included hunting, crafts and pottery, and maternity care skills.

- **Video Coverage and Video Shows:** This service was not demanded regularly and it was hard to compute the income it generated per day or even per month. It did not enjoy a big customer base but was used once in a while.

Source: Mugume (2006)

less networks also comprised of new potential partners identified in the 2006 site visits. The final list of partners included schools, hospitals, small businesses, and local government offices. In total, sixteen partners were connected including nine partners at the Kachewekano and Kabale telecentres based in Kabale, four partners at the Nabweru telecentre, and three partners at the Lira CPAR telecentre. In all cases, the partners contributed to Internet costs on an equal basis independent of their usage, and the subscription to the Internet Service Provider (ISP) was managed by the telecentre manager. The telecentre managers were also specially trained by the CWRC to provide technical support to the networks.

In establishing the community wireless networks, anticipated outcomes included the increased use and uptake of the Internet as a source for information gathering, exchange, and dissemination due to the reduced costs and convenience of access. It was also anticipated that sustainability of the telecentres would be enhanced, and in particular, the Internet facilities due to sharing of the costs of maintenance. Risks related to the initiative as identified by both the CWRC and the telecentres included:

- Willingness for potential partners to continuing meeting their financial obligations towards the community wireless networks—including failure or delay to pay their Internet access contribution, and unwillingness to contribute towards salaries of the ICT staff in charge of maintaining the network.
- Relatively high cost of network equipment (for example, about US$200 for the access points), compounded by the limited availability of such equipment locally. For the project, most of the equipment was sourced from out of Uganda.
- Limited skills at the telecentres, and in rural areas generally, to manage and maintain the networks. While the project did provide a number of trainings and hosts CWRC Annual Meetings to review the situation at the telecentres and their networks, the telecentres also have to grapple with retention of skilled ICT staff.

2.4. Research

The CWRC research activities involve supervision, mentoring and training of students from the Department of Electrical and Computer Engineering. Students have been a mainstay in this initiative having been involved in conducting the preliminary technical inspection surveys and assessment of partners' willingness to cooperate in a shared Internet access initiative, to actual implementation of the community wireless networks, through to basic technical support to the telecentre managers. Through this all, the CWRC and students greatly benefited from technical support provided by IT+46. Students have since also undertaken research projects related to issues of concern for the telecentre-based networks, for example, design of a low-cost cantenna – an antenna from locally available materials and cans as well as set up of a Community Wireless Network (CWN)-based voice network.

The first engagement of students was in 2006 when three third year students conducted the technical readiness and partner assessments of six IDRC-supported telecentres. The assessments were to gain a better understanding of the local conditions for each of the telecentres and their potential partners, which would influence the final selection of telecentres to participate in the CWRC project based on technical feasibility and economic viability. The students conducted the assessments as part of their third year industrial training period with the CWRC, and hence the assessments were an integral part of their undergraduate training.

In addition, the CWRC hosts an annual meeting involving the CWRC staff and students, telecentre managers and other key stakeholders. Key stakeholders are invited to share particular expertise, for example, some of the invitees have included the communications sector regulator and a telecommunications services provider. The annual CWRC meetings are an avenue for highlighting key technical and management issues that need to be addressed. It is also an opportunity to learn and share about the experiences from the telecentres as well as the research directions of the CWRC. For example, the 2010 Annual CWRC Workshop that was held under the theme: "Rural connectivity for community development" had key issues emerging such as:

- Telecentres and the community wireless networks need to provide the kind of services demanded by and relevant for the communities.
- There is need to create appropriate business models for sustainability of telecentres and, in turn, the community wireless networks.
- There is need to address the availability of equipment required for the community wireless networks.

Indeed, an area of increasing significance is the type of business models that telecentres can use for successful operations of the community wireless networks. Consequently, the CWRC and telecentres partnered with the Economics Department at Kabale University to assess the impact of the community wireless initiative as well as to propose business models for the networks.

The community wireless networks have brought forth a number of research challenges including the following:

- The partners are not sensitized for instance on the ways of managing bandwidth to reduce on the cost of bandwidth. Much more bandwidth than had been anticipated is currently consumed, with one community wireless network consuming up to ten times more traffic—due to largely Web-based traffic. This increase in bandwidth usage is due to the high demand that was generated by the wireless networking initiative—leading to more users and usage-hours.
- Breakdown of equipment is a major challenge for these wireless networks. This could be due the harsh environment or lightning. For example, the central hub at one of the telecentres was destroyed due to lightning, rendering the network dormant. Worse still, there is limited local supply of the wireless equipment required and at costs that the communities can afford.
- The network partners are currently not convinced about paying the same flat rates. The differences in user behavior mean that their bandwidth usage is different; hence, they should pay different rates. This presents an administrative challenge to the telecentre manager to explain why all network partners are paying the same flat rate—in spite of the differences in utilization of the network.

- The remuneration of the telecentre managers who maintain the networks is still not clear. They depend on "handouts" from the partners. This of course is not a sustainable way of running these networks.

The CWRC is involved in research initiatives to solve some of these challenges. The CWRC also contributes to capacity building in the Department of Electrical and Computer Engineering through providing industrial training opportunities to second and third year undergraduate students, as already highlighted. The objective of industrial training is for students to gain practical knowledge to reinforce theoretical concepts learnt from class as well as to expose them to the day-to-day engineering profession. These students are equipped with skills and hands-on opportunities in wireless and in Internet Protocol (IP) networking. Twelve out of twenty-four industrial training students have gone on to do their final year research projects under the supervision of the CWRC and three students have further stayed on as Research Assistants with the CWRC after completing their undergraduate studies.

Through their research projects, students associated with the CWRC have been able to test and demonstrate applications on the community wireless networks such as Voice Over IP (VoIP), traffic shaping, and bandwidth network management. In all, over thirty undergraduate and five graduate students have undertaken industrial training and/or research activities related to the community wireless networks—including students that did not undertake their industrial training with the CWRC. Where we have been short is on widely publishing the various research works arising out of the students' efforts. However, this too is set to change especially with inauguration in 2011 of the Uganda National Conference on Communications (NCC). The NCC is an annual event that seeks to provide an avenue to build capacity and strengthen the academic and industry communities in communications—particularly in the research of locally relevant solutions.

The CWRC has also been able to attract additional support to address a number of bandwidth-related challenges for community wireless networks. So far, this has been through a grant from the Millennium Science Initiative (MSI) of the Uganda National Council of Science and Technology (UNCST) and World Bank as well as through a grant from the Presidential Initiative Grant (PI) to CEDAT. The MSI and PI projects involve supervision, mentoring and training of graduate and undergraduate students. The projects are addressing questions like:

- How can traffic shaping be implemented in a "static"—as opposed to an adaptive wireless environment so as to ensure latency and congestion management as well as fairness within the network? Instead of having the same "static" capacity for all partners, could bandwidth capacity be availed according to usage required?

- In the case of adaptive bandwidth schemes, what key factors should be actively monitored in order to determine and quantify the "extra" bandwidth utilized by a user? How can this information then be used to develop a billing model that charges partners based on bandwidth consumption?

- What is the impact of the propagation environment on the viability of adaptive bandwidth management?

- What impact does cognitive/smart radio technology have on the number of users that can be hosted on a single cooperative network employing an adaptive bandwidth strategy?

- What policy recommendations can be provided to improve the capacity of wireless technologies in the provision of diversified services and value addition—particularly for rural and underserved areas?

The primary considerations in the development of community wireless networks are not only technical in nature, but also include socio-economic considerations. This means that there is also need to study of what use the Internet services have been put to in the different networks and the impact that the Internet has had in the different communities. There is also need to test and propose appropriate business models for the telecentre-based community wireless networks. Figure 5 shows a billboard at the Kabale Telecentre highlighting available services. However, over time, the telecentre is facing ever more stiff competition from other ICT service providers who are increasing in number within Kabale town—particularly from those that are more centrally located within the business center of the town.

Under the Presidential Initiative for Science and Technology funding awarded to the College of Engineering, Design, Art, and Technology (CEDAT), the CWRC successfully applied for a grant to examine the impact of public access to ICTs with a focus on the CWRC community wireless networks. To this end, a partnership with Kabale University in South Western Uganda was developed in order to obtain a socio-economic expert to join the team in reviewing the wireless networks in Kabale, Nabweru and Nakaseke. Following the review, a dissemination workshop was held to draw contributions for a way forward for the community wireless networks. Participants in the workshop represented a wide range of stakeholders including telecentre users, community and local government leaders, civil society, and academia. The impact of public access to ICTs study had obtained the following (CWRC, 2011):

- The telecentres had eased their ways of communication and had also facilitated easy access to information.

- Users had gained better farming practices information which had made them better informed.

Figure 5. Billboard at Kabale Telecentre highlighting available services (Source: CWRC)

- Users had improved their computer literacy and their typing skills.
- Users had improved their chances of employment with some being employed by the telecentres while others who had gained typing skills had got employment elsewhere.
- Students had been able to access information for their research.

The public access impact study also found that users wanted the telecentres to spread their services further within the communities—and certainly the community wireless networks would serve the need to spread Internet access within the community. It is indeed telling that the reasons advanced by the study respondents on why they wanted a spread of the telecentre services indicate that all the needs could be met online – again strengthening the case for the telecentre-based community wireless networks. These reasons included:

- Communities need more information on farming skills, agricultural products, good crop yielding varieties.
- Communities need improved communications within the communities as well as improved Internet services.

- Need to support students do their coursework well.
- Need to source for employment opportunities over the Internet.
- Need to share information about the communities on a national and global stage with a view of encouraging more investment and so by increasing employment opportunities within the community.

In spite of the perceived value of the telecentres and the community wireless networks, the telecentres are struggling to remain afloat with the Kabale community wireless network not in operation presently. The public access impact study did identify a number of barriers affecting the sustainability of the telecentres including:

- Location of the telecentres was found to be remote and not readily accessible. When they were the only ICT providers, users would commute to access the services. However, with mushrooming kiosks and other ICT service providers, location, location, location is becoming a key issue.
- Communities were not sensitized on the availability and services offered by the telecentres, particularly those further within the communities.

- Telecentres continued to suffer from lack of reliable energy supply as well as lack of skilled personnel to operate and maintain the centers.
- Telecentres needed to come up with innovative service packages to remain relevant within their communities, for example, one telecentre offering computer training packages to government institutions. Given high levels of illiteracy in rural areas, the telecentres could also provide information assistants to support users access information online.

It should also be noted that, as highlighted in Table 2, Uganda has witnessed a significant fall in ICT prices since the community wireless networks were established. Furthermore, Internet access options have also increased with advances in technology. For example, one can readily obtain an Internet dongle (wireless USB modem) for a one-time fee of 85,000 Uganda Shillings (US$ 34) and a 30-day Internet bundle of 500MB for 25,000 Uganda Shillings (US$ 10). The drop from monthly Internet access charges of US$ 250 to US$ 10 means that telecentres need to come up with attractive service packages to maintain and grow partners within the coverage area of their community wireless networks. Two of the active telecentre networks have maintained their advantage as preferred ICT service providers by offering technical service packages to their partners in addition to the Internet access. In addition, by purchasing bandwidth in bulk, the telecentres are in position to offer lower costs per unit bandwidth to their community wireless partners and for the telecentre users. Being located in rural areas means that even with advances in technology and growth of ICT service providers—particularly in major towns, there is still room to explore appropriate business models that can enable the growth of telecentre-based community wireless networks in a sustainable manner.

Indeed, in 2009, the CWRC participated in the Wireless Africa (WA) project that was aimed at implementing low cost, affordable technologies and applications that result in the high use, potential revenue and/or dramatic cost-savings (WA, 2010). The WA initiative gave the CWRC an opportunity to carry out demand side studies to establish the needs of users at the telecentres. It was anticipated that this will result into the implementation of appropriate business models and Value-Added Services (VAS) for sustainability of telecentres. Using one of the telecentres as a pilot, and following training and receipt of some networking equipment from the Wireless Africa initiative, it was planned to test a wireless VoIP service across the telecentre-based community wireless network.

After successful VoIP tests in the laboratory, the CWRC was ready to do the installations by June 2009 but the major challenge was resources (CWRC, 2009). These mainly included the VoIP equipment that should be purchased, and other equipment to facilitate the implementation. The CWRC has a limited financial muscle and the network partners were not willing to contribute to a project whose benefits they were yet to perceive. In addition to this, the CWRC sourced for local availability of VoIP equipment. At the time, all the required equipment was locally available except the Analog Telephone Adapters (ATAs). The CWRC had an option to implement the VoIP solution with only soft phones but since this was the first installation, it was decided to pursue hardware (in this case the ATA). Hardware, the use of analog phones or VOIP phones, was expected to make a bigger impact in making the technology acceptable to the community as opposed to the soft phones on computers. Eventually, the CWRC was successful in sourcing for some ATAs from abroad and a demonstration of the VoIP solution was successfully conducted. Still, due to limited community sensitization and awareness of the potential savings of communication among themselves via the VoIP network instead of over the commercial telecommunication networks, the CWRC was not able to marshal enough resources to rollout the VoIP service beyond the demonstration phase.

As an emerging program, the CWRC seeks to build on our own quality of expertise in community wireless networks by linkages with established programs in wireless networking and sustainable ICT models. For example, with the MSI support, two graduate students have been exposed to advanced research environments. The objective of this exposure was to enable students acquaint themselves with experiments/simulations using advanced equipment and tools as well as to be exposed and interact with other researchers in telecommunication systems. In 2010, the first graduate student spent ten weeks at the Sentech Chair in Broadband and Wireless Multimedia Communications (BWMC) Lab at the University of Pretoria, South Africa. In 2011, the second graduate student was hosted by the Meraka Institute at the Council for Scientific and Industrial Research (CSIR) in South Africa. In December 2011, the CWRC also hosted a Professor from the University of Navarra in Spain to provide specialized training on the use of spectrum and network analyzers with a view to establishment of a wireless communications testbed at the Department of Electrical and Computer Engineering.

3. CONCLUDING REMARKS

Telecentres have long played a key role in availing access to ICTs and in supporting the provision of universal access. The initiative of the community wireless networks builds upon the community approach of telecentres by expanding the telecentres' outreach in terms of Internet access and by providing another form of entrepreneurial activity for the telecentres. Already, the telecentres were offering a much needed variety of services including Internet access, printing, typing, scanning, and photocopying.

The initiative to establish the telecentre-based community wireless networks has also resulted in skills transfer from the university to the community as the telecentre managers have benefited from training and hands-on experience in wireless networking. The CWRC staff and students have also been on hand, as needed, to go to the telecentres and assist in diagnosis and resolution of technical problems encountered by the community wireless networks. In addition, an electronic mailing list was established to which the telecentre managers as well as CWRC staff and students are subscribed. The list facilitates ongoing interaction among the telecentre managers as well as between the CWRC and the telecentre managers.

From the CWRC perspective, the networks have also provided fertile ground for a number of studies and research opportunities for our undergraduate and graduate students. In addition, to continued efforts to study and deploy new technical services and applications that can add value to the community wireless networks, there is also need to continue to recognize the key role such networks can play in empowering rural communities in Uganda.

Going forward, and as a university-based research group, the CWRC needs to have further community engagement beyond only the telecentre managers so as to maintain a grasp on the changing situation and needs of the rural communities. This would also complement the efforts of the telecentre managers in raising awareness about the role of ICTs in community development. Clearly, there is also need for further review and piloting of a variety of business models in order to obtain an optimal solution at each of the telecentres. With falling prices and new technologies, the telecentre-based community wireless networks need to continually innovate in order to remain relevant to both the telecentres and the partners that together comprise the networks.

REFERENCES

Abdelaal, A., Ali, H., & Khazanchi, D. (2009). The role of social capital in the creation of community wireless networks. In *Proceedings of the 42nd Hawaii International Conference on System Sciences 2009 (HICSS 2009)*, (pp. 5-8). IEEE Press.

Abdelaal, A., & Ali, H. H. (2009). Community wireless networks: Emerging wireless commons for digital inclusion. In *Proceedings of the IEEE International Symposium on Technology and Society (ISTAS 2009)*, (pp. 1-9). IEEE Press.

Community Wireless Resource Center. (2009). *Status report on wireless Africa initiative in Uganda. Technical Report*. Kampala, Uganda: Makerere University.

Community Wireless Resource Center. (2011). *Examining the impact of public access to information and communications technologies (ICTs): Selected IDRC-funded telecentres and the associated wireless networks in rural Uganda. Technical Report*. Kampala, Uganda: Makerere University.

International Telecommunication Union. (2009). *Information society statistical profiles 2009 – Africa*. Retrieved from http://www.itu.int

International Telecommunication Union. (2011). *Measuring the information society*. Retrieved from http://www.itu.int

Ishmael, J., Bury, S., Pezaros, D., & Race, N. (2008). Deploying rural community wireless mesh networks. *IEEE Internet Computing, 12*(4), 22–29. doi:10.1109/MIC.2008.76

Mugume, K. E. (2006). *Telecentre assessment surveys: Nakaseke and Buwama multipurpose community telecentres. Technical Report*. Kampala, Uganda: Makerere University.

National Conference on Communications. (2011). *Website*. Retrieved from http://www.cedat.mak.ac.ug/ncc

Organisation for Economic Co-Operation and Development. (2009). *African economic outlook 2009*. Geneva, Switzerland: OECD.

Rural Communications Development Fund. (2012). *Website*. Retrieved from http://bit.ly/ucc_rcdf

Siochru, S. O., & Girard, B. (2005). *Community-based networks and innovative technologies: New models to serve and empower the poor*. Geneva, Switzerland: UNDP.

UbuntuNet Alliance. (2009). *Overview of fibre infrastructure opportunities in the UbuntuNet region*. Kampala, Uganda: UbuntuNet Alliance.

Wireless Africa Alliance. (2009). *Website*. Retrieved from http://www.wireless-africa.org

Wireless Networking in the Developing World. (2006). *A practical guide to planning and building low-cost telecommunications infrastructure*. Kampala, Uganda: Limehouse Book Sprint Team.

Women of Uganda Network. (2006). *Uganda country-based research, policy support and advocacy partnerships for pro-poor ICT*. Kampala, Uganda: Women of Uganda Network.

KEY TERMS AND DEFINITIONS

Bandwidth: A measure of the amount of data that can be accessed or transferred in a given time period.

Community Engagement: Collaboration between a university and the community for enhanced learning, research and exchange.

Community Wireless Network: A wireless communication network that is shared and managed by a community.

ICT Price Basket: A measure comprised of the prices of fixed telephone, mobile cellular and fixed broadband Internet services in a country.

Internet Point of Presence: A physical location from which Internet can be accessed or be further distributed within a community.

Renewable Energy: Energy from natural resources such as solar energy, wind energy, bioenergy and geothermal energy.

Telecentre: A public space that makes available access to a variety of ICTs including computers and Internet.

Universal Access: Policies and programs availing personal and/or public access to ICTs for the benefit of all citizens irrespective of social or economic status.

ENDNOTES

[1] Gini Index – Measure of distribution of individual or household income/expenditure relative to an equal distribution; PPP – Purchasing Power Parity; GDP – Gross Domestic Product.

Chapter 10
The Case of Chapleau Network:
Why Community Wireless Networks Fail?

Sylvie Albert
Laurentian University, Canada

ABSTRACT

Remote and underserved communities do not attract telecommunication companies because of their low income, remote location, and limited capacity. This chapter discusses the challenges small communities face when developing their own Wi-Fi network, even when an investment is made. In particular, this chapter examines the technical, social, and economic challenges faced by the community of Chapleau (Ontario, Canada) while building its Wi-Fi network. The project adopted a public-private partnership in which Bell Canada and Nortel Networks funded its pilot phase. However, the project failed because of unclear and divergent goals, lack of sustainable applications, and insufficient technical skills on the part of the community. Using a change management framework, the chapter identifies key lessons learned and success factors required for public-private partnerships.

BACKGROUND

Chapleau is a small town in Ontario, Canada, with a population of 2,600 people. It is located in a remote area, which is 800 kilometers northwest of Toronto. It is recognized as a destination for hunters and anglers, naturalists and eco-tourists, and was built to provide lumber and act as a rail-

way hub. Its surrounding area is rich in mining resources such as gold, copper, and nickel. The leaders of the community planned to provide ubiquitous and wireless communications for the purpose of improving the overall economic opportunities in the area. They hoped that ubiquitous and affordable Internet access would diversify and

DOI: 10.4018/978-1-4666-2997-4.ch010

reinvigorate the community through integrating it into the information economy.

In 2004, a group of community members worked to find a solution for its outdated telecommunication infrastructure and sought government funds for a large telecommunication project that could leapfrog current technology and provide Fiber-To-The-Homes (FTTH) and Fiber-To-The-Businesses (FTTB) connectivity. Chapleau is currently served by party lines, a limited dial-up infrastructure, and an overcrowded fiber backhaul. In contrast, surrounding cities are served by an advanced national telecommunication superhighway that uses fiber cables. Although unsuccessful in its bid to the government to install fiber throughout the community, the visionary work of community leaders garnered the attention of Bell and Nortel. The small community could not provide a business case for the 200-350 kilometers of fiber cable that would connect Chapleau to the nearest city or for its local loop needs.

Traditional rural communities are restricted in the provision of broadband infrastructure, as it is financially difficult for Information and Communication Technology (ICT) companies to cover the cost of extending the infrastructure to rural areas (Youtie, 2000). Chapleau leaders felt that, as in many rural communities, they were continually last in line when it came to technology rollouts. For instance, Sudbury (a community of 160,000 located three hours south of Chapleau) has had a number of telecommunications services for more than five years (such as smart phones and several improvements in broadband) that are still not available in Chapleau. In addition, when technology is brought to Chapleau, it is not always up to date. For example, Chapleau is still using a digital 1x cellular system, a technology that cannot support smart phone applications.

Generally, many leaders in rural and remote communities see ICTs as an important enabler of economic and social development, and view broadband as a basic infrastructure similar to roads and sewers (Albert, et al., 2010; Fernback, 2005). Communities that are left out of important

developmental infrastructures face a growing digital divide (Castels, 2006), and this concern leads remote communities to seek assistance in deploying ICTs infrastructure. The Chapleau community was well aware of the possibility of being technologically left behind. In addition, residents were eager to be included in the information society, but were skeptical on the capabilities of ICTs to solve larger economic problems. ICTs are perceived as a tool and a force for change, but seldom as a full-fledged solution to economic woes. Real answers are incumbent on the actions of individuals and organizations, not the technologies that support them, but this level of engagement by stakeholders takes time and well rooted organizations.

Infrastructure projects can be complex, capital intensive, include long gestation periods and involve multiple risks to the project participants (Agrawal, et al., 2011, p. 52). This level of complexity encourages stakeholders to seek partners for the purpose of sharing risks and costs, especially when resources are limited and where institutions are looking for 'value for money' and accountability (Agrawal, et al., 2011; Clarke & Healy, 1999; Demirag & Khadaroo, 2011; Kwak, et al., 2009). Although attractive from a resource point of view, researchers of Public-Private Partnerships (PPP) point to the complexity that arises from partnered projects, including increased risks in governance, pronounced problems in managing agreements, communication and expectations related problems, and challenges inherent in asymmetric abilities. Still, PPPs remain an important solution for many communities and private sector organizations, especially in times of resource constraints (Fischer, 2010).

Bell and Nortel extended their collaborative partnership with Chapleau (called Project Chapleau) to assist it to overcome the daunting challenge of building its own wireless network. The intention was to develop a test-bed demonstration network using WiFi within an 18-month period of time. In addition, the project was expected to provide the skills, knowledge, and resources needed to eliminate some of Chapleau's infrastructure concerns and showcase a number of social and economic development applications. As a result, the Chapleau project began the dialogue on how the community could explore opportunities to co-build economic foundations using ICTs.

LITERATURE REVIEW

Castells (2006), known for his work on the 'network society,' described an information society as one that fosters the use of ICT to allow greater openness, adaptability and flexibility. Albert et al. (2009) proposed a model for building networked communities in which individuals, organizations and communities are able to achieve economic and social goals, have choices and freedom of expression to share ideas, and to innovate regardless of location. Many communities, and Chapleau was among them, are interested in working towards the opportunities afforded by a planned networked community. Among the requirements for a networked community (Albert, et al., 2009) is access to affordable and upgradable broadband, ICTs applications, and support systems that promote collaboration, leadership and community engagement. Access to broadband infrastructure is increasingly considered a basic right not dissimilar to roads and municipal services, and when it is not available it becomes an agenda item for municipal representatives. Cables (copper, coaxial, fiber) are important primary instruments in the provision of broadband, in addition to wireless technology such as satellite, microwave, WiFi, etc. However, a number of authors have also identified the relative importance of wireless infrastructures in the provision of broadband to communities as a supplementary tool that can help provide universal and flexible modes of access and to guard against abuse from single and larger providers of basic services (Lehr, et al., 2006; Middleton, et al., 2008; Middleton & Crow, 2008).

Based on data from two successful wireless projects in India, Surana et al. (2008) identified a number of challenges in surviving wireless networks beyond the pilot stage, most of which are technical in nature. As they note, the real challenge is 'to find solutions that are sustainable and low-cost at all levels of the system' (p. 119). Sustainability includes a number of factors such as cost-benefit, community engagement, technical affordability and scalability, and change management processes (Albert, et al., 2010). Adding multiple stakeholders and partners as part of program governance adds more complexity and affects sustainability.

The introduction of broadband services and new infrastructures into communities is a change process that must be managed. Pettigrew (1987, 1992) is among the many authors who have provided models and frameworks to manage change. Pettigrew argues that events are best studied in their setting where there is an interaction between actors, events, and environment. The Pettigrew framework, which is used in this case study, includes three interacting components (usually acting as a cycle): the content of change (what), the context of change (why), and the process of change (how). An investigation of the interplay of these three components reveals patterns of change that can lead to a better understanding of the phenomenon studied (See Figure 1).

Figure 1. Pettigrew change management framework

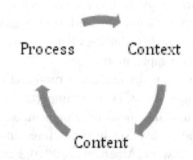

Pade and Sewry (2011) bring specificity on planning a change in ICTs projects, as they outline that organizations need to go beyond measuring the traditional impact (such as number of jobs created) on the target population in ICTs projects. The authors provide eight areas of assessment that are needed to help define expectations and plan future steps toward a successful measurement system. These are strategic in nature and require planners to think about the change management process outlined by Pettigrew. For example, advanced planning regarding what will be measured should consider using an evaluation approach that will be 'multidisciplinary, comparative, participatory, and multicultural to contribute to research on different elements associated with the contribution of ICTs to rural development' (p. 365). This requires that community ICTs planners think in terms of why change is needed, what requires change, and how the change process will occur and be measured. This leads us to conclude that a successful change process requires time and expertise. The Pade article points out that evaluation is often set aside or is an afterthought in the development of innovative projects, especially in pilot projects with short timeframes. Poor planning at the onset affects the sustainability of projects in the long run. The Pettigrew framework assumes that communities and its project partners will take the time to plan and agree on context, content and processes in introducing change, and the Pade and Sewry study outlines the importance of setting an adequate program measurement system at the onset that also requires careful discussion, planning, and agreement. Both sets of authors point to a need to better understand where projects are heading and how the partnership will work through content and processes toward agreed-upon goals.

The Chapleau project can be studied in terms of its change management process, but also in terms of its governance, which was a public-private partnership. Fischer (2010) classified 17 projects in Europe under three different types of public-private cooperation: Public-Private Collaboration (PPC), Public-Private Partnership (PPP), and Public-Private Joint Ventures (PPJV). The types vary in the duration of the agreement, in the structuring of the partnership and in the sharing of risks. Given our experience, the Chapleau project appears to fit into the PPP definition: "the key characteristic of this kind of PPP is the transfer for a limited period of time of integrated services relating to the planning, construction, financing, maintenance and operation (in a life-cycle approach) of public infrastructure... the main objective is to generate efficiency gains in the provision of services... to bridge liquidity bottlenecks on the part of the public partners."

Kwak et al. (2009) offered a definition of public-private partnerships based on the differences in expectations from project partners. Under the continuum shown in Figure 2, the authors detail an array of expectations on the responsibilities of each partner. With respect to the Chapleau case, the private sector partners were building a pilot project for the sake of evaluating its business model. Their intention was a design-build-operate model according to which the project would be handed back to the public sector partner (the municipality of Chapleau) upon completion of the trial. Conversely, and in the eyes of the municipality, their hopes were for sufficient broadband infrastructure to realize economic and social development objectives in a build-operate-transfer format in which the municipality would have a concession period to assume the transfer. A build-own-operate model was also considered acceptable by Chapleau, and in this form the assets would remain in the hands of the public sector partners. As explained by several motivation theorists, congruence occurs when interests are aligned (Schaffer, 2007) and motivation occurs when the objective is seen as achievable and valued (Vroom, 1964). The expectations of the Chapleau PPP were not fully aligned and the impact of this will be seen through the various challenges discussed in this chapter.

Figure 2. Continuum of types of PPP (Kwak, et al., 2009, p. 54)

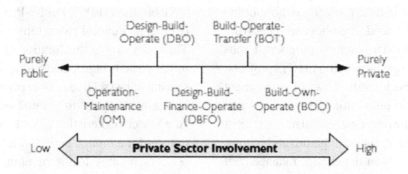

Chapleau had already done some planning regarding opportunities afforded by a broadband economy and had started to educate itself, but this knowledge acquisition process was focused on developmental objectives rather than on technical rollout. Expectations were similar but not necessarily aligned—Bell primarily saw this project as a method for piloting technology and finding its profitability (business role). Nortel was interested in identifying applications that could be duplicated to meet the social and economic needs of rural communities (social responsibility role built into a business-level strategy). Chapleau was interested in acquiring broadband infrastructure to increase future economic and social development potential, though what these were are unclear, but the potential of creating an 'information society' promised opportunities that could not be ignored. Nortel's vision was slightly more multidimensional and aligned to Chapleau's than Bell's was, as Nortel was interested in finding a model that was replicable in many other communities internationally, and this model would include the development of applications and a model for engagement.

The Bell technology trial was attractive and appealed to the vision of community leaders. The difficulty was in understanding how the various pieces fit in and how it could be done in 18 months. In reality, the private sector plans were short term while the community hoped for a more

extensive, longer-term involvement. It became clear that although Nortel spoke about long-term achievements, the company was unwilling to stay beyond 18 months (viewing its involvement as a mentor that would point the community in the right direction), and Chapleau would have to be prepared to take over at that point. It should also be noted that community development experts in small and rural communities work in 3-5 year timeframes to accomplish any significant objective. Technology projects are no different; it can take long lead times to convince key stakeholders to buy in and develop applications, especially when the decision makers are not located in the community. The Chapleau project represented the beginning stages of a grander plan to diversify the economy of the small community, and Bell and Nortel represented an idyllic partner that could yield substantial longer-term results if they could be convinced to stay a while longer.

CASE DESCRIPTION

The Network Infrastructure

Bell Canada and Nortel Networks deployed municipal Wi-Fi (Wireless Mesh, WLAN 2300), which included Nortel Multimedia Solutions: IP Telephony, CallPilot, Multimedia Communication Server (MCS) 5100, Nortel Ethernet Routing

Switches, Nortel Optical Solutions, and Nortel Security Solutions (VPN). The Wi-Fi operated on a 1-km radius on a fiber ring, covering the majority of the populated areas of Chapleau, except for residents who live slightly outside the municipality on lakes, tourist outfitters removed from town, and First Nations communities. Residents directly in the municipality now had mobile, flexible, and reliable access to Internet-based services, and they were provided, as part of the installation, with an online encyclopedia, adventure and strategy games, educational software, and a downloadable music store. In addition to the expanded optical and broadband network, four of the five schools and the local hospital installed wireless Local Area Network (LAN) solutions and upgraded their Wide Area Network (WAN) connectivity (Nortel, 2011).

The design and deployment of the network were the responsibility of the Bell/Nortel team. They identified the technical needs for updating the backhaul and expanding a new fiber infrastructure. The local loop was wireless with a small fiber ring provided through town to serve the needs of larger bandwidth users. Pricing for residential users was below the current DSL rate in northern Ontario but slightly above the dial-up rate, which meant that consumers needed to be somewhat convinced about the utility of broadband.

All projects are faced with resource restrictions and in this case Chapleau was asked to make choices between further investments in either infrastructure or applications. The required trade-offs were not easy to decide upon. One of Chapleau's priorities is to develop its tourism infrastructure and outfitters were eager to utilize the broadband infrastructure to access online resources. In particular, tourist outfitters were reporting that 70% of all bookings came from online sources. In addition, the emerging online economy requires publishing more videos and other rich content on websites. Therefore, broadband access was viewed as a necessity but it was not possible to meet the needs of many tourism operators and outlying

residents due to distance. At the same time, the government was eager to find solutions to serve First Nations communities. There was a First Nations community situated within the community of Chapleau and another located a few kilometers away from the downtown core and was not within the limits of the Wi-Fi project. The government was willing to help finance the infrastructure to ensure that First Nations would not be left out and WiFi expansion was considered, but in the end, Bell chose DSL for this leg of the project. This decision changed the landscape of the Wi-Fi project, as there would now be two broadband infrastructures, with DSL promising to be slightly more expandable and reliable.

The issue of buy-in was discussed in planning meetings amongst the partners in terms of the appeal of the infrastructure and its services to businesses and households. Buy-in for broadband was managed through incentive programs such as allowing people to test the service for a period of time at a much lower cost (or for free). Buy-in for business applications meant changes to the strategic structure of businesses as well as the need to develop websites and generally gain trust in its value. Some businesses needed to be convinced by people they trusted (supporting research from Fernback, 2005) and wanted to know that the broadband technology and its applications would help and not create another layer of challenges. A small team was deployed to link businesses with testimonials and mentors, and to assist in website development.

Before the construction of WiFi began, the project was announced during a yearly home show, which was followed by media announcements. These generated expectations as citizens began to ask more questions regarding whether the infrastructure would reach their home or business, and when. Chapleau residents had not had good experiences in the past with projects such as the introduction of cellular service, which had long wait periods. As a result, their level of patience was low and technical problems were met with skepticism. Nevertheless, Bell/Nortel labored to fix problems quickly despite the long distances crews had to travel to reach Chapleau. This expectation issue affected the perception of residents on WiFi's value.

Working on Content and Applications

Within an 18-month timeline, Chapleau spent one third of the time dealing with basic infrastructure and formulating a structure and agreements to go forward, and had approximately 12 months to explore the development of content applications. The applications, which are the life of a telecommunication network (Albert, et al., 2009), would be complicated and the community was eager to find solutions for sustainability. These solutions were assumed to be 'killer' applications, or those that promote use and generate revenues. Sector meetings were organized to identify new and innovative ideas. Stakeholders who were already using the technology could see the opportunities and suggested potential pilot projects or extensions of existing services. Meetings with less experienced sectors did not generate many new ideas but served to educate attendees on future potential. Chapleau leaders investigated popular applications used in international projects as a method to stimulate local innovative thinking. The community, having failed to discover a 'killer' application settled on a combination of applications supported by niche groups (such as in e-health, e-learning, and so forth). More applications required more resources (financial and human) and time to develop them. Bell and Nortel stepped in to develop partnerships in various disciplines, meeting with public and not-for-profit organizations that had been singled out as potential key contributors to the project (many of these were located outside of Chapleau in larger centers). It became clear that many of these partnerships needed 3-5 years to see true change and buy-in, including growing the pool of volunteers and leaders.

Among the applications considered were those in e-health, e-learning, e-commerce and a community portal. The community hoped that it could also seek longer-term projects such as a call center, data warehousing, a tourism technology/wildlife interpretive center focused on e-delivery, a wildlife research center equipped with GPS tracking technology, online gaming competitions online, tele-work, and more. The timeframe did not permit a detailed evaluation of feasibility of these projects. In addition, project partners were required to make judgments on the projects that generate the highest value. The following projects were chosen:

1. **A Community Portal:** The economic development office collaborated with municipal staff to identify information that should be placed on the portal and then communicated with local organizations to seek their support in collecting data. The majority of stakeholders were supportive but had difficulty assigning resources to its development. To meet deadlines, the economic development organization was asked to write a majority of the content needed for the portal, aggregating bits of information provided by e-mail or fax. The small team was concerned that local organizations would not be able to update the information going forward. Many organizations in smaller communities were already swamped with day-to-day affairs and were often assigned a wide range issues, leaving little time to work on new projects.

2. **An E-Learning Project:** Through its school boards, the provincial Ministry of Education had showcased a number of distance learning opportunities and Chapleau was eager to take part in this initiative since it was having difficulty recruiting teachers for several courses. A number of presentations were made to school boards, some with early buy-in because the latter were already invested in other communities. Others needed

more time to investigate the feasibility of their involvement, or acquire the necessary knowledge before they could be engaged.

3. **A Tele-Health Facility:** The provincial Ministry of Health and a number of regional health networks were approached to initiate tele-medicine and home tele-health projects. Diabetes and hypertension disease management trials were developed, as was local patient online access to specialized health (such as meeting health specialists in larger centers through video-conferencing at the Chapleau hospital).Positive outcomes were realized from these initiatives.

4. **An Online Production System:** The Steering Committee agreed to develop a creative production facility to encourage video and audio productions that could be posted on the Web. The school boards bought into the idea and offered programs to youth and adults on using the technology. When community organizations were challenged to find content, a number of suggestions were made by Bell and Nortel staff such as interviewing seniors who had lived through various forestry practices or were filming the work of researchers who were following patterns of wildlife. Bell and Nortel made a substantial investment in the capital infrastructure and staffing the production center, and the community was asked to find a solution for sustaining the project once the partnership came to an end.

Nortel Learn IT helped develop technology skills in an applied, learner-centric approach that allowed students to create content to demonstrate their ability to apply the knowledge they gained and their capacity to communicate using ICT. According to Nortel, Learn IT "begins with the preparation of Chapleau teachers and trainers to effectively use technology for educational purposes. After their professional development, educators are provided with lesson plans and other training materials that embed the use of technology through hands-on projects. Students use their technical skills to demonstrate their proficiency in the subject matter" (Chapleau Portal, 2011).

Sustainable forms of buy-in and engagement are needed for the development of long-term technology projects and for achieving innovation and synergy. In their evaluation of three wireless projects in Canada, and Albert and LeBrasseur (2010) in a review of engagement in intelligent communities worldwide, Middleton and Crow (2008) pointed out that wider engagement is critical for the development of a more integrated community network. Frohlich et al. (2009) proposed that ICT and literacy skills are important to engagement, and ICT solutions need to cater to social, infrastructural, and political needs as well as be optimized for the benefit of each community (p. 33). By building the capabilities of local citizens, a community is more apt to realize stronger engagement and strengthen its ability to develop innovation and synergy, a process that takes time, education, and communication. Sustainability is a difficult puzzle for small communities and finding diverse applications and users becomes critical to create viability in the long term. Strategic questions need to be asked about what sorts of diversified offerings make sense given current resource capabilities and needs.

Partnership and Finances

As in any partnership, the first step was to understand the extent of the collaboration. This was an asymmetric three-way partnership, which has the tendency to increase the level of complexity, especially in implementation (Bourgeois & Brodwin, 1984). Asymmetry in size and knowledge are known to have repercussions on the vision, governance, culture, and managing communication (Minshall, et al., 2010). Minshall et al. have suggested that asymmetric partnerships can work but need clear actionable plans to resolve inequities, for example, partners need to identify best practices and consider how they will involve all partners to minimize asymmetries. However, Minshall also identified that a partnered project can quickly reach an impasse if the intention is to increase the level of readiness of the smaller partner and one or all of the partners are unwilling or unable to make the investment to do so.

The Fischer (2010) study identified classes of projects that tend to affect outcomes. The Chapleau project is a class B-PPP project, or those that have 'potential for development, but lack sufficient private funding' (p. 109). The Chapleau project was fully funded by Bell and Nortel, and they also offered the equipment and software at no cost as well as the team to manage the trial. Chapleau provided volunteers, community leadership and agreed to sustain some applications such as the community portal. The provincial government provided funds to extend the physical network to First Nations communities. Non-profit organizations (such as schools and hospitals) and businesses were provided with a business case to join the network (such as purchasing fiber connectivity to enhance programming and value for dollar). The project was sustainable as long as all of the resources worked collaboratively, but was less so once some of these key resources left the partnership.

Chapleau was unprepared for the size of the array and the speed at which projects would occur. The Bell/Nortel partnership was well organized, but as in most of these types of ventures, private sector companies tend to be hypersensitive in their protection of intellectual property. Bell and Nortel were on a timeline and were also careful in unveiling their plans and taking time to ensure that they could deliver on promises. Budget details were closely guarded, which sometimes made it difficult for community stakeholders to understand the extent of the project. Minshall et al. (2010) outlined that when there is one partner who is inexperienced with low levels of readiness and a larger partner with an extensive history and a complex organization, the smaller partner

needs more support such as a point of contact to cushion against the larger partner's stronger and slower organizational structure. Some firms have a clear structure while others do not and someone is needed to help navigate the smaller partner to the right resources.

The Minshall article suggests that in cases of asymmetry, an element that may be as important as an efficient point of contact is the availability of a mentor to bridge the gap. Minshall suggested considering choosing someone from an experienced community to fill this role. At the early juncture of this pilot project, there were few comparable examples to provide viable models. Most of the available cases were similar to Fredericton, which had a population of 85,000 (30 times that of Chapleau) and financed its wireless infrastructure by leveraging the investments of large public and non-profit organizations that had purchased fiber in city conduits (ICF files – Fredericton, 2008; Middleton & Crow, 2008; Powel & Shade, 2006). Chapleau was not in a position to fashion its own wireless involvement as Fredericton did (the latter being largely an infrastructure project) and therefore, Nortel staff filled the mentorship position with support from Bell.

Governance and Management

Among the success stories of this project is the fact that Nortel and Bell assigned well-positioned executives to the Chapleau project team and these folks and their team visited the community on a regular basis. As such, the Bell/Nortel team was able to develop an environment of trust with Chapleau and quickly suggested an effective governance structure to stimulate community stakeholder involvement. This included a steering committee and a number of sub-committees responsible for government liaison, communication, research, education, health, and economic development. Although recommendations were made from sub-committees to the steering committee, most of the decisions had financial resource constraints and therefore Bell/Nortel had significant power to influence the chosen solution.

An important challenge throughout all phases of the project was an adequate supply of volunteer resources to sit on committees and perform the necessary work. Although small in size, the community appeared to have a healthy number of volunteers readily available at the Economic Development Corporation (12 people), municipal council and staff (8-10 people), and in its telecommunication committee (6 people). However, the same individuals typically represented half of the membership of these committees, which meant that instead of 26-28 people, there were only 12-14 individuals potentially[1] available to sit on committees. Gaining outside interest from other sectors meant identifying people who did not typically volunteer for this sort of assignment, which would change the structure and culture of committees. It also meant dealing with trust issues and obtaining buy-in to the vision, and also required training new collaborators. These collaboration-building issues take time and are difficult to implement in 12-18 months.

Communication

Chapleau chose to hold sector meetings to discuss the vision of the project and encourage new stakeholders to get involved and provide ideas. Brainstorming sessions were held and citizens were encouraged to write in their ideas to the economic development office. The meetings were well attended and concepts were accepted but it was difficult to encourage community stakeholders to move into action. The Chapleau project team was trying to sell a vision that was not well understood and did not have immediate buy-in from stakeholders. Community stakeholders were not coming to the project with problems, and it was left to the Project Team to find out how the project could help the community, encourage innovation, and lead to new economic opportunities. This proved to be a slow process given that people need time to understand change, including how the telecommunication project would affect them, how it could provide them with new enhanced opportunities, and whether they were prepared to engage.

Introducing a new technology and a number of demonstration applications often means major change for businesses, organizations, and residents. Despite information and training, real change happens when people see a benefit and understand how they can innovate with the technology. Symbolic interaction (Blumer, 1969) posits that people will react to social stimulation based on the meaning it has for them and they will derive meaning from social interaction. The continued communication of project information and the interpretation of meaning from trusted sources such as peers reinforces behavior (study by Fernback, 2005, in distressed neighborhoods of urban communities). Community image and attitudes toward ICT are shaped by individual sources of information such 'neighbors, local merchants, educational institutions, church and the agents of the social services available to them' (Fernback, 2005, p. 486). The relationship between residents and The Chapleau project was fragile—many changes were proposed in education, health, business, and social interactions that required buy-in or engagement. Deriving meaning from the wide range of options was perhaps overwhelming to some and challenging to most, and would need more time than was allotted during the short trial. In the end, a great number of residents and businesses bought into ICT programs and a number of projects (such as the portal, online education, and online health) are still operating today.

Communication was handled by a stakeholder with the knowledge to address each group involved the project. For example, the skills required to communicate with media are different than the skills needed to communicate with government officials, industry sectors or potential volunteers. Bell and Nortel provided much expertise in communication with national media and worked with The Chapleau project committees to manage the communication agenda in all target markets. Chapleau utilized the local newspaper to educate citizens, keep them informed of new developments, and relay information about the portal to register people for services and training programs. It was also used to conduct a survey in order to accumulate a baseline for evaluating results. The mayor was involved in all media launches and releases, meetings with government leaders, and special advisories to the community. Bell/Nortel public relations personnel were involved in government funding programs and national releases. Individual committees prepared their own communications to the sectors they were targeting and each committee included members of the overall steering committee for continuity. The project was continually looking for ways to communicate its messages and evaluate the results. Bell/Nortel established an action register and each committee reviewed the progress of all groups on a weekly basis, which helped keep people on track. During the one-hour weekly call, each person involved in an action reported their progress and made decisions on new action items to be added to the register, and a Bell/Nortel representative provided an executive overview of the activities of other committees when relevant. The same process was followed for other regular committees outside of the register, including the economic development committee and the portal committee.

Politics and expectations need to be managed within the community as part of an effective communication strategy. People tend to build silos and empires, and to build stronger economic foundations many communities require a multi-organizational, multi-sector collaborative approach (Albert, et al., 2009). Some people and organizations may feel threatened and wonder how a new project will affect current operations. Widespread communication efforts (such as diffusion through media or websites) are never enough to dispel these feelings of fear and apprehension. In intelligent community formation, Albert et al. (2009) found that community leaders regularly need to take time to contact potential stakeholders directly and give them the time to accept transformation and agree to become an active participant in the change effort. The Chapleau project utilized its leaders as often as it could to influence decision makers within the key organizations identified. However, the issue of authority became a challenge with decision makers in partnering organizations that were not located in Chapleau. Essentially, the Chapleau agenda was competing against a wider and stronger array of interests and a more pressing list of challenges. These leaders had to balance Chapleau's requirements against their portfolio and prioritize their activities and commitments, and Chapleau had to be prepared to build a good business case for their project.

BENEFITS OF THE PROJECT AND LESSONS LEARNED

In the end, the pilot project ran its course and the wireless equipment was removed, as the community had become more comfortable with the DSL technology that had been installed during the WiFi trial. The pilot was still considered a win for all sides—it allowed the private sector partners to learn some important lessons about their technology and about partnerships with small municipalities, and it allowed a rural community to mobilize their resources and think critically about ICTs opportunities. Whether the community would consider the removal of the WiFi today as the right thing to do is another question. Among the more notable wins of this project at the time were the following:

Community Engagement and Awareness

Education formed an important engagement mechanism for Chapleau as it did for the LaGrange case (Youtie, 2000). In Chapleau, lessons were provided on the community portal, classes offered in schools and community centers, and Nortel provided their Learn IT program. First-run attempts at building engagement included town hall meetings, presentations at community events, the use of media, students training seniors, evening classes, and one-on-one selling and training sessions with businesses. Building awareness took time and there were a number of environmental constraints. For example, to become part of larger tele-health or e-learning initiatives in Chapleau, you needed the approval of bodies outside of the community. If these organizations were well versed with online applications and felt that Chapleau was an effective way to manage a regional initiative, approvals were provided more readily (perhaps within a few months or more). However, when the IT department, often cited as a first point of entry, needed to convince management, which then had to obtain approval from a board (as was the case with some school boards and health organizations) that had to make an investment (whether in the short or long run), patience was required and the only option was to slowly work projects with stakeholders (over a year's time or longer). The purpose of a demonstration project is to learn and therefore the vision was clear in its intent, but the implementation process needs details (who, what, when, and how much), which were not readily available due to a lack of experience and time, and this fact impacted organizational engagement in e-projects.

Increasing Technical Knowledge and Broadband Access

The rollout of WiFi and fiber sped up the deployment of fiber on the backhaul and in-town and between regional organizations (such as school boards and hospitals) that saw The Chapleau project as an opportunity to rollout broadband and take part in a nationally recognized project. The rollout also encouraged a regional management initiative (that was provincially funded) so that servers and data were relocated to the region, creating some

jobs and decreasing costs to municipalities. It also encouraged people to think about the technology and what it could do. Businesses moved from dial-up to DSL and developed websites. Chapleau also indicated that they learned about what was going on in the rest of the world through the talent sent by Bell and Nortel.

Chapleau leaders learned much about ICT infrastructures and were in a better position to make decisions about its future in the knowledge economy. This is evidenced by the many presentations made by Chapleau municipal leaders at conferences and media events. For example, the Treasurer of the Town of Chapleau identified the following rollout successes (Chapleau Portal, 2007):

1. Providing a local support contact.
2. Bundling the wireless network with other services such as a community portal.
3. Avoiding "over design" and "under design."
4. Determining the required level of service considering the needed applications.
5. Creating contracts for local access point support (installation, replacement, and maintenance by local energy services company).

On the other hand, it was evident that small municipalities are not often equipped with ICTs technical capabilities sufficient to run a telecommunication infrastructure as reported by Youtie (2000) in a USA case study on the LaGrange community. Municipal initiatives providing telecom services lack sufficient expertise to keep up with rapid technological change unless the technical infrastructure can be managed by a utility department that has specialists to answer production requirements. In a purely private sector partnership, the firm with more experience is expected to make engineers or other technical staff available to bridge smaller partners' knowledge gaps. Private partnerships do not usually involve partners with little knowledge, and alliances are formed because one partner has a value to provide, often in the form of complementary knowledge. Although Bell and Nortel made substantial staff available for the project, the knowledge gap was extensive and would have required a more permanent solution such as investing in full-time technical personnel. Bell was not interested in extending its network management centre beyond 18 months and Chapleau was going to have a problem finding a suitable pool of personnel to fill the void.

The DSL solution was well known and could be managed and supported externally, taking the weight off the municipality's shoulders. The negative part of the technology removal was in lost innovation, diminished interest in the 'networked community' concept, loss of competition, and future loss of investment, as having WiFi infrastructure might have spurred on a more competitive landscape, encouraged more thought regarding building a networked community and achieved more investment into the community.

If Chapleau had retained the Wi-Fi infrastructure, it would have become the operator of all aspects of the technology, and the leaders of the time felt that this was not within their core strengths. Technical/IT capabilities did not exist in the community and Chapleau would have had to provision everything from support help lines, installation, and maintenance to purchasing and leasing equipment for the backhaul. The fact that DSL had been rolled out meant an immediate competitor and if nothing else had been available, it may have influenced the decision to keep the Wi-Fi equipment under a municipal wing. Chapleau looked at the landscape and could see that carriers were focusing on cellular and using current copper technologies. This exacerbated concerns over the long-term value of rural Wi-Fi when so many other technology solutions were looming. Chapleau was planning a system for commercial, industrial, and residential users, each with diverse needs, making technology management even more multifaceted and fraught with risks. Strategic success is all about execution. It is a little easier when you are in the business and understand the

key success factors and how to establish incentives and control mechanisms to properly implement your plan, and is substantially harder when the needs of very different parties must be taken into consideration and there is no substantial history of how to manage program needs.

Managing the Change Process

Using the Pettigrew change framework, Figure 3 outlines the elements that were positive and negative in the public-private partnership. For example, the partners had similar economic goals, which is part of the context of the change management framework and is therefore identified as a positive (+) in the diagram. On the other hand, partners had varying visions and development goals, with some alignment and some misalignment, and therefore a positive and negative relationship is outlined in the figure. The phenomenon that we are trying to understand is the dynamics involved within the public-private partnership and whether the change process was successfully achieved. At the heart of the project are people who work hard to realize their collective aspirations but each has a different context: what they are striving to achieve, their own conception of what the project is able to achieve, and constraints dictated by a process that is time and resource limited. The framework could arguably demonstrate a number of wins and a number of losses. It is meant to provide an overview of all factors that were discussed in this study which were either positive or negative or perhaps both, and which influenced the success of the project.

The Pettigrew framework below is amended to reflect three influencing factors found in this study: the impact of time, resources, and capabilities. Each of these factors overarches and influences all three components of the model in one way or another. As was outlined in this chapter, time had a negative influence, as an 18-month project timeline meant it was difficult to achieve the desired change. Capabilities had a positive

Figure 3. Change management framework

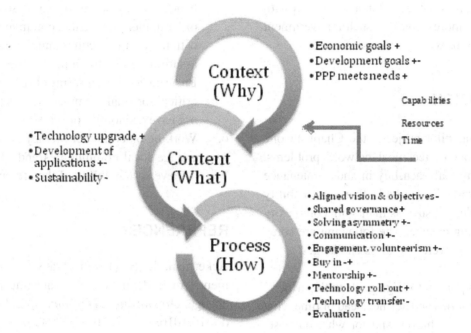

and a negative influence on the project, as Bell and Nortel certainly brought much in terms of capabilities and allowed the process to flow, but capabilities did not have time to filter to the community, thereby creating a negative influence. Resources had a positive influence in the short term and a negative influence in the long term.

The Chapleau Treasurer also identified additional success/challenge factors deemed necessary for the sustainability and repeatability of Project Chapleau, including: a) municipal engagement, which includes identifying stakeholders, creating business model framework, and analyzing municipal strategy; b) exploring partnerships or joint ownership; c) network cost analysis; and d) identifying operational and financial objectives. Over-design was identified as particularly important so that small municipalities do not fall behind in their cellular, fiber or broadband services to home and businesses. Economic development change is a large challenge and takes time, but communities also need to pay attention to sustaining quality of life and basic infrastructure, and

telecommunications is a critical aspect of this. When infrastructure falls behind it is more costly to update, and if population numbers are dwindling then consumers bear a larger proportion of the cost through higher taxes for major overhauls. Chapleau is one of many rural communities that need to upgrade costly municipal infrastructures in a time of declining revenues. This hampers their ability to make investments and the community becomes less attractive to development.

The partnership had a facilitating role in helping to shape and negotiate meaning in the community, helping develop a shared vision of ICTs use, and in creating developmental opportunities (supporting research by Fernback [2005] on the benefits of PPP in urban regeneration). Hall (2003) claimed that communities can build social capital by altering self-image and dominant social structures. This proposition is demonstrated each year in intelligent or smart cities (ICF, 2012) that do exactly that, using ICTs to fashion a new image and build collaborative models for e-health, e-learning, e-entertainment and so forth. Some rural communities are able to demonstrate this level of progress but are often also blessed with substantially more resources (such as government subsidies) or proximity to larger centers.

CONCLUSION

As in similar pilot projects, the Chapleau one had a number of technical network problems, including physical reach, buy-in, and maintenance of the network. It also had other sustainability related challenges such as managing complexity and governing change. The following are some suggestions to overcome these challenges:

1. It is important to align goals with partners' expectations and define ahead of time what would be achieved and for what purpose. If goals are misaligned, unrealistic, or too broad, it is difficult to agree upon or achieve them in the given timeframe.
2. There is a need to appoint the right people to manage bureaucracy and deal with the organizational issues of partners.
3. Public-private partnerships should have a mentor who can bridge the gap between large and small partners along with public and private ones, as needed. The Chapleau project was well supported by the team assigned by Bell and Nortel. However, it could have potentially benefited from a municipal mentor to help better define achievable objectives and work on sustainability given rural community constraints.
4. Developing a plan to manage technical competency is important if transfers of technology is contemplated. Again, the right resources and approaches must be used so the receiving partner's needs are met. In addition, the community should build an adequate pool of human resources that may go beyond regular staff and include volunteers.

5. The project pan should include efficient benchmarking and information sourcing in order to identify feasible community applications, promote community engagement, assure buy-in from community stakeholders. Ensuring a sufficient array of applications is critical for small communities, as they will not have economies of scale.
6. Working on computer literacy is critical to engage local communities and, therefore, achieve sustainability of the project.

REFERENCES

Ackerman, L. S. (1982). Transition management: An in-depth look at managing complex change. *Organizational Dynamics*, *82*(1), 46–66. doi:10.1016/0090-2616(82)90042-0

Agrawal, R., Gupta, A., & Gupta, M. C. (2011). Financing of PPP infrastructure projects in India: Constraints and recommendations. *The IUP Journal of Infrastructure*, *9*(1), 54–57.

Albert, S., Flournoy, D., & LeBrasseur, R. (2009). *Networked communities: Strategies for digital collaboration*. Hershey, PA: IGI Global. doi:10.4018/978-1-59904-771-3

Albert, S., & Lebrasseur, R. (2010). Citizen engagement in the networked community. *The International Journal of Knowledge, Culture, and Change Management*, *10*(4), 55–67.

Blumer, H. (1969). *Symbolic interactionism: Perspective and method*. Englewood Cliffs, NJ: Prentice Hall.

Bourgeois, L. J., & Brodwin, D. R. (1984). Strategic implementation: Five approaches to an elusive phenomenon. *Strategic Management Journal*, *5*, 241–264. doi:10.1002/smj.4250050305

Castells, M. (2006). *The network society: From knowledge to policy*. Washington, DC: Johns Hopkins Center for Transatlantic Relations.

Chapleau Portal. (2007). *The Chapleau project highlighted in Nortel e-seminar*. Retrieved December 26, 2011, from http://www.chapleau.ca/portal/en/connectingchapleau?paf gear_id=1000 025&itemId=3100038&returnUrl=%2Fportal%2 Fen%2Fconnectingchapleau%3Bjsessionid%3DJ IYQRVKO2BQBNTRPH3XHLRQ

Chapleau Portal. (2011). *Nortel learn IT*. Retrieved December 26, 2011, from http://www.chapleau.ca/portal/en/connectingchapleau/nortellearnit

Clarke, P., & Healy, K. (2003). Investigating aspects of public private partnerships in Ireland. *The Irish Journal of Management, 24*(2), 20–30.

Demirag, I., & Khadaroo, I. (2011). Accountability and value for money: A theoretical framework for the relationship in public-private partnerships. *Journal of Management & Governance, 15*(2), 271–296. doi:10.1007/s10997-009-9109-6

Fernback, J. (2005). Information technology, networks, and community voices – Social inclusion for urban regeneration. *Information Communication and Society, 8*(4), 482–502. doi:10.1080/13691180500418402

Fisher, K. (2010). ACT4PPP – A transnational initiative to promote public-private cooperation in urban development. *European Public Private Partnership Law Review, 5*(2), 106–111.

Frohlich, D. M., Bhat, R., & Jones, M. (2009). *Democracy, design, and development in community content creation: Lessons from the StoryBank project*. Los Angeles, CA: University of Southern California.

Hall, P. M. (2003). Interactionism, social organization and social processes: Looking back and moving ahead. *Symbolic Interaction, 26*(1), 33–55. doi:10.1525/si.2003.26.1.33

ICF. (2012). *Intelligent community forum: Smart 21*. Retrieved December 26, 2011, from https://www.intelligentcommunity.org/index.php?src=gendocs&ref=Smart21&category=Events&link=Smart21

ICF – Fredericton. (2008). *Top 7 intelligent community award*. Retrieved August 25, 2008, from http://www.intelligentcommunity.org/index.php?src=gendocs&ref=Top7_ 2008_Frederict

Kwak, Y. H., Chih, Y., & Ibbs, C. W. (2009). Towards a comprehensive understanding of public private partnerships for infrastructure development. *California Management Review, 51*(2), 51–78. doi:10.2307/41166480

Lehr, W., Sirbu, M., & Gillett, S. (2006). Wireless is changing the policy calculus for municipal broadband. *Government Information Quarterly, 23*, 435–453. doi:10.1016/j.giq.2006.08.001

Lewin, K. (1948). *Resolving social conflicts: Selected papers on group dynamics*. New York, NY: Harper & Row.

Lewin, K., & Grabbe, P. (1945). Conduct, knowledge and acceptance of new values. *The Journal of Social Issues*, 2.

Middleton, C., Clement, A., Crow, B., & Longford, G. (2008). *ICT infrastructure as public infrastructure: Connecting communities to the knowledge-based economy & society*. Final Report of the Community Wireless Infrastructure Research Project. Retrieved January 26, 2012, from http://www.cwirp.ca/files/CWIRP_Final_report.pdf

Middleton, C., & Crow, B. (2008). Building wi-fi networks for communities: Three Canadian cases. *Canadian Journal of Communication, 33*, 419–441.

Minshall, T., Mortara, L., Valli, R., & Probert, D. (2010). Making Asymmetric partnerships work. In *Research Technology Management* (pp. 53–63). Industrial Research Institute Inc.

Nortel. (2005). *Bell Canada and Nortel deliver advanced broadband services and applications to northern Ontario community*. Retrieved December 27, 2011, from http://www2.nortel.com/go/news_detail.jsp?cat_id=-8055&oid=100190608

Nortel. (2011). *Chapleau case study*. Retrieved December 26, 2011, from http://www2.nortel.com/go/news_detail.jsp?cat_id=-9252&oid=100204737&locale=en-US&NT_promo

Pade-Khene, C., & Swery, D. (2011). Toward a comprehensive evaluation framework for ICT for development evaluation – An analysis of evaluation frameworks. In *Proceedings of the European Conference on Information Management & Evaluation*. IEEE.

Parikh, T. S. (2009). Engineering rural development. *Communications of the ACM, 52*(1), 54–63. doi:10.1145/1435417.1435433

Pettigrew, A. (1987). Context and action in the transformation of the firm. *Journal of Management Studies, 24*(6), 649–670. doi:10.1111/j.1467-6486.1987.tb00467.x

Pettigrew, A. (1992). The character and significance of strategy process research. *Strategic Management Journal, 13*, 5–16. doi:10.1002/smj.4250130903

Powell, A. (2011). *#Fail: What we learn from failed tech projects*. Retrieved January 24, 2012, from http://lse.academia.edu/AlisonPowell/Talks/49255/_FAIL_What_we_learn_from_failed_community_technology_projects

Powell, A., & Shade, L. R. (2006). Going wi-fi in Canada: Municipal and community initiatives. *Government Information Quarterly, 23*(3/4), 381–403. doi:10.1016/j.giq.2006.09.001

Pritchard, W. (2004). *Wireless networks: Opportunities and challenges for foothill college*. White Paper. Retrieved from http://fhdafiles.fhda.edu/downloads/etsfhda/WirelessWhitePaper.pdf

Ramirez, R. (2007). Appreciating the contribution of broadband ICT with rural and remote communities: Stepping stones toward an alternative paradigm. *The Information Society, 23*, 85–94. doi:10.1080/01972240701224044

Schaffer, B. S. (2007). The nature of goal congruence in organizations. *Super Vision, 68*(8), 13–17.

Surana, S., Patra, R., Nedevshi, S., Ramos, M., Subramanian, L., Ben-David, Y., & Brewer, E. (2008). Beyond pilots: Keeping rural wireless networks alive. In *Proceedings of NSDI 2008: 5th USENIX Symposium on Networked Systems Design and Implementation*. Retrieved January 26, 2012, from http://www.usenix.org/event/nsdi08/tech/full_papers/surana/surana.pdf

Ubilium. (2012). *Project Chapleau*. Retrieved June 13, 2012 from http://www.ubilium.com/chapleau.html

Vroom, V. C. (1964). *Work and motivation*. New York, NY: Wiley.

Youtie, J. (2000). Field of dream revisited: Economic development and telecommunications in LaGrange, Georgia. *Economic Development Quarterly, 14*(2), 146–153. doi:10.1177/089124240001400202

ENDNOTES

[1] Availability does not entail interest and one can generally reduce the number of interested parties by at least 50-75% in any given committee.

Chapter 11
Challenges and Opportunities of ICTs for Rural and Remote Areas

Yasuhiko Kawasumi
ITU Association of Japan, Japan

ABSTRACT

Broadband Internet access is important for rural and remote areas to access e-commerce, e-government, e-learning, e-healthcare, Internet telephony, and other online resources. This chapter discusses the main opportunities and challenges of developing telecommunication infrastructures for rural and remote areas. In addition, affordable high-speed Internet access is important for communication (voice, data, Internet, etc.), community empowerment, job search and career development, and weather and climate monitoring. Expanding Internet access to rural areas, in particular, faces a number of challenges, such as lack of sustainable and affordable power supply, limited funding opportunities, and selecting a suitable technology. The authors discuss these issues using anecdotic evidence from a number of projects and case studies developed in the last 30 years by International Telecommunication Union (ITU). They conclude the chapter with recommendations of successful practices and policy guidelines.

INTRODUCTION

The International Telecommunication Union (ITU)[1] is the agency of the United Nations responsible for Information and Communication Technologies (ICTs). In particular, it strives to provide ICTs to underserved and rural communi-

ties all over the world, allocates radio spectrum and satellite orbits worldwide, and develops network standards to ensure seamless interconnection. ITU is committed to include all people in the information society and to support the rights of people to communicate. In other words, one of its objectives is to remedy the imbalance of telecommunications

DOI: 10.4018/978-1-4666-2997-4.ch011

between "haves and have-nots" in both developed and developing countries. For that purpose, ITU established the Telecommunication Development Sector (ITU-D) which is responsible for:

1. Assisting member countries in accessing and mobilizing technical, human and financial resources needed for the implementation of ICTs and promoting access to their service;
2. Working on bridging the digital divide;
3. Promoting the extensions of the benefits of ICTs to all human beings; and
4. Developing and managing programs that facilitate information flow geared to the specific needs of developing societies.

Its efforts go back to almost 30 years. In particular, the United Nations General Assembly chose 1983 as the "World Communications Year." This was when the ITU established the Independent Commission for World-Wide Telecommunications Development chaired, by Sir Donald Maitland. Later it is called Maitland Commission, which issued the famous "The Missing Link Report" or "The Maitland Report" in January 1985 (Maitland Commission, 1985). The report identified the communications gap between developed and developing countries. The target set in the report was that by the early part of the 21st century virtually the whole humankind should be brought within the easy reach of a telephone and of all the benefits it can bring. Since then, higher investment in telecommunication has been allocated, coupled with the emergence of new technologies and innovative strategies, as well as the general understanding of the socio-economic effects of communications infrastructure, have led to a remarkable degree of telecommunication development observed in most of developing countries throughout the 1990s. Whereas the goal set in the Maitland Report is deemed to be a realistic and achievable target, the progress of digital technologies, and the proliferation of

Internet related services and applications have brought to us new challenges.

A new goal namely to connect all humankind on this planet was set by 2015 through two phases of the World Summit on Information Society (WSIS) 2003 in Geneva and 2005 in Tunis. In the meantime the World Telecommunication Development Conference in 1998 (WTDC-98, Valletta, Malta) decided to create a Focus Group on topic 7 (later it is called FG7) to identify new technologies designed to fulfill the needs of developing countries and to tackle the issue, particularly in rural and remote areas. The final report entitled "New technologies for rural applications" was published in February, 2001 by the ITU (ITU-D Focus Group 7, 2001). The report highlighted emerging technologies suitable for use in rural and remote areas. It also identified potential applications for tele-education, tele-medicine, development of small businesses, emergency support, disaster relief and environmental monitoring, etc. It concluded that wireless networks, combined with packet-based Internet Protocol (IP) networks can be used in rural and remote areas of developing countries because of its cost effectiveness, fast roll-out time and capability of affordable and sustainable multimedia services. In addition, the report includes recommendations such as creating a handbook for renewable energy and small-scale power systems for rural ICTs. ITU's challenge to tackle the rural ICTs development issues was succeeded by the ITU-D SG2 Rapporteur's group on Q10 "Telecommunications/ICTs for rural and remote areas" since 2002 (ITU-D SG2, 2004). During three study cycles (each one is four years), it conducted global survey on the status of telecommunications in rural and remote areas. It also collected case studies, issued analytical reports, established library of cases on the website of ITU-D[2].

The following are the key issues identified from the 60 case studies:

1. Though demand on voice communications is met by the rapid penetration of mobile phone services in rural and remote areas of developing countries, needs for multimedia services over the broadband infrastructure are emerging.

2. Broadband wireless infrastructure is highly demanded and its implementation is planned in most of developing countries in their rural and remote areas.

3. The development of ICTs Infrastructure is the key driver for empowerment of the economy of rural and remote areas in developing countries. This issue should be recognized by policy and decision makers.

Social Settings of Rural Areas

Development experts believe that deploying ICTs in rural and remote areas will significantly empower their dwellers. This is because Internet access enables villages, community centers, schools, and individuals to access online resources and services. In other words, rural economy will be vitalized remarkably by various e-applications and services. Migration of population to urban areas may be suppressed and hopefully contribute to the eradication of poverty and hunger since people migrated to urban areas encounter unemployment or jobless situation with consequent poverty and hunger. Many job opportunities are created because of the deployment and operation of ICT infrastructure and community centers in rural and remote areas.

According to the report of global survey on the status of rural and remote areas conducted by ITU-D SG2 Rapporteur's Group on the question for telecommunications of rural and remote areas, the following are the findings about the reality of rural and remote areas:

1. Lack or inadequate basic enabling infrastructure such as regular electricity supply;

2. Absence of telecommunications infrastructure;

3. The relative high cost of telecommunication infrastructure if existed;

4. Geographic access problems due to harsh terrain, poor quality of roads and transportation network and remoteness of some rural communities;

5. High cost of physical access and equipment installation due to any combination of the above geographically related issues;

6. Low income, lack of disposable income and relative poverty of rural population;

7. Low population density of target populations. These target populations usually live in small rural populations or in sparsely populated communities that are geographically away from one another;

8. High degrees of illiteracy in some rural areas;

9. Low levels of awareness (if any) of the benefits of modern ICTs, leading to low current demand in some areas;

10. Higher rate of vandalism and theft of infrastructure components such as cables and solar panels in rural areas;

11. Unproven feasibility of telecommunication services in rural areas;

12. Overall lack of funding (both public and private); and

13. Lack of energy sources.

Because of the above difficulties, migration of population from rural areas to urban areas is globally a remarkable phenomenon as shown in the Figure 1.

According to the ITU-D analytical report, many population groups are digitally left behind in rural areas. These groups include children and young people (under 16 years old), elderly people (older than 55 years old) and women. It also shows that males aged between 15-55 years old leave rural areas and migrate to urban ones seeking more attractive job opportunities and better life

Figure 1. World urbanization prospects of the UN population division

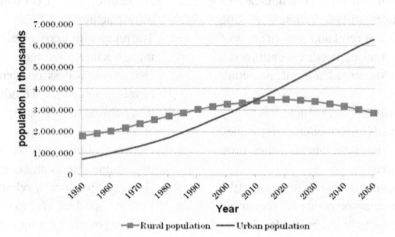

quality. In 2011, about half of the world population (around 3.5 billion) was living in rural areas, as shown in the Figure 1. This proportion was less than a billion in 1950 worldwide. The UN also predicts that rural population will be less than three billion in 2050 while urban population will exceed 6 billion. In addition, migration of rural residents to urban areas continues in every region of the world. Meanwhile, the migration of population is slow in the case of rocky areas and Small Islands in Developing States (SIDS) because they consist of many outer or remote small islands. This is because of the difficulty and cost of leaving rural and remote areas. In the case of Nepal, cross point of rural and urban population is ex-

pected after the 2050, as shown in Figure 2. However, in African and Asian regions cross point of rural and urban population is prospected around 2030, according to the UN Population Division.

Moreover, living standards of rural dwellers is lower than that of urban residents because of the limited economic activities and opportunities in rural areas. In addition, livelihood in rural and remote areas is based on the agriculture life style with low cash income. Furthermore, in rural areas there is limited information about market prices of products and lack of ICTs services. This makes their lives even more difficult compared to those who live in urban areas.

Figure 2. Urbanization prospects in Nepal by UN population division

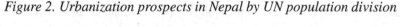

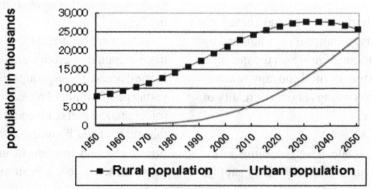

Information Needs for Rural and Remote Areas

Using ICTs to empower rural dwellers has become an urgent need in every developing country in order for them to improve their quality of life. As shown in Figure 1, half of world population lives in rural and remote areas. The capital investment for developing ICTs infrastructure in rural areas is considered rather modest compared to other large-scale infrastructures such as transportation, electricity, healthcare services, etc. On the other hand, the effectiveness of the investment on ICTs in rural and remote areas is expected to be remarkable according to recent case studies. Currently, policy makers in most of developing countries consider the development of ICTs infrastructure in rural areas a top priority. This is evidenced in their national master plan shown in the contributed papers submitted to the ITU-D meetings and the presentation by policy makers at ITU events. They particularly give priorities to e-education, e-health and e-government services. The WSIS Tunis 2005, and agencies of UN identified some other e-applications. The WSIS Geneva 2003 (Declaration of principles, and Plan of actions) and WSIS Tunis 2005 (Tunis Agenda for the information society) clearly identified a roadmap for achieving the information society by 2015 (WSIS, 2003; WSIS, 2005). Its purpose is to harness the potential of ICTs to promote the UN Millennium Development Goals. These goals are:

1. Eradicating extreme poverty and hunger;
2. Achieving universal primary education;
3. Promoting gender equality and empower women;
4. Reducing child mortality;
5. Improving maternal health;
6. Combating HIV/AIDS, malaria and other diseases;
7. Ensuring environmental sustainability; and
8. Developing a global partnership for development.

CASE STUDIES

The following is a brief discussion of selected case studies from the ITU-D case library:

The Case of Republic of Marshall Islands

The discussed data about this case is extracted from the presentations of the APT Development forum and ITU case library, and from our own observation while participating in the development of the project.

Marshall Islands (RMI) have 29 atolls, in which each atoll has numerous islands on it, as shown in Figure 3. Its population is about 60 thousands and its area is 181 km^2 with population density of 326 pers./km^2. We implemented the Mejit Island project for the purpose of educating the society about the benefits and services of ICTs they were deprived for couple of decades. Another purpose is to build a network for the government can use to provide a wide range of electronic service and connect all Marshall Islands.

The Mejit Island is one of the many under developed islands in the Marshall Islands that has 80 households (about 300-400 inhabitants), living on 1.86 km^2. The inhabitants have limited economic opportunities, poor entertainment facilities, and lacks of means of livelihood. Most of them only rely on their daily sustenance from crops and the ocean resources.

Currently, GSM mobile phone services are provided in Majuro (the Capital of RMI), Ebeye, Jaluit, Wotje, Kili, and Rongelap atolls/Island of the above map. In addition, Telecenters are built on Enewetak, Mejetto & Santo. Information services are offered in all of the aforementioned islands and will be extended to other 20 islands. APT funded the project proposal for Mejit Island and the project has started providing services in 2012. Currently, the residents of the Mejit Island are using the HF radio to communicate to other islands, especially Majuro, for commercial or personal related com-

Figure 3. Map of Republic of Marshall Islands (RMI)

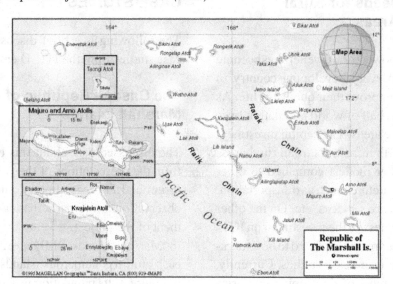

munications. Mainly, there is a huge shortage of computer access in the schools in Mejit. Thus, this limited access to computers deprives students from current technologies and electronic facilities used in e-learning. Figure 4 shows the configuration of the used satellite network.

Thanks to this project, the leaders of the society and government officials have become aware of the usefulness of ICTs for conducting the daily life affairs of individuals. Through ICTs, Marshallese can preserve the culture and stimulat-

ing small businesses in the area. This project is expected to drive the installation of telecenters in other locations that are not being served by local telecommunications operator. Telecenter is another connectivity alternative for islands and rural areas that are digitally disconnected from the Internet and the rest of the world. They enable people to share communications equipment as well as personal computers.

The project provides the following services:

Figure 4. Configuration of the satellite link connecting Majuro and Mejit islands

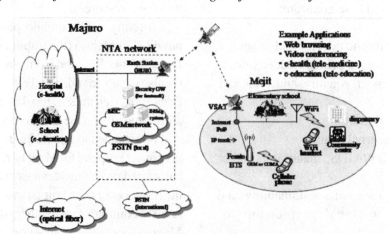

Public Communications

Internet access would be available at the telecenter and can be accessed by "WiFi internet access scratch card" for those who have their own laptops such as individuals for private or business use and tourists.

Computer Laboratory

With a separate acquisition for laptops, the computer laboratory will be available for e-learning such as basic computer to facilitate the delivery of training exercises and course assignments. Teachers will be trained for the usage of the facilities. As for the culture preservation, the educational material can be posted or archived at Youtube. com or any legitimate video streaming hosts that can be utilized for e-learning in other Islands or in the future. The material can also be fed to the Marshalls Broadcasting Cable Company (MBC) or Marshall Island Air TV (MHTV).

Femtocell Base Station

Femtocell base station is a low-power and compact cellular base station. With the state-of-the-art femtocell technology, cellular access is available in outer islands. Domestic and international telephone calls have become available through satellite link and integrated system consisting of a C-band earth station system and a DAMA14 system.

Healthcare

The network is also used for delivering healthcare and providing tele-consultation from a hospital in Majuro through Internet at affiliated health posts.

Key Partners and Their Contributions

1. **The RMI Government:** Its responsibilities are to build the telecenters with its own local manpower and locally purchased construction materials thus giving employment to locals.

2. **The Mejit Local Government and Mejit Senator:** Their responsibility is to find ways and means for shipping equipments from the Majuro port to the Mejit Island (400 km in distance). They should also assist in transporting Japanese experts and local experts from Majuro to the Mejit Island.

3. **The Ministry of Transportation and Communications:** It is responsible for establishing training facilities. In addition, it is responsible for ensuring the sustainability of operation of the systems after delivery and to utilize the results of the project appropriately for enhancing the development of ICTs infrastructure in other Islands. Recently, other concerned ministries and organizations signed a memorandum of agreement to contribute and/or benefit from the project. The Ministry of Transportation would also create subsidiary agreements with other stakeholders to ensure the sustainability of the project. In addition, it is supposed to facilitate encourage the community that ICT will greatly help in putting up small-scale businesses.

4. **The Marshall Islands National Telecom Authority:** It should purchase bulk bandwidth and run related promotion programs. The telecenters can purchase a scratch card for 9 dollars for a value of 10 dollars or similar purchase arrangement with MINTA.

The Case of Nepal Wireless

The purpose of Nepal Wireless is to serve the Pokhara-Mustang area in the Himalayan Mountain using WiMAX. This data discussed in this case is extracted from APT development forum presentation and the ITU case library and from our own observation while participating in the development of the project.

Nepal is mostly a rocky land with most of the highest peaks of the world including the Mount Everest. This means that digging for expanding landline networks would be so expensive. The population of Nepal is 25.3 million people scattered throughout mountains, hills, valleys, and plains. The Gross Domestic Product (GDP) per capita is 311 dollar in 2006, with an average growth rate of 2.9 percent in the past three years. The poverty rate was 31 percent in 2004. Agriculture is still the main economic activity in Nepal and it employs more than 80 percent of the population and provides 38 percent of GDP. The lack of physical infrastructure (e.g. transportation, power, etc.) and insufficient ICTs are key reasons for the low economic performance of the country. Nepal does not have many industries. Therefore, most of the young men have to seek job opportunities abroad to make money for themselves and their extended families. That is why remittances have been the main source of foreign currency. Tourism is also another major source for foreign currency.

The telecommunication connectivity penetration rate nationwide reached 6.48% in 2005; for landlines it was 2.46% and for mobile telephones, 4.03%. This rate is one of the lowest in the world. Investment in ICTs is very much skewed in favor of major cities. Therefore, the rural coverage is very low with a penetration rate of only 0.06%. More than 40% of rural districts of Nepal have almost no telecommunication services. Incumbent operator Nepal Telecom (NT) has planned to rollout 3.5 million CDMA line for voice and data services. Although CDMA is capable of contributing to the improvement of rural telephone services, it is not compatible with high-speed data services.

The project was implemented to enhance the Nepal Wireless Project and was funded by Asia-Pacific Telecommunity program. The project is serving villages in remote areas, which are far from cities. Actually, the villages have been chosen in Mustang districts in the northern side of the Himalayas and near Tibetan plateaus. The health clinics and government schools which are connected do not have enough qualified health workers as well as qualified school teachers. These districts are producing lot of agro-products such as apples, potatoes and vegetables, however, farmers are not able to receive reasonable prices and access market for their products.

The Nepal Wireless project was started from grass root level in 2001 to provide mainly communication services for the people living in the remote villages of Myagdi district. Now it is providing educational support to the students and teachers, health services to the villagers through telemedicine, local e-commerce service through local Web server, and VoIP phone service to the villagers. Because the project has penetrated remote areas where no service providers have reached, the benefit that the villagers are getting cannot be just measured, particularly when we consider its financial success. Moreover, it is not simply duplicating services that are offered elsewhere but finding innovative ways to optimize the benefits of the wireless technology. Other than providing some basic services to villagers, it is helping to save both time and money of the people because the villagers do not need to go to the cities anymore to seek the services that the project is providing.

As noted before, before the wireless network was built in the region, there was no telecommunication infrastructure. The nearest telephone and fax service that was available that time from Nangi village was six hours walk downhill to a small town called Beni. In addition, the nearest Internet service was available in Pokhara city, which is another four hours bus ride from Beni. There was no modern communication system between the villages as well. Villagers used to walk by themselves to other villages to communicate with their relatives.

In the enhanced project, network backbone is deploying WiMAX equivalent equipment (5.8GHz) with extension from PoPs to villages by WiFi (2.4GHz). Figure 5 shows the WiMax network in Nepal Wireless.

Asia-Pacific Telecommunity (APT) has funded this project for the following reasons:

1. To maximize the benefits of the technology to the rural population;
2. To optimize the benefits of ICTs for rural communities;

Figure 5. The WiMAX network of Nepal wireless

3. To facilitate the delivery of tele-education and tele-training, sharing of online educational content, and conducting video conferencing between teachers and students;

4. To provide tele-medicine service through online or offline consultation using the Internet and Intranets;

5. To enable rural communities to use advanced communication tools such as e-mail, Internet Telephony, chatting venues, and social media;

6. To give community members to benefit from information sharing tools, webpages, bulletin boards, group discussion venues, research works; and

7. To promote e-Business, Internet banking transaction, remittance services, etc.

The power generation facilities (solar pane and windmill power generator) are installed at the hilltop relay station at about 3000m high. Transporting materials and installation by local volunteer staff were hard work to be performed only under the appropriate weather condition.

The network has used one single backbone to connect all the proposed eight villages and connected them to the nearest city as well. It has

deployed WiMAX radios for backhaul and access points. However, Wi-Fi radios were deployed for last mile connections in two villages. The reason WiMAX is deployed because of its long range and high data throughput capacity. Also the reason Wi-Fi is deployed for the last mile connectivity in two places is because the distance of those two villages from the relay stations are closer and it has lower cost than WiMAX.

E-Applications

The project provides the following applications:

The leaders of the project built a content and data Linux server in Pokhara, which connects all other villages to the server. All the applications such as content management, VoIP, local e-commerce and money transfer, and e-learning applications are installed in the server.

1. **Education and Training:** The project increases educational opportunities in the rural communities by creating tele-teaching and tele-training program and by making e-learning materials accessible to students, teachers, and villagers through e-libraries created by different organizations. It also

provides training in each village to overcome the shortage of skilled human resources. Project leaders train up to minimum two people from each village.

2. **Healthcare:** It provides quality tele-medicine and tele-education services in those villages, high quality and already proven video conferencing equipment and new medical technologies such as digital vital sign monitoring system, electro-cardiograph and echo-graph, etc. To connect the rural health clinics and health workers to the city hospitals in order to provide quality medical assistances through telemedicine program and make available the healthcare in the rural communities by virtually bringing medical doctors.

3. **Communication:** Each of the five villages will have one Internet enabled communication centers having four computers, one multi-function printer, one digital camera, one VoIP phone, and one color printer. These communication centers will provide e-mail/Internet, VoIP phone, bulletin boards, and other online social forums. It helps villagers to communicate using the Nepali language.

4. **E-Commerce:** The network also enables villagers to buy and sell their products in the local market through local intranet. It also facilitates online money transfer services.

5. **Job Creation:** One of the key purposes of the project is to generate jobs in the rural areas for young generation through ICT related services such as communication centers, VoIP phone services, remittance services, and virtual ATM machine and the operation and maintenance of the facilities.

6. **Weather and Climate Monitoring:** The network helps researchers to collect data for the researchers on climate change monitoring projects and provide real time weather information of the air route between Jomsom and Pokhara for airlines companies. The data obtained by field servers are made visible on the Internet.

More than 60 case studies are collected since 2002 by the questionnaire format, which are posted on the ITU-D case library. These case studies are analyzed by the ITU-D SG2 rapporteur's group (ITU-D SG2, 2004). Variety of case studies is analyzed for identifying successful practices. 20 case studies collected during the study cycle 2006-2010. There was 8 cases from Asia and pacific, 6 cases from Americas, 4 cases from Africa and 2 cases from Europe. It is important to note that no case studies were collected from Arab countries.

The collected case studies show that transmission media deployed in each country is different depending on the geographical landscape. In addition, a combination of media is also found in the cases. Provision of broadband connection is demanded to meet the emerging multimedia applications in every region.

To provide multimedia applications and services, last mile access line should also be broadened to accommodate IP platform for the emerging services of the Next Generation Networks (NGN). Wireless Local Loop (WLL) has become popular for its decreasing cost and the increasing rollout observed in the collected case studies (see Figure 6).

Another study cycle started since 2010 for the purpose of collecting additional case studies. Currently, the rapporteurs group is analyzing these case studies to identify emerging technologies, applications and services deployed over the networks extended in rural and remote areas.

EMERGING OPPORTUNITIES

Suitable Technology

Choosing a specific technology should be based on given criteria that have been identified considering the environment and the social settings of selected sites in the target countries. Examples of these criteria include cost effectiveness, short time for rollout, capability, affordability, and sustainability of services. Currently, broadband

Figure 6. New trends to respond to the needs of various applications and services

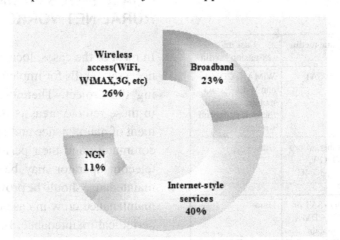

wireless technologies are highly demanded in rural and remote areas of developing countries. A minimum speed requirement of 256/512Kbps bit rate should be taken into account for the provision of multimedia services such as video or image transmission. These multimedia services are necessary for e-conference, e-consultation, and e-learning, etc.

Mobile technologies are penetrating even in rural and remote areas, which overcome the barriers facing landline infrastructure. However, the 2.5G technology is still prevailing in developing countries, which is limiting the transmission speed (bit rate) for Internet services and multi-media services.

According to the statistics of ITU as of 2011, global mobile phone users reached 5.9 billion. The penetration of mobile broadband users is only 1.2 billion (one 5th). However, this penetration is only four percent in Africa compared to one percent for fixed broadband penetration. In the meantime, the 3G, CDMA, and WiMAX broadband wireless technologies are highly demanded for connectivity in rural and remote areas. WiFi is deployed for the last mile connectivity in many countries. Use of satellite communications combined with VSAT (Very Small Aperture Antenna) is also another option for remote areas, geographically harsh sites, and Islands. The satellite channel could be

redistributed or extended more using the WiFi technology. There are case studies deployed Over Power Line Grand Wire (OPGW) coupled with fiber optic or fiber backbone network, in addition to WiMAX, CDMA and WiFi extensions. In the case of satellite communication, the cost of transponder lease is the bottle neck. Sharing transponders among a group of countries is a solution for reducing cost. Use of frequencies is another key issue that needs to be solved. It depends on the frequency management and policy of the country, deployed wireless technology, and satellite transponders. C-band and Ku-band transponders are common. Inmarsat or other satellite services provides voice and data (Internet) services globally such as INMARSAT BGAN services, etc. Table 1 shows the backbone media suitable for different types of countries and compatible last mile extensions. For instance, large and vast countries are best served using satellite backbone. This backbone could be extended using WiMax, WiFi, and 3G/4G cellular network.

The deployment of Long Term Evolution (LTE) technologies for mobile services is increasing in the studied countries. LTE has the capability to provide high data speed compared to fiber optic cables. However, it may not be cost effective for implementation in rural and remote areas.

Table 1. Technology options for transmission media

Type of countries	Backbone media	Last mile extension media
Vast/large size countries	Satellite/VSAT	WiMAX, CDMA, and WiFi, 3G/4G mobile, Nanocell, Picocell Femtocell wireless.
Small/middle size countries	Optic Fiber, Micro wave, OPGW, WiMAX, 3G/4G mobile	Ditto
Land lock countries	Satellite/VSAT or WiMAX, CDMA, 3G/4G mobile	Ditto
Islands and outer islands	Satellite/VSAT	Ditto

E-Applications for Rural Areas

Table 2 presents the e-applications identified in the Tunis 2005 Agenda for the information society.

As mentioned in the section above, three major applications are e-education, e-health, and e-government (administrative services) to be provided in rural and remote areas by ICTs. Video- conferencing is useful for delivering e-learning, e-health, and consolation. It saves travel costs and facilitates sharing of human resources. However, the broadband connection for video transmission over IP-based platforms needs an Internet speed between 256 Kbps or 512 Kbps. There are other options of e-services observed in the collected case studies such as e-commerce (on line shopping, electronic money remittance, e-banking, etc.), electronic library, e-services for disabled people, etc. Table 2 shows a list of e-applications and possible facilitators recommended in Tunis World Summit for Information Society.

CHALLENGES FACING RURAL NETWORKS

In most of the cases, local communities lack the necessary skills for implementing and maintaining the project. Therefore, providing training in these remote areas is the key issue. Assignment of maintenance and operation staff in rural community and their periodical training by the telecom operator may be required. High-level maintenance should be provided by the operator's maintenance crew in case of serious failure and periodical maintenance. Partnership agreement/ MOU between community/telecom operators and ISPs is preferred for the sustainability of network facilities and customer premise equipment. Rural community participation is also recommended by investment or by provision of human resources, spaces to accommodate ICT facilities and/or community center buildings or houses. Revenue from the provision of services at the community center will be shared by the community investors and used for the operational expenditure. To promote awareness of the community users, school teachers, health workers, etc. of telecommunication/ ICT the program should be prepared by the community in collaboration with local administrators and telecom operators and concerned agencies.

Table 2. Action line for e-applications extracted from Tunis 2005 agenda

ICT Applications	Possible moderators/ facilitators
E-government	UNDP/ITU
E-business	WTO/UNCTAD/ITU/UPU
E-learning	UNESCO/ITU/UNIDO
E-health	WHO/ITU
E-employment	ILO/ITU
E-environment	WHO/WMO/UNEP/UN-Habitat/ITU/ICAO
E-agriculture	FAO/ITU
E-science	UNESCO/ITU/UNCTAD

Limited Funding Opportunities

Provision of ICTs services in rural and remote areas is not commercially viable in any case for the telecom operators or service providers. The mechanism providing subsidies for capital and operational expenditure should be sought for the sustainable infrastructure development projects in rural and remote areas. Community participation with their partial investment financially or in-kind from the initial stage of project formulation is highly recommendable. Public Private Partnership (PPP) scheme is found in some of the cases. CSR scheme is promoted in some countries together with universal service fund to raise the fund from private sector. Service licensing fee and frequency licensing fee may also be the sources of development fund. Figure 7 shows the funding methods for the projects observed in the collected case studies by the ITU-D rapporteur's group.

The Power Supply Problem

Most of rural and remote areas of developing countries are not well equipped with public or commercial power supplies. The cost of power supply for small-scale ICTs projects in areas that does not have power grids could be as much as 80% of the initial cost of the project. Even if power supply is available, it is usually limited or unreliable for several hours or so per day. Community mini-power grids are implemented in some cases to feed ICTs facilities in addition to commercial grids. Small community network needs energy from 10 to 12 kWh of energy (ITU-D SG2 Rapporteur, 2006). This need could be supported by electric generators with capacity ranging from tens of Watts up to 2 to 3 kilowatts kW. A number of standalone power generators are available in the market to provide power for small-scale last-mile ICTs infrastructure. They include solar power, mini-hydro power, and windmill power. Some projects use a combination of these sources, depending on the surrounding conditions, to provide reliable energy sources.

Figure 7. Funding sources of the projects

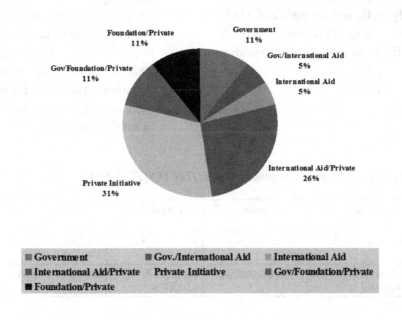

Table 3 compares the capital cost and operating cost of different power sources.

In addition, project managers should consider using low-power ICTs equipment in order for them to save power. This equipment includes laptops, low-power desktops, LCDs screens, and ink jet printers. Even if grid power is available, low-power equipment could be useful when we use a back-up electric generator or a battery in the case the grid is unreliable and subject to regular power outages. The cost of a back-up power supply usually increases with the capacity of the battery bank. In general, the less power the facilities consume, the less expensive it will be to supply any outage that may arise. For detailed information about options for power supply for rural communities, we refer readers to ITU-D SG2 Rapporteur (2006).

RECOMMENDATIONS

The ITU-D SG2 suggests that ITU-D 19 recommendations for rural telecommunications (ITU-D 19, 2010). It suggests recommendations and guideline for the member countries, sector members and associate. The Recommendations are compiled from the past recommendations of the previous study cycles and the works of the SG2 on the development of rural communications. It is also the compilation of analysis of case studies for the past study cycles.

ITU-D 19 recommends:

1. That developing countries should include provision of telecommunications/ICTs in rural and remote areas in their national development plans;
2. Choosing a suitable technology is a key success factor. Therefore, it is important to assess all available technologies in the market taking into consideration the regulatory environment, geographical conditions, climate, costs (both capital and operational costs), maintainability, operability, sustainability, etc., based on the results of the site survey;
3. That community access to ICTs facilities and services is particularly important in rural and remote areas. Engaging local entrepreneurs and local communities is important for the financial and operational sustainability of adopted business models. The facilities, where necessary, should also be supported by a suitable program for Universal Service Funds. Such a program is essential instrument for the digital inclusion of rural communities;
4. That post offices have a communicative presence in the lives of the population in rural areas and their use as vehicles for provision of telecommunication/ICTs should be encouraged;
5. That local institutions, such as village committees should be involved in planning and implementing ICT facilities;
6. That enhancing local technical expertise and adoption is important for successful implementation of ICT services and applications in rural and remote areas. Attention should

Table 3. Energy costs of off-grid ICTs installations (ITU-D SG2, 2006)

	Grid extension	Solar	Windmill	Mini-hydro	Diesel/Gas Generator
Capital cost	U$4,000 to 10,000	U$ 12,000 to 20,000/ kW	U$ 2,000 to 8,000/ kW	U$ 1,000 to 4,000/ kW	U$ 1,000/kW
Operating cost	U$ 80 to 120 per 1,000kWh	U$ 5 per 1,000kWh	U$ 10 per 1,000kWh	U$20 per 1,000kWh	U$250 per 1,000kWh

be paid to training, exchange of information, creation of shared maintenance facilities in order to achieve sustainability and viability;

7. That shifting to broadband technology should be encouraged;

8. That keeping even technologically obsolete equipment in good working condition through effective preventive maintenance programme is an essential part of making telecommunications in rural areas viable and should be encouraged, while guarding against making developing countries a dumping ground for obsolete technologies;

9. That it is important to take steps to ensure continued reliability of equipment in rural environments such as developing an appropriate maintenance and operation strategy and encouraging training for technical staff;

10. That given that lack of energy supply is a major bottleneck in the provision of telecommunications/ICTs in rural and remote areas, renewable energy sources should be used whenever feasible taking into consideration the environmental problems; and

11. That partnership among governments, industry, local agencies and international organizations is desirable in the development of low cost ICT infrastructure, including renewable energy sources and terminals for the provision of telecommunications/ICTs in rural and remote areas and should be pursued.

The recommendations of ITU D-19 on rural ICTs were approved during the World Telecommunication Development Conference in Hyderabad, India (WTDC-10).

In addition, Focus Group 7 (2001) recommend promoting the development of low-cost appliances for rural communities; creating a handbook for renewable and small-scale power systems for ICTS; Increasing collaboration with micro-finance organizations to fund ICTs-based businesses and applications; Conducting pilot projects for wireless access infrastructure for multimedia applications; and promoting emerging and new technologies for rural applications.

We also encourage developing countries to include the provision of ICTs in rural and remote areas in their national development master plans. To better serve these communities, developing countries can call for financial support for these projects from international funding institutions and development agencies. The other measures to fund the development projects in rural and remote areas in many developing countries are establishing regulatory framework for funding the development of infrastructure and service provision in rural and remote areas such as Universal Service Obligation (USO) and Universal Service Fund (USF). For instance, one percent of the telecommunication turnover gained by operators or service providers could be set aside for funds managed by the national regulators for subsidizing or funding the development of rural infrastructures and services. Percentage to be set aside for the funds or framework is in variety depending on the countries' national policies. In a country, universal service fee is charged on each fixed telephone or mobile phone number by the amount of 10 US cents per month. Funding policy for infrastructures and services is also different from country to country, such as funding infrastructure development for basic services, or covering telephony, Internet access or broadcasting services to remote areas.

CONCLUSION

Although more than ten years are spent on studying issues of developing rural communications development in ITU-D SG2, there are still under served or unconnected areas in many developing countries. ITU is promoting the "connect the unconnected" initiative by cost effective technologies, namely broadband wireless technologies to provide Internet services to every corner of this planet by 2015.

In order to promote the development of rural infrastructure and expand service provision, lack of funds is the key challenge. Another key challenge is the need for system integration to formulate into the solutions and projects. It is also important to recruit experienced experts who possess sufficient knowledge about the state of the art technology and applications. Expertise is the first step to start these projects. It is also important to form a steering committee to promote the project, co-ordinate concerned ministries, contact agencies, and attract community members. In addition, building partnerships with schools, health clinics, farmers, and other local entities is a critical success factor. Moreover, a backup power supply is very important because of the unreliability of power supply in rural areas. In addition, project managers should consider low-power equipment as cost of off-grid power supply could reach 80 percent of the total cost of the project.

With the rapid growth of mobile phone and Internet service penetration, the impact of tele-communications/ICTs on the socio-economic development is remarkable. More than half of the world population (7 billion) which is so called the Base of Pyramid (BOP) is now formulating the most promising economic market. Given the growing growth rate of GDP in developing countries, the market of the BOP will grow and this call for more investment in the ICTs sector. However, there is a concern about the commercial viability of investing in rural and remote areas. Concept of social business is advocated by the academia. We also recommend subsidizing rural Internet for needy communities.

Broadband commission established by ITU and UNESCO in response to UN Secretary-General's call issued the report titled "Achieving Digital Inclusion for All" in September, 2012. The report examines the possible ways to define "broad-band"; for example, as minimum upstream and/or downstream transmission speeds or according to the technology used or the type of service that it can deliver. However, countries differ in their definitions of broadband, and, as technologies advance, the minimum defined speeds are also likely to increase rapidly. The broadband commission, therefore, decided to focus on considering broadband based on a set of core concept, such as an always-on service (not needing the user to make a new connection to a server each time), and high capacity: able to carry lots of data per second, rather than at particular speed. The practical result is that broadband enables the combined provision of voice, data, and video at the same time.

REFERENCES

Broadband Commission of ITU and UNESCO. (2012). *The state f broadband 2012: Achieving digital inclusion for all*. Retrieved from http://www.broadbandcommission.org

International Telecommunication Union. (2005). *WSIS outcome documents*. Retrieved from http://www.itu.int

ITU-D 19. (2010). *Recommendation ITU-D 19: Telecommunication for rural and remote areas*. Retrieved on September 19, 2012, from http://www.itu.int/dms_pubrec/itu-d/rec/d/D-REC-D.19-201003-I!!PDF-E.pdf

ITU-D Focus Group 7. (2001). *New technologies for rural applications*. Retrieved from http://www.itu.int/ITU-D/fg7/pdf/FG_7-e.pdf

ITU-D Focus Group 7. (2001). *New technologies for rural applications*. Retrieved September 19, 2012, from http://www.itu.int/ITU-D/fg7/pdf/FG_7-e.pdf

ITU-D SG2. (2004). *Rapporteur for telecommunications for rural and remote areas: Analysis of replies to the questionnaire on rural communications*. Retrieved from http://www.itu.int

ITU-D SG2. (2010). *Recommendation D-19 telecommunication for rural and remote areas.* Retrieved from http://www.itu.int

ITU-D SG2 Rapporteur. (2006). *Analysis of case studies on successful practices in telecommunications for rural and remote areas.* Retrieved September 19, 2012, from http://www.itu.int/pub/D-STG-SG02.10.1-2006/en

ITU-D SG2 Rapporteur. (2009). *For telecommunications for rural and remote areas: Analysis of case studies on successful practices in telecommunications for rural and remote areas.* Retrieved from http://www.itu.int

Kawasumi, Y. (2002, March). Challenges for rural communications development. *ITU News.*

Kawasumi, Y. (2005). *Maitland+20-Fixing the missing link, focus on rural connectivity.* London, UK: The Anima Center.

Maitland Commission. (1985). *The missing link.* Retrieved on 9/19/2012 from http://www.itu.int/osg/spu/sfo/missinglink/The_Missing_Ling_A4-E.pdf

WSIS. (2003). *Declaration of principles and plan of actions.* Retrieved on September 19, 2012, from http://www.itu.int/wsis/docs/promotional/brochure-dop-poa.pdf

WSIS. (2005). *Tunis agenda for the information society.* Retrieved on September 2012, from http://www.itu.int/wsis/docs2/tunis/off/6rev1.html

KEY TERMS AND DEFINITIONS

Asia Pacific Telecommunity (APT): APT was founded on the joint initiatives of the United Nations Economic and Social Commission for Asia and the Pacific (UNESCAP) and the International Telecommunication Union (ITU). The APT was established in Bangkok in July 1979. The APT is an Intergovernmental Organization operates in conjunction with telecom service providers, manufacturers of communications equipment, and research and development organizations active in the field of communication, information, and innovation technologies. The APT covers 38 member countries, with 4 associate members and 130 affiliate members. Through its various programs and activities, APT has made a significant contribution to the development growth in ICT sectors.

Base of the Pyramid (BOP): Lower income population is now said to be more than half of the total world population. It is also called bottom of the pyramid to indicate undeveloped world market in general. International Finance Corporation (IFC) and World Resources Institute (WRI) defined that population of annual income less than U$3,000 should be called BOP in 2007. The population in this group is now 72% or about 40 billion and their purchasing power is estimated U$ 5 trillion as of 2007. This large market is called world attention as new emerging market of the developing world by the developed world since 2009.

Broadband Global Area Network (BGAN): BGAN is a global Satellite Internet Network with telephony using portable terminals. The terminals are normally used to connect a laptop computer to broadband Internet in remote locations, although as long as line-of-sight to the satellite exists, the terminal can be used anywhere. The value of BGAN terminals is that unlike other satellite Internet services which require bulky and heavy satellite dishes to connect, a BGAN terminal is about the size of a laptop and thus can be carried easily. The network is provided by Inmarsat and uses three geostationary satellites to provide almost global coverage.

C-Band and Ku Band (Frequency Bands): The C band is a name given to certain portions of the electromagnetic spectrum, including wavelengths of microwaves that are used for long-distance radio telecommunications. The IEEE C-band (4 GHz to 8 GHz)—and its slight variations—contains frequency ranges that are used for many satellite communications transmissions, some Wi-Fi devices, some cordless telephones, and some weather radar systems. For satellite communications, the microwave frequencies of the C-band perform better under adverse weather conditions in comparison with Ku band (11.2 GHz to 14.5 GHz) microwave frequencies, which are used by another large set of communication satellites. The adverse weather conditions, collectively referred to as rain fade, all have to do with moisture in the air, including rain and snow.

Corporate Social Responsibility (CSR): CSR is a form of corporate self-regulation integrated into a business model whereby a business monitors and ensures its active compliance with the spirit of the law, ethical standards, and international norms. The goal of CSR is to embrace responsibility for the company's actions and encourage a positive impact through its activities on the environment, consumers, employees, communities, stakeholders. CSR is titled to aid an organization's mission as well as a guide to what the company stands for and will uphold to its consumers.

Demand Assignment Multiple Access (DAMA): DAMA apparatus was developed for satellite communication system using K (sub u) band of private communication satellite. The apparatus was introduced for the purpose of following functions: (1) keeping up communication trunk during emergency and disaster; (2) economically constructible user's own satellite telephone system; (3) securing traffics during emergency and disaster due to setting user's own communication trunk; and (4) effective utilization of satellite communication band by using communication trunk during telephone call only.

Inmarsat: Was originally founded in 1979, as the International Maritime Satellite Organization (Inmarsat), and now it is a British satellite telecommunications company, offering global, mobile services. It provides telephony and data services to users worldwide, via portable or mobile terminals, which communicate to ground stations through eleven geostationary telecommunications satellites. Inmarsat's network provides communications services to a range of governments, aid agencies, media outlets, and businesses with a need to communicate in remote regions or where there is no reliable terrestrial network.

International Telecommunication Union (ITU): ITU is the United Nations specialized agency for Information and Communication Technologies – ICTs. It is headquartered in Geneva and was established in 1865 in Paris. ITU has three main areas of activities organized in 'Sectors' namely ITU-R, ITU-T, and ITU-D which work through conferences and meetings to allocate global radio spectrum and satellite orbits, develop the technical standards that ensure networks and technologies seamlessly interconnect, and strive to improve access to ICTs to underserved communities worldwide.

Internet Protocol (IP): IP is the principal communications protocol used for relaying datagrams (also known as network packets) across an internetwork using the Internet Protocol Suite. Responsible for routing packets across network boundaries, it is the primary protocol that establishes the Internet.

Internet Service Provider (ISP): ISP is an organization that provides access to the Internet. Internet service providers can be either community-owned and non-profit, or privately owned and for-profit.

Next Generation Networks (NGN): NGN is a packet-based network able to provide Telecommunication Services to users and able to make use of multiple broadband, QoS-enabled transport technologies and in which service-related functions are independent of the underlying transport-related technologies.

Public Private Partnership (PPP): PPP describes a government service or private business venture, which is funded and operated through a partnership of government and one or more private sector companies. PPP involves a contract between a public sector authority and a private party, in which the private party provides a public service or project and assumes substantial financial, technical, and operational risk in the project. In some types of PPP, the cost of using the service is borne exclusively by the users of the service and not by the taxpayer. In other types (notably the private finance initiative), capital investment is made by the private sector on the strength of a contract with government to provide agreed services and the cost of providing the service is borne wholly or in part by the government.

Small Island Developing States (SIDS): SIDS were identified as a special group during the 1992 Earth Summit and subsequently a number of internationally agreed development goals have been formulated to specifically address SIDS vulnerabilities and to build resistance and sustainability. Currently, fifty-one small island developing States and territories are known collectively as Small Island Developing States (SIDS).

Social Business: The concept of social business was first defined by Nobel Peace Prize laureate Prof. Muhammad Yunus and is described in his books Creating a world without poverty—Social Business and the future of capitalism and Building Social Business—The new kind of capitalism that serves humanity's most pressing needs. A number of organizations with which he is involved actively promote and incubate social businesses. These include the Yunus Centre in Bangladesh, the Yunus Social Business Centre University of Florence, the Grameen Creative Lab in Germany, and Social Business Earth. In Yunus' definition, a social business is a non-loss, non-dividend company designed to address a social objective within the highly regulated marketplace of today. It is distinct from a non-profit because the business should seek to generate a modest profit but this will be used to expand the company's reach, improve the product or service or in other ways to subsidize the social mission.

Universal Service Fund (USF): USF was created by the United States Federal Communications Commission (FCC) in 1997 to meet Congressional universal service goals as mandated by the Telecommunications Act of 1996. The 1996 Act states that all providers of telecommunications services should contribute to federal universal service in some equitable and nondiscriminatory manner; there should be specific, predictable, and sufficient Federal and State mechanisms to preserve and advance universal service; all schools, classrooms, health care providers, and libraries should, generally, have access to advanced telecommunications services; and finally, that the Federal-State Joint Board and the FCC should determine those other principles that, consistent with the 1996 Act, are necessary to protect the public interest. As of the third quarter of 2012, the USF fee, which changes quarterly, equals 15.7 percent of a telecom company's interstate and end-user revenues.

Universal Service Obligation (USO): USO is an economic, legal, and business term used mostly in regulated industries, referring to the practice of providing a baseline level of services to every resident of a country. An example of this concept is found in the US Telecommunications Act of 1996, whose goals are 1) To promote the availability of quality services at just, reasonable, and affordable rates; 2) To increase access to advanced telecommunications services throughout the Nation; 3) To advance the availability of such services to all consumers, including those in low income; and rural, insular, and high cost areas at rates that are reasonably comparable to those charged in urban areas. In many countries now, they are promoting universal service to underserved areas by charging operators (incumbent operator in particular) obligation to provide universal services in conjunction with USF scheme.

Very Small Aperture Antenna (VSAT): VSAT is a 1.2 -1.8 m small diameter antenna for satellite communication system for a cost effective solution for providing communications network connecting a large number of geographically dispersed sites.

Wireless Fidelity (Wi-Fi): Wi-Fi it is wireless LAN technology certified by Wi-Fi Alliance to guarantee compatibility of wireless LAN equipments of different manufacture. Communication standard is developed as IEEE 802.11series standards such as 802.11/802.11b.

World Summit for Information Society (WSIS): The World Summit on the Information Society (WSIS) was held in two phases. The first phase took place in Geneva hosted by the Government of Switzerland from 10 to 12 December 2003, and the second phase took place in Tunis hosted by the Government of Tunisia, from 16 to 18 November 2005. The outcome documents are Geneva declaration of principles, Geneva plan of action, Tunis commitment, and Tunis agenda for the information society.

World Telecommunication Development Conference (WTDC): The International Telecommunication Union, through the Telecommunication Development Bureau (BDT), organizes a World Telecommunication Development Conference (WTDC) every four years. The Telecommunication Development Conferences serve as forums for free discussion by all concerned with the Development Sector. In addition, they review the numerous programmes and projects of the Sector and BDT.

ENDNOTES

1 http://www.itu.int/en/Pages/default.aspx
2 http://www.itu.int/ITU-D/ict/cs/

Chapter 12
Best Practices in Designing Low–Cost Community Wireless Networks

Tomas Dulik
Tomas Bata University, Czech Republic

Michal Bliznak
Tomas Bata University, Czech Republic

Roman Jasek
Tomas Bata University, Czech Republic

ABSTRACT

The Czech Republic (CR) has been ranked the 1st among the countries of the European Union (EU) countries in the growth rate of broadband access. The Internet penetration rate has increased by 48 percent between 2005 and 2011. This high growth rate is driven by the entry of new operators and the proliferation of Community Wireless Networks (CWNs). The CR holds the first place in EU in the number of newly entered operators. There are 1150 companies providing Internet access in 601 Czech towns and 5645 villages. In addition, a number of community wireless networks have emerged as an alternative of these commercial Internet Service Providers (ISPs). Their main purpose is to increase the affordability and penetration of broadband Internet in the country. This chapter discusses the contribution of CWNs to the proliferation and affordability of broadband access in the CR, focusing on the reasons for their success and popularity. Their key success factors include obtaining a non-profit status, engaging academics, and cooperating with government entities. They formed the CZFree.net forum for experts and volunteers to exchange information and best practices with respect to new technologies, design considerations, and technical and social issues. It also articulates on technology options and best practices for building low-cost CWNs. Furthermore, the chapter discusses the role of the Netural czFree eXchange association in aggregating their technical, financial, and personal resources of individual CWNs. Thanks to this association and the CZFree.net forum, CWNs in the CR have become influential competitors in the local telecommunication industry.

DOI: 10.4018/978-1-4666-2997-4.ch012

BACKGROUND

The Czech Republic (CR) is currently ranked the 1st among all the 27 EU countries in the growth rate of broadband access (Eurostat, 2012). The number of Czech households with Internet connection has risen by 48 percent (from 19 to 67) in the period between 2005 and 2011. In addition, the number of households with broadband connection has increased by 58 percent (from 5 to 63) in the same period. Most of the growth is driven by the entry of new operators that have dominated about 67 percent of the fixed broadband market (European Commission, 2012). However, the most unique aspect of the CR case is its long-term leadership in the provision of Wireless Local Loops (WLL). In 2009, 50 percent of the 1.3 million WLL fixed broadband lines installed in the whole 27 EU countries was in the CR (European Commission, 2010a), although its 10 million citizens represent only 2 percent of the EU population.

As in many other countries, the Internet in the CR has begun as an academic initiative (Kirstein, 2004). The academic Internet access has been started officially in February 1992 (CESNET, 2012). In the following few years, all major Czech universities had been connected for the purpose of providing Internet access in their campuses for their students and employees. In 1996, these academic networks were covered by a newly formed legal association called CESNET. Around that time, students could use the Internet also in some dormitories. In addition, they could use a dial-up connection to access their university networks from home. However, the latter option was very expensive if long-distance connections had to be used.

The first commercial Internet in the CR was provided by COnet in 1994. However, their services could not be offered openly because of the monopoly license of Eurotel, which controlled the provision of data services. Eurotel was a joint venture company owned by the ex-state driven company SPT Telecom and the American joint venture Atlantic West (Smithfarm, Lemmio, Jklamo, & Voženílek, 2011).

State guaranteed monopoly has major negative impacts and the situation in CR was a typical example. This is because when COnet and CESNET had started their commercial Internet offerings, Eurotel was not providing *any* Internet access services (Peterka, 2001), only the voice tariffs on its NMT mobile network. Therefore, the Internet was mostly locked inside the academic walls until 1995. Afterward, the monopoly for data services was terminated when SPT Telecom purchased NexTel, which was the data division of Eurotel (Peterka, 2001).

Since then, commercial ISPs have started to enter the market. In 1998, there were more than 150 ISPs (Peterka, 2001). Typical small ISPs of that time used the Plain Old Telephone Service (POTS) and the Integrated Services Digital Network (ISDN) modems for both uplink and last-mile connections, as the e-carrier leased lines were very expensive. However, there were already reasonably priced P2P microwave links produced by several Czech companies. These links could be used for building larger network backbones without the burden of leasing lines.

The last-mile ISDN dial-up connection was not affordable for most households because of the high price of phone services. Phone services were suffering monopoly as well. In the 1995-1999 era, the Czech households did not have viable or affordable connectivity alternatives. In particular, the P2P microwave links were too expensive and the 10/100 Mbit/s Ethernet had short range. Cheap fiber optic technologies emerged many years later.

The situation in 1998 was such that the Internet was already perceived as a standard service in many Czech companies. Yet, households were still dependent on dial-up connections that could not be used for long periods because of the old charging schemes, which were adequate for voice calls but expensive for Internet connections.

This caused permanent frustration among Internet users, who blamed the government for protecting the monopoly of SPT Telecom. The frustration quickly transformed into public protests and boycotts (Zandl, 2001) when SPT Telecom published even more expensive voice calls tariffs in 1998. The result was the Internet99 dial-up tariff (Peterka, 2001). This tariff was better than the previous voice tariffs, but not satisfactory.

Meanwhile, the telecommunication monopolies were collapsing in the whole Europe, but with much greater dynamics in the western countries. For instance, France Telecom started to offer the ADSL in 1999. At that time, there were already 43000 households with Internet broadband cable-TV access provided by various competing companies (Grange, 1999).

At the same time, SPT Telecom in the CR was still relying solely on its ISDN and dial-up offers. late as in 2002. Moreover, it started to offer ADSL services in 2002, which was delayed because of competing companies, which protested against the missing local loop unbundling policy. This protest made the telecommunication regulator ČTÚ stop SPT Telecom's ADSL service until an unbundling policy came in effect (Peterka, 2002).

It is worth mentioning that a free society could overcome the problems facing them in an innovative manner. Fortunately, this was the case of the CR. During the following three years (from 2000 to 2003), a huge wave of new operators entered the market using wireless links in license-free bands. The CR became the European leader in Wireless Local Loop (WLL) for Internet access (European Commission, 2009, 2010a, 2010b).

The Boom of Community Networks

The boom of CWNs in the CR at the beginning of the 21st century was catalyzed by the following factors:

1. The monopoly of SPT Telecom of local loops and its inability to launch affordable ADSL services, while 802.11 devices were getting more affordable and more reliable;

2. The increasing popularity and usage of the Internet among the public which led to high demand for affordable high-speed Internet access;

3. The emergence of a group of Czech companies, whose business was producing and/ or selling affordable microwave hardware, quickly adapted their offerings to fit the new market of wireless networking equipment;

4. The general licenses GP-01/1994 and GP-02/1994 (ČTÚ, 1997) have been updated in 2000 by GL-12/R/2000 and GL-14/R/2000 (ČTÚ, 2000) to allow toll-free use of the 2.4GHz frequency and 10GHz bands;

5. The advocacy of the civil society which has a 150 years long tradition of civic participation and involvement of public affairs;

6. The increased positive perception about non-profit organizations and the frustration from the economic disasters in 1990s, after privatization of coupon (Pithart, et al., 2006), evasions of multi-billion tax (Fijnaut & Paoli, 2004), frauds in asset tunneling (Johnson, Porta, Silanes, & Shleifer, 2000) and various other criminal activities committed by big businesses which were close to government officials. These economic conditions left many of the previously flourishing regions in ashes. Since then, the word "entrepreneur" has a bit of pejorative connotation among the Czech people. Non-profit organizations were perceived positively because of active participation of ordinary local people. This participation was taken as a sign of expected public support and sustainability of community-centered projects.

It is difficult to determine when the first community network in the CR started. Back in 1990s, the most typical scenario were small coax or UTP Ethernet networks created by few friends who wanted to share their data, but not the Internet connection. Sharing the Internet connection, at this time, did not provide any advantage with the dial-up and ISDN time-based accounting committed by SPT Telecom.

The situation changed as soon as commercial ISPs started business with flat rate tariffs starting from 25 USD per month for a 64-128kbit/s wireless link. Their prices targeted business users, so they were expensive for a common household user[1], but not that expensive for a group of households. The only problem was the short range of UTP Ethernet, which was limited only for a group of two to three neighbors.

In 2000, affordable wireless LAN devices for the 2.4GHz band appeared in the market. In the beginning, second-hand WaveLAN PC cards were available for fraction of their original price. Sooner, a flood of reliable IEEE 802.11 devices with Orinoco/Proxim appeared in the market with prices less than 50 USD which was affordable for most households. Although these devices were not designed for building external fixed wireless access networks, it was possible to use them for this purpose using cheap homemade external antennas such as the bi-quad or "canntenna."

The advocates of CWNs are scattered all over the country. These advocates include:

1. Amateur radio community members who got interested in new way of radio communication;
2. University students who got used to high speed Internet in their campuses and had the abilities, skills and competence to design and administer the newly built networks;
3. Computer professionals who are able to help the end users to solve their problems; and
4. End users who are eager to explore, troubleshoot and manage their networks themselves.

In 2001, they established a common communication channel for sharing their experiences which is the CZFree.net online discussion forum.

At that time, community members already had everything needed for getting cheap Internet access

to everyone. This includes inexpensive technology, motivated experts and advocates, a platform to exchange their knowledge and a country full of anxious end-users.

Therefore, the boom of CWNs has started. Sooner, it was possible to get Internet access for only 8 to 10 USD. This rate was about three to four times less than the average commercial flat-rate tariffs of that time. There were even extreme cases like the UnArt.cz network with membership fees only two USD per month.

The outcomes of this shared model included affordable Internet access for all and exponential growth of CWNs membership base. In 2005, CWNs in the CR had become so popular that the "free community" label was a marketing hype exploited even by many commercial ISPs.

ISSUES OF THE CZECH NETWORKS

As with other countries (Abdelaal & Ali, 2012; Forlano, Powell, Shaffer, & Lennett, 2011), the leaders of the Czech community networks were not just typical IT geeks who wanted to provide their IT skils and services to their families and neighbors. In addition, the community networking was not "yet another service for fee." It was a movement with long-term mission whose leaders were aware of their roles and objectives. However, they faced the following legal, economic, and social challenges:

Tax Legislations

One of the critical challenges of CWNs is dealing with tax legislations if they charge members fees for Internet access. Except few anarchistic groups, most of the Czech CWNs always try to comply with the current tax legislations. As CWNs was completely a new phenomenon representing a small group of the society that works outside the scope of law, the legal status of non-profit Internet solutions was not clearly defined. Therefore, CWNs had to take over methods and principles used by other Czech non-profit groups and associations while avoiding practices that have been already penalized as non-desirable or illegal for a typical non-profit group or organization.

The Czech telecommunication law allows non-profit Internet service simply by not prohibiting it. The law clearly states that it is defined only for commercial telecommunication operators. According to the law, commercial operators are obliged to register at the Czech telecommunication regulator (ČTÚ), and then comply with its various requirements. The most tedious requirement is filling and sending monthly statistical reports. Non-profit operators are spared of all these duties.

More stringent requirements are given by the tax legislation. Indeed, CWNs must collect some money for building the network, upgrading its technology and backbones, paying for the Internet connectivity, etc. It might be possible to minimize the taxable incomes by distributing the costs and incomes between members of the community network, and this path was definitely followed by many Czech CWNs in the early years. However, as the community grows, the distributed economy administration overheads might grow beyond acceptable limits. In addition, within this model, it is impossible to fund bigger infrastructure investments, e.g. for laying underground fiber optic cables, building radio towers, etc. For such projects, money must be accumulated on a single bank account for a long period of time and obviously, these incomes are taxable.

As soon as the standard accounting model with single cash register and/or bank account is chosen, the aggregate income will easily cross the revenue threshold for the tax payment, which is currently 1000 USD/year for occasional/irregular activities of physical persons in the CR. If this threshold is crossed, the informal community network group should decide how to cope with this situation. It could either ignore the legislation and risk the tax audit (and pay the fines afterwards) or start acting as a standard for profit entity.

Non-Profit Status

Obtaining a non-profit status is very beneficial for CWNs. Non-profit civil associations have long tradition in the CR (Anderle, 2005). The first law regulating non-profit civic associations was issued in 1843. After the collapse of the communist regime in 1989, key civil associations were quickly re-created and many new ones were founded. The new civil law of 1990 provided simple and fast requirements for registering new associations. The tax law of 1992 provided tax reliefs for non-profit organizations—the member fees are not subject to VAT and organizations can even create profit up to 13000 USD, which is free of the income tax.

The civil association law of 1990 lacks important features like clarification of the process of central registration of associations and their statutory representatives, definition of the allowable non-profit and for-profit activities etc. Although the need for better law has been broadly discussed for more than 20 years, the progress towards a new law was very slow. For various reasons, many politicians were refusing the improvements for various political reasons (Pehe, 1994). The new civil society law was finally approved in February 2012, but it will take effect as late as in January 2014.

By 2001, some networks already had a substantial number of members and thus non-negligible cash flow. From tax legislation perspective, they crossed the taxable revenue threshold 1000 USD/year. In addition, they had also other problems such as:

1. Lack of consensus in their decision making processes, which paralyzed them; and
2. The emergence of the "parasite phenomenon," when some of the active members used the community network infrastructure for making their personal profits.

The various organizational issues convinced community leaders that registering for a non-profit status is unavoidable. At that point, many groups started their transformation and they had two choices: either to become a non-profit association or a normal commercial ISP.

The decision between these two options was driven by philosophical or/and practical reasons. Leaving the philosophy aside, practical reasons for obtaining non-profit status could be:

1. Better perception of non-profit organizations among the population and the local government;
2. Much better tolerance of network outages by the members who were aware of non-profit nature of the network;
3. Simpler accounting and tax benefits; and
4. Easier motivation of the members for voluntary work.

In bigger networks, however, it is usually hard or impossible to estimate or manage the amount of time and money that active members already invested in the network. These groups were then unable to start a commercial company because it was hard for them to negotiate their shares.

The organizational structure of the newly established community networking associations was implied by the civil association law: any association must have an executive board (board of directors), an audit committee (or single auditor) and the general assembly. Some associations have also established a technical committee to be responsible for the technological upgrade of the network.

In the beginning, most of the newly established associations were run on purely amateur basis and loosely affiliation. In other words, they had no employees and everything they did had to be done by volunteers during their free time.

However, in many cases, as communities grew, the voluntary-based model started to be impractical. Therefore, more sustainable models had to be adopted. Some associations hired full-time employees, part-time people, or freelancers.

Many CWNS have undertaken the government tax audits. In all cases, tax officers were cross-

checking the articles of the association with its real practices and procedures. In general, they inquired the costs and outcomes of the non-profit activities such as research and development, cultural events, any evidence of cooperation with other non-profit organizations, etc.

These audits helped the communities to polish their declared mission, improve their knowledge, and understand the legislative system. They collectively specified the following set of rules and regulations that a Czech CWN association should follow:

1. The primary objectives of any non-profit association must not coincide with for-profit organizations. Therefore, the article of atypical association should clearly declare its primary mission as a community cause such as "improving the life of the local community," "improving the economic conditions of the local region," etc.
2. The primary activities of the association must be non-profit, typically: "building and improving the community network," "supporting research and development of network technologies," "providing training and organizing cultural and sport events," etc.
3. The association should avoid providing paid services. In other words, the fixed and operation costs should be covered through membership fee.
4. The membership fee must not be similar to the fees paid for for-profit services. Therefore, there should be only one type of membership fee provided by the association. The association should not charge any Internet tariffs.
5. If the association provides a paid service, (e.g. charging fees for VOIP calls or IPTV), the profit from such a service must be treated the same way as in for-profit companies. The Czech legislation allows the non-profit organizations to do for-profit activities. However, incomes from profitable services

that goes beyond 35000 EUR/year trigger permanent VAT payment duty. Although it might seem that this is not a big problem, the VAT payments make the bookkeeping more difficult. Generally, it is not wise to go this way without a full-time accountant whose salary would be a real overhead; therefore, most of the Czech CWNs avoid such activities.

Collaboration with Academicians

Another factor that contributed to the development of low-cost CWNs in the CR is employing students and researchers to develop necessary software and hardware, to manage the various networking systems, and troubleshoot them. As mentioned earlier, university students and alumni were one of the key actors in the Czech CWNs movement. Their engagement in the movement was often connected to their schoolwork. Typically, they have developed various tools as school projects and these tools were then used for their community as well. There were also many graduate and undergraduate projects inspired by these CWNs. For examples of these projects, we refer readers to Haleš (2005), Jaroš (2005), Sporek (2005), Vilímovský (2005), Bula (2007), Fryšták (2007), Sviták (2007), Matúšů (2008), Turek (2009), Vágner (2011), Šlinz (2012), and Dulík (2012).

In addition, students used CWNs to test and publish their projects results (see Grygárek, 2012 for examples). The students have been keeping this practice in their CWNs as well—most often on the community wiki pages. This is much different from commercial ISPs, which do not share or publish any knowledge at all.

Moreover, free educational seminars and workshops for the members have been organized by most of CWNs. The topics cover currently used and emerging technologies, technological and organizational challenges and problems. They also address issues from other IT disciplines unrelated to networking, like digital signal and image processing.

Few communities produce paper publications on regular basis—namely, Czela.net bulletin "Včela" (Procházka, 2012) and the "UnArt Revue" magazine (O.s. UnArt, 2009), whose printed form is distributed to the mailboxes of all members.

Engaging Government Authorities

The development and sustainability of such societal projects, and non-profit organizations in general, require the support of both local and federal governments. However, the attitude of the government authorities towards CWNs varies greatly from case to case. In case of local municipal councils, the network often brings Internet to towns or villages where there are no other local ISPs. Local councils happily supported this type of networks by:

1. Providing sites for building the wireless APs for free;
2. Allowing to lay the underground HDPE ducts/pipes for fiber optic cables; and
3. Helping with the promotion of the initiative.

Commercial ISPs, in a similar situation, may obtain a similar support, but if it had to compete with a community network, a non-profit organization would probably win the council sympathies. On the other hand, CWNs were mostly unable to build relations with higher government levels - e.g., with counties or even national institutions, where lobbies and big business competition do not leave much space for amateur movements.

Our vision is to adopt successful practices with respect to cooperation between government entities and non-profit organizations or armature groups. The following are few examples:

- Czech firefighters are composed of both professional organization and amateur teams, which are irreplaceable in rural areas and also in cities during extraordinary disaster events like floods, extensive fires etc.

- Most social services for elderly and disabled people are run by non-profit organizations, although these are mostly fully professionalized.
- Nature protection organizations are both professional and amateur/non-profit entities. The professional ones are mostly responsible for natural parks, while the non-profit ones usually take care of their local environment.

Although we are not aware of any current or future plans for cooperation between CWNs and counties or state, it is possible that such cooperation can arise in the future thanks to the growing growth and maturity of many CWNs. In addition, most of the new generation of politicians advocate the civil society and have an anti-corruption agenda. They have raised the hopes that they will not be under the influence of telecommunication lobbies.

Competition in the Telecommunication Industry

It is important for community leaders to consider and plan for a strong competition from commercial ISPs and even from other CWNs. Most of CWNs did not expect such competition. This changed as soon as their networks acquired significant number of connected households in some regions. One of the eminent examples of fierce competition was experienced by Pilsfree.net in 2009 (Sova, 2009). Specifically, its pro-active community building approach had provoked the leaders of the local UPC Cable Company who started deceptive advertising campaigns. Similar misleading campaigns from local competitors were experienced by most of CWNs.

The defensive tactics in these cases focused mostly on informing the members by all available channels about the maligned competition actions. We are not aware of any legal actions and law suits invoked by the affected communities. In other words, CWNs have avoided legal actions and kept it as a last option.

However, the competition has also a positive aspect. In particular, it has forced communities to constantly improve their networks and promote their mission in better ways. One of the leaders said "If there was no competition, we would have to make it up."

THE NEUTRAL CZFREE EXCHANGE PROJECT

Since the beginning of this movement, its leaders were attempting to develop a platform of cooperation and find ways to aggregate their scarce resources. Until 2006, there was a shortage in human resources. Therefore, starting a sustainable countrywide platform was not viable.

Of course, limited cooperation opportunities between CWNs have already existed. They were mainly based on exchanging experience and know-how through the CZFree.net forum. In addition, there were also few country-wide projects that provide critical services such as IP address coordination (RIPE.czfree.net) and service mapping (mapa.czfree.net) for affiliated networks. Some neighboring networks like Klfree.net, Unhfree and some communities in Prague were already cooperating at closer level. Their cooperation included building shared wireless backup links, sharing Internet connections, and exchanging experts and/or volunteers.

In 2006, the leaders of several CWNs were already fully convinced about the necessity and advantages of countrywide cooperation and were determined to establish an organizational framework for it. They founded the Netural czFree eXchange (NFX). NFX is an association of legal entities, which became an umbrella organization for various countrywide projects. These projects include:

- Building common Points-Of-Presence (POPs) connected to both the NIX.cz (Czech Internet eXchange) and to international uplink providers;

- Buying capacity for international connectivity as a whole bundle for better prices, saving yet more money by using the aggregation effect;
- Building a common IPTV and VOIP infrastructure; and
- Sharing the best practices and know-how during regular meetings.

Currently, NFX has 20 members—17 non-profit associations and 3 commercial ISPs. They serve about 50000 households in total, covering most of the CR regions—see the map in Figure 1. The members of NFXinclude: Bubakov: 2496, CZF-Praha: 1310, Czela.net: 757, Evkanet: 2000, Gavanet: 390, HKfree: 3443, Jablonka: 757, JM-Net: 1045, KHnet: 2605, Klfree.net: 5827, Křivonet: 616, Freenet Liberec: 452, mh2net: 584, PilsFree: 16037, PVfree: 1965, Steadynet: 500, Sychrov.net: 3000, UnArt: 4807, UNHfree: 1814.

The peek download traffic (from NIX.cz + abroad) is currently about 10 Gbit/s and it grows at the rate of 1Gbit/s per year.

The NFX is represented by a board of directors, who are driven and controlled by the resolutions of the quarterly general assembly meetings. The general assembly is composed of the member associations, each member association has one voting right carried by a duly appointed representative or proxy. The resolutions are passed on a show of hands by a simple majority.

The projects and activities of NFX are organized and accounted in workgroups. A typical workgroup consists of member associations interested in this specific activity. Workgroups have a flat structure where each member has one representative.

In 2009, NFX has established a workgroup for Research and Development. This workgroup immediately started to sponsor various development activities and research projects, including enhancements of the ath5k drivers, development of special firmware for ubiquity devices, developing new map for CZFree, and developing networking hardware.

The NFX membership fee is currently 420 USD a month. The membership fees are used for covering the regular costs—typically, the NIX.cz fees, the site rental fees for the 2 racks used as NFX POPs, the RIPE membership fees, the Cisco support fees, etc.

Figure 1. Map of the Czech Republic community networks associated in NFX

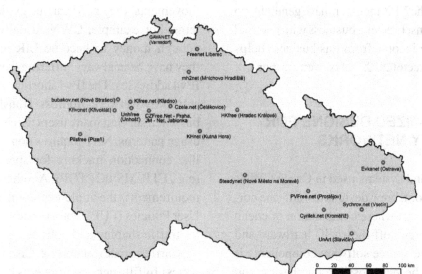

As for the non-regular costs like building connectivity capacity, these were charged and billed to each of the member associations proportionally to their usage of the commodity (i.e., the connectivity capacity).

In 2011, NFX decided to establish its commercial subsidiary—the Freetel s.r.o. The reasons for this step were the following:

- NFX decided to start its own networking hardware sales for its members, exploiting yet another area of mass volume purchases for getting better prices and thus saving more money;
- Aside of foreign connectivity, NFX succeeded in arranging bargains for leasing national fiber optic backbones; and
- The growing amount of invoiced incomes would trigger the VAT payment duty for NFX, which would be disadvantageous for its members who are not VAT payers. Therefore, all the above mentioned business activities were transferred to Freetel s.r.o., which functions as a service organization for the non-profit NFX association and its members.

Thanks to these activities, Freetel offerings became an interesting option even for some commercial ISPs. The NFX members have generally no objections against Freetel's business with external subjects, as the income from this business helps reducing the Freetel/NFX costs even further.

COST-OPTIMIZED DESIGNS FOR COMMUNITY NETWORKS

Technologies and systems used in Czech CWNs evolved from homebrew designs towards the current state of the art where combination of cheap COTS (commercial off-the-shelf) hardware and open and closed source software components is optimized for the best possible purchasing and operating costs.

As for producing hardware on their own, generally speaking, Czech CWNs always try to develop do-it-yourself solutions in areas where there are no low-cost alternatives. However, they usually do not have enough resources or motivation for competing with network equipment manufacturers—e.g. by starting a mass production of hardware that was already available in the market. Instead, they have always been supporting the manufacturers by examining their products thoroughly in laboratories or field tests, see better documented examples in Bula (2007), Vágner (2011), and Dulík (2012). Many of the Czech manufacturers appreciate this cooperation and use CWNs for beta-testing their products. The deployed product samples are often offered to the testing community for fraction of their retail price or even for free.

The technologies used by Czech CWNs are discussed in greater details in the following sections.

Low Cost Routers

Every network needs at least one router. Simple boxes like Linksys BEFSR41 and BEFW11 were available already in 2001 and they were used by some CWNs for a while. Yet, because of their limited features and frequent freezes, they had to be replaced by more advanced ones.

Although these CWNs are basically amateur movements, they need carrier-grade routing systems. For example, CWNs usually do not have enough money to become LIR and, therefore, they have been always suffering from shortage of IPv4 addresses. The IPv4 shortage is traditionally solved by Network Address Translation (NAT), but when serving many users with many different usage patterns, NAT requires non-trivial features like connection tracking for specific protocols (e.g., FTP, MS POPTOP). Another carrier-grade requirement is the implementation of various Fair User Policies (FUP) requiring advanced software for traffic shaping and policing.

Carrier-grade routers (e.g. Cisco Catalyst 6500 series) fulfill such requirements but they are too expensive for CWNs. Fortunately, Linux and BSD systems have these features have long time ago before the boom of community networks.

Therefore, developing and deploying Linux or BSD routers of their own is the most favorite option for computer geeks in all CWNs in the country.

As mentioned above, first generation routing boxes made by Linksys and other vendors were not suitable for deployment in CWNs because of their limited features, simplicity, and instability. Therefore, most communities started building routers from ordinary desktop PCs.

As shown in Žáček (2007), even old computers with 266 MHz CPU clock can achieve about 50Mbit/s routing performance and that was just enough back in 2001 when people were absolutely happy with 64 kbit/s Internet. An old, useless computer can be found in almost every household, so the initial cost of a new wireless node is composed only of the wireless card(s), cables, and antennas.

However, a router built from an old computer has the following disadvantages:

- Higher operating costs. An ordinary desktop PC takes at least 100W of input power so it consumes electricity worth of 175 USD[2] per year. This can be reduced by factor of 20 by using an embedded PC platform like WRAP (PC Engines GmbH, 2007), Alix (PC Engines GmbH, 2008), or non-x86 architectures like Linksys WRT54G (Wikipedia, 2012), or Mikrotik Routerboard (Mikrotik Inc., 2003). The reduced power consumption also prolongs UPS runtime on batteries during power outage.
- Hardware watchdog which resets the system when it freezes is not available on ordinary desktop mainboards. A workaround solution is using an external watchdog, but this is not so elegant as above-mentioned embedded routers with integrated watchdogs.
- The management/monitoring IPMI module is not present on desktop mainboards.

- An ordinary desktop PC requires a lot of space in the rack/cabinet. Therefore, it is difficult or impossible to place it directly on the radio tower or antenna mast.

For these reasons, the Czech CWNs converged to the current state of the art: they use embedded router platforms for nodes with traffic up to 100 Mbit/s. When a higher throughput is needed, they deploy either L3 switches or routers built from low power Intel Atom mainboards, preferably with IPMI management/monitoring modules, for example (Super Micro Computer, 2009).

The most frequently used routing software in Czech CWNs was Linux Debian. In 2006-2007, the "CZFree Debian" project was even distributed preinstalled on compact flash cards by some e-shops. There were also non-Linux developments like the custom router distribution based on FreeBSD, which is still used by the Praha12.net.

The development of Linux routers in CR has a significant backing in the local kernel developers. E.g., the most frequently used QoS queuing discipline used in most Linux-based networking products is HTB (Devera, 2002), which was developed by a Czech developer. The Linux wireless driver for Atheros chipset—the ath5k—has two contributors from the Czech community (Jiří Slabý and Lukáš Turek). And there are many other developers who helped to debug and document various Linux functionalities useful for CWNs, like the OSPF and BGP routing systems.

Around 2006, some communities started deploying the commercial StarOS and Mikrotik RouterOS systems, which are also based on Linux kernel. The most frequently cited reason was user-friendly GUI and few features not available in Linux-based systems such as MAC telnet. As the time went, some communities have switched to single-vendor devices and consequently, they have started struggling with vendor-lock-in disillusions: the firmware bugs were not being fixed for long periods of time, the accumulation of bugs caused network outages and outraged both users and network administrators. Currently, most communities converged to infrastructure composed of devices produced by multiple vendors. The advantage is that a single vendor bug or a security flaw will not put down the whole network.

With the recent availability of reasonably priced 10Gbit switches and fiber optic SFP+ modules, some NFX members started upgrading their core backbones to the 10Gbit/s technology. In this area, the price difference between 10Gbit Linux router + 10Gbit L2 switch and a complete single-box Cisco Catalyst 4948E or HP5800 L3 switch is so negligible that there is currently minimal motivation for spending time on experiments with 10Gbit Linux routers. This trend could change if/when the 10Gbit becomes standard NIC on computer mainboards and 10Gbit L2 switches become as cheap as 1Gbit switches today.

Using License-Free Bands

The oldest wireless networking hardware whose deployment was reported in the CZFree.net communities were second-hand WaveLAN cards (Robison, 2012). Their production started in 1988, but for many years, they were too expensive for a generic household. In 1989, the WaveLAN vendor NCR/COMTEN[3] contributed the design to the IEEE 802 LAN/MAN Standards Committee (Robison, 2012) and therefore the WaveLAN cards were already 802.11-compliant when the standard was released in 1998/99.

In 2001, there were many 802.11-compliant cards and boxes ready on the market. The CZFree.net community tested most of them and selected the ZCOM XI-626 PCI card as the recommended product. Besides reasonable price and very good sensitivity, the paramount advantage of this card was the Proxim chipset supporting AP functionality implemented in the Host AP Linux driver (Malinen, 2012).

Around 2005, the wireless networking hardware landscape was filled by all kinds of cards

and boxes using chipsets from all major silicon vendors, but only few of them provided the AP functions supported in a Linux driver. Cheap products based on the Proxim chipsets quietly disappeared from the market and were replaced by Atheros 802.11 a/b/g products.

In 2005, the Czech national telecommunication regulator (ČTÚ) made the 5GHz band available for free by increasing the maximum allowed TX power to 1W EIRP, which allowed mid-range 802.11a radio links to be built. The new 11 non-overlapping 20MHz channels in the 5.4-5.7 GHz band brought fresh wind in all CWNs, because the 2.4 GHz band was exhausted and suffered from heavy interferences.

Aside of using cheap 802.11 wireless cards, some CWNs were experimenting also with other solutions such as Motorola Canopy, Proxim Tsunami or Trango system products, all based on proprietary protocols, but as their price/performance ratio was not that convincing, the 802.11-compatible gear stayed the most preferred solution after all.

In 2009, the Ubiquity Inc. started production of 802.11n products with TDMA MAC, which brought a substantial increase in the local wireless loop performance. After extended laboratory and field tests, the Ubiquity products were adopted by most of the NFX members and are currently the most preferred solution for the local wireless loops.

Inexpensive Wireless Links for Backbones

In 1993, Czech companies Konwes s.r.o., Miracle and SVM started producing full duplex point-to-point (P2P) 10 GHz wireless links (Novák, 2009) with 4-8 Mbit/s capacity (Miracle Group, 2001). At that time, the 10 GHz band was not license-free in the CR, but the companies were successful in lobbying for setting it free (Sova, 2010) for P2P full-duplex links. The result was the general license GP-02/1994 published in 1994. This was a major success factor for the whole Czech telecom industry, because the 10 GHz band is ideal for long-distance links. For example, with 120cm dish antennas, it is possible to build 35 km link with BER=10^{-6} and 20 dBm signal-to-noise reserve. Such a link will work reliably even through a heavy rain or snow. Because the link is full duplex, its behavior is similar to fiber optics Ethernet, but it costs only fraction of the fiber optics Ethernet link of the same length.

In 1994, another company—Alcoma—entered the 10 GHz market. One of its developers then based the Summit Company in 1999. Yet another incomer to the crowded 10GHz market was the Ath system, which also produces components of microwave links (e.g., modems) for other manufacturers. In 2006, another company KPE came with 10GHz products of its own. The last competitor on the Czech 10GHz market is Racom, which started producing 170 Mbit/s links in 2009.

The prices for the 10 GHz links have always been in the 5000-10000 USD range, but that was acceptable considering the advantages of this WLAN technology such as its full duplex, low access control overhead, minimal delay, great directionality and capability of placing many links with same channels to a single co-location site / radio tower.

The competition between the manufactures did not bring the prices lower, but forced them into constant innovation cycles bringing new and more capable products. For instance, in 2005, it was possible to buy a 16Mbits link with long-range antennas for 5000 USD.

Aside from the Czech-made microwave equipment, more expensive foreign products made by companies like Ericsson, NEC, Ceragon, Nera etc. were also deployed by CWNs for the licensed frequency bands (11 GHz, 13 GHz, and 17 GHz) in cases where:

- The 10 GHz band was already saturated at the deployment place;
- A 10 GHz link range was not enough; and
- A greater link capacity was needed more than what was provided on the 10 GHz market.

Most recently, some CWNs started to deploy the license-free millimeter waves links from Siklu, GigaBeam etc. The physical properties of the free space medium at 60-80 GHz frequencies limit the range of radio links to only 2-3 km, but it still seems to be a good alternative to fiber optics for metropolitan area networks where laying the fiber optic cables is prohibitively expensive.

Low-Cost Microwave Antennas

In 2001, there were no cheap external antennas for the 2.4 GHz band available on the market. Making a simple client-side antenna is not a rocket science; therefore, most new CWNs were producing antennas by themselves.

The most frequently used designs for sector antennas were bi-quad (Marshall, 2001) and tin can waveguide antenna (Rehm, 2007), as they were easy to make. There were some experiments with slotted waveguide antennas (Marshall, 2002), but the complexity of their construction prevented significant dissemination of this design.

As for omnidirectional antennas, some CWNs experimented with the designs using coax segments (Wireless Gumph, 2002; Martell, 2009; Moon, Songatikamas, & Wall, 2009) and coils (Lin, 2002; Pot, 2005). However, the omnidirectional antennas were cheaply available right in 2002, so the time spent by making them at home did not pay off. Moreover, using the omnidirectional antennas on an exterior AP with CSMA/CA medium access layer usually exacerbates the hidden node problem (Kapadia, Patel, & Jhaveri, 2010). Therefore, when building a new AP, the omnidirectional antennas are recommended as the option of last choice—the preferred solution is using sector antennas.

Directional antennas were also produced using several designs: the offset parabolic antennas with waveguide can feed (Morton, 2010) or biquad feed (Marshall, 2001; Dobričić, 2007) were the most frequently used ones.

However, the demand for quality antennas excited developers and manufacturers of microwave equipment and since 2002, the market got progressively saturated with hi-quality antennas whose parameters are guaranteed and stable in time. When the 5 GHz band was freed in 2005, there were already many COTS (commercial off the shelf) antennas with reasonable prices, therefore making 5GHz homebrew antennas with unknown parameters (gain, impedance, VSWR) did not make sense. In fact, since 2005, we are not aware of any Czech community network that produces homebrew antennas for the 2.4GHz or even 5 GHz bands.

Wireless Optical Links

Back in 2001, when most CWNs depended solely on the 802.11b half-duplex links, many of the communities considered the laser wireless links as the most prospective way forward. Thus, the most CZFree hardware related developments happened in projects like Ronja (Kulhavý, 2010) and Crusader (Myslík, 2008).

Ronja was the first complete open source, open hardware, "do-it-yourself" full duplex 10Mbit/s laser link introduced by Karel Kulhavý in 2002. It was then built and deployed by many CWNs around the world. The design had many odds and quirks, mainly caused by the internal electronics which was not integrated on a single PCB. Therefore, there was a good motivation for better and more repeatable solution, which led to the Crusader project. The Crusader was 10/100Mbit/s laser optical link developed by a team led by Vladimír Myslík. The development was successfully finished in 2004. The electronics was completely fitted on a PCB, which allowed mass production. However, there were problems with mechanical construction: the laser wireless links require sturdy but precise housing and me-

chanical support and this was difficult to produce cheaply in mass-production, not even mentioning the "do-it-yourself" scenario. When the 802.11n wireless links with throughput near to 100 Mbit/s started appearing in 2008, it was clear that the Crusader project is coming to a dead end.

The most important lesson learned from Ronja and Crussader were the deployment pitfalls: any side of the wireless optical link must not face the sun. Otherwise, the sun projected inside the unit through the lens will blind overheat or even burn the unit. In addition, fog, rain, and snow usually stop the transmission. Thus, it was necessary to back up the laser link by radio wireless link.

With the current state of the art, most communities do not deploy new optical wireless links anymore, because of the problems mentioned above. The more convenient solution for high availability is having two wireless radio links and running OSPF or spanning tree protocols above.

Whatever problems the Ronja and Crusader projects experienced, they helped the communities to achieve greater capacities with reasonable cost and to explore the pros and cons of wireless optical technology. The communities also learned how to support a common development project.

Fiber Optics

It is hard to find the first written note about an extensive deployment of fiber optics in a Czech community network, but one of the first CZFree. net forum threads on that topic was started by Cz-ela.net members in 2004 (Vondráček, 2004). These discussions inspired the rest of other communities to re-evaluate their perception of fiber optics, which they considered expensive and unrealistic option for building their networks, but suddenly, they found the prices of fiber optics comparable to metallic Ethernet.

The communities activated their members who were civil engineers and made projects for laying underground optical cables. After the projects were approved by the authorities, most communities hired a friendly construction company who dug the trenches for the HDPE pipes. Some of the communities, like UnArt.cz, made the excavation works themselves by organizing brigades attempting to lower the overall project costs as much as possible.

Around 2009, KLFree.net started exploring the FTTH GPON technology, which promised to lower the fiber optics installation cost and complexity even further. They decided to buy the technology and started deploying it massively. They are satisfied with the decision, although they admit the price point is almost the same like with standard FTTB. The GPON has these advantages:

- The households are electrically isolated from the rest of the network,
- The household units support IPTV and VOIP,
- There are no active devices needed along the infrastructure,
- In a block of flats, there is no need to route cable bundles from the Ethernet switch to each flat. Instead, there is single optical riser cable, which usually fits in the gooseneck tubes used for TV antenna cabling or door bells.

We are currently not aware of any other community network who would invest into the GPON technology, but this will probably change in the near future, because the prices are falling every other month. Most of the communities are still waiting for the turnaround point when FTTH GPON becomes more cost-effective than the classic FTTB and then they switch to it progressively in the process of regular network devices upgrades.

Open Source Software

Every community network needs system software for purposes such as automation of billing and accounting of members' payments. It is also required to keep records of the IP addresses assigned to the members. By 2001, there were no free open-source CRM/ERP/ software available and thus most communities started developing systems of their own. Most of these systems were not open-sourced because of the code immaturity and lack of developers with free time needed for supporting installations at third parties. The only open source system created by the CZFree.net movement which is freely available to anybody is Freenetis.org (Kliment, et al., 2007). Another option for NFX members are the BlueAngel and RedDevil systems—probably the most advanced network management systems we have ever seen, because it automates the network management almost completely. There were plans to put together the features of Freenetis and BlueAngel in a new software project. However, this project is suspended until fund is secured and/or time is allocated for it.

Aside from developing information systems of their own, some CWNs considered systems targeted at commercial ISPs. In the CR, the most popular system is ISPadmin (NET service solution, s.r.o., 2012), but there are also other systems like Kompex (Alfeus s.r.o., 2012), CIBS (Orange & Green, 2012), etc. There is also an open source LMS (Mozer, et al., 2012) made in Poland. All these systems could be used for non-profit subjects, but a lot of tweaking and development would be necessary.

Another area of development is monitoring of the networks. Again, some smaller projects like CaLStats (Krupa, 2012a), WeWiMo (Krupa, 2012b), VisualOSPF (Turek, 2012), Mapstats (Klír, 2012), etc. have been created by community members, but most often, other open source systems are used for monitoring the networks—e.g. Nagios, Cacti, or Zabbix to name a few.

There are also some closed-source proprietary single-vendor systems used—typically Mikrotik Dude (Mikrotik Inc., 2012) or Ubiquity AirControl (Ubiquiti Networks, Inc., 2012).

LESSONS LEARNED AND FUTURE PLANS

During the last 10 years, the members of Czech CWNs have tried all possible and impossible technologies in order to build and maintain reliable networks with the lowest possible cost. Many successful projects can be considered as patterns for other countries. The following are some examples:

- The Tier2 network operator established by the CWNs as the national association NFX, which helps to achieve the best possible Internet connectivity prices by using the aggregation approach;
- The fiber optics projects realized in Czech CWNs. The extent of these projects is much larger than what the local commercial ISPs usually do. Moreover, CWNs have built fiber optic cables even in rural areas, that would never be covered by commercial ISPs; and
- The open source software systems developed for CWNs such as the FreenetIS software.

However, there were also some significant failures. For instance, a lot of human energy was invested in developing, producing, deploying, and maintaining wireless optical links. However, the long-term experience from several years of operation did not match the expecations. Eventually, the wireless optical links are consecutively replaced by wireless radio links.

In addition, community members have built experience with large-scale deployment of products made by some well-established vendors. A typical problem of many low cost products is the capacitor plague (Wikipedia, 2004) which cannot be detected in laboratories or field tests, but will ruin the network reliability after a long period of time.

The Czech telecommunication industry is influenced by the innovation cycles in the IT sector. However, while a five year old desktop computer is still perfectly sufficient for most of today's tasks, a 5 years old wireless link or a network switch is almost useless. For example: in 2007 UnArt association bought a 150Mbit/s Ceragon FibeAir 1500 link to use it as a backbone. The link had to be upgraded to 300 Mbit/s in 2010 and then to 400 Mbit/s in 2011. In 2012, the 400Mbit/s link was running over the top and UnArt decided to lease a 10Gbit/s fiber optics line. This is a typical example of permanent investments which must be done for keeping the network capacity adequate to the growing needs of end users. In addition, all operators are decreasing their prices and their profits are diminishing because of strong competition from CWNs and even from their end-users. Simply, lay users can set up a low-cost access point on the roof and share it with their neighbors.

All Czech ISPs are aware of these trends and those who do not have enough resources to struggle with the competition would eventually sell their business to a bigger operator. For several years, the most acquisition-hungry company has been the incumbent operator O2 Telefonica. Their strategy in the ISP buyouts looks quite random – the ISP size does not matter, neither the quality of its network (e.g. the share of the fiber optics and wireless links). Acquisition contract negotiations details have not leaked in any single case and thus we cannot speculate about the strategy of O2. From the customers' point of view, the acquisitions has been almost invisible—the acquired companies are kept running with their original employees, tariffs and sometimes the original infrastructure, if it was sound and reliable.

Working on sustainability, the members of NFX have improved their strategy and included more members who have common interests. In other words, the NFX group is getting big enough to resist the market pressures. In addition, with the help of its commercial subsidiary Freetel, NFX is even big enough to create pressures of its own.

Despite the success of these networks, there are many challenges that need to be addressed. Although the NFX members have very similar interests, not all members are aware of these interests. In addition, a commercial operator of the same size can easily adopt and deploy a chosen technology. This is not easy in the case of NFX because every member association has different priorities and possibilities. Thus, the resource aggregation savings are not optimized as they could be in a single-headed organization.

NFX has the broadest spectrum of IT professionals from all Czech regions and its democratic decision making process provides a good guarantee that the aggregated money are spent only for future-proof projects or technologies and with minimal overheads.

The NFX leaders are planning many ambitious projects for the near future. For achieving even better foreign connectivity prices, they want to establish peering partnerships with similar projects abroad. They also have a plan to strengthen the research and development activities, particularly in the area of fiber optics, GIS, community applications, quality of service, and security. Another challenging task is to improve the government and structure of NFX

CONCLUSION

CWNs in the CR have contributed to the current state of the art of low-cost CWNs solutions. In addition to the development of systems of their own, they have tested and published their identified pitfalls dead ends, which should be avoided. They finally converged to a rather small set of proven reliable set of hardware and software solutions.

The driver of developing low cost networks is the hard work of many enthusiasts who have been spending several years of their free time in developing the missing pieces or testing low cost COTS networking products to find the few "needles"—devices usable in a large-scale network—in the "haystack" of useless garbage. As the CR is the most important European market for outdoor wireless hardware especially in the license-free bands. Our recommendations and test

results helped vendors to strengthen the capability, security or quality of many commercial hardware products in the market. This is closely related to another important role of CWNs—the incubating one. As CWNs typically use many redundant data links, they are great platforms for testing beta samples of new products. Network equipment manufacturers in CR are well aware of this and most of them appreciate close cooperation with a community network exactly for this purpose.

The Czech CWNs have played an important role in proliferation of the Internet between the public in the country. In addition, they are strongly committed to various public good activities. Typical examples are the development of open source systems and building the capacity of community members. They have raised a significant number of experts, who acquired their hands-on experience while solving real-world problems without the risks of impending penalties they would face up in a commercial ISP if something goes wrong. Many of active members were students, who developed their diploma or dissertation projects within these networks.

With recent developments in NFX, we believe that the future of Czech non-profit CWNs is brighter. This is mainly because NFX finally reached critical mass of stable resources needed for a long-term sustainability and success.

However, replicating such projects in other countries depends on the specific social stings of these countries. Finally, we are seeking cooperation opportunities with similar projects worldwide to leverage this movement.

ACKNOWLEDGMENT

This work is supported by the European Regional Development Fund under the project CEBIA-Tech No. CZ.1.05/2.1.00/03.0089. We want to thank NFX.cz members for their valuable and creative feedback, namely Petr Lázňovský, Daniel Tencl, Pavel Kocourek, and Adam Přibyl.

REFERENCES

Abdelaal, A. M., & Ali, H. H. (2012). Measuring social capital in the domain of community wireless networks. In *Proceedings of the Hawaii International Conference on System Sciences,* (pp. 4850–4859). Los Alamitos, CA: IEEE Computer Society.

Alfeus, o. (2012). *Systém komplex.* Retrieved April 12, 2012, from http://www.alfeus.sk/index.php/komplex/

Bula, J. (2007). *Measuring the RF parameters of WiFi devices.* (BSc. thesis). FAI UTB ve Zlíně.

CESNET. (2012). *Czech internet celebrates its 20th anniversary today.* Retrieved February 27, 2012, from http://www.ces.net/doc/press/2012/pr120213.html

ČTÚ. (1997). Gerální povolení GP - 01/1994. *Telekomunikační věstník, 1.* Retrieved from http://www.ctu.cz/cs/download/vseobecna-opravneni/archiv/tv_mimoradna_castka_01_1997.pdf

ČTÚ. (2000). Gerální licence GL - 12/R/2000. *Telekomunikační věstník, 9,* 15–16.

Devera, M. (2002). *Hierachical token bucket theory.* Retrieved August 30, 2012, from http://luxik.cdi.cz/~devik/qos/htb/manual/theory.htm

Dobričić, D. (2007). Efficient feed for offset parabolic antennas for 2.4 GHz. *antenneX, 128.* Retrieved from http://www.qsl.net/yu1aw/ANT_VHF/fid24ghz.pdf

Dulík, T. (2012). *Methods for interference mitigation in wireless networks.* (Doctoral Thesis). FAI TBU in Zlin. Retrieved from http://zamestnanci.fai.utb.cz/~dulik/dissertation/

European Commission. (2009). *14th report on the implementation of the telecommunications regulatory package - 2008.* Retrieved March 5, 2012, from http://ec.europa.eu/information_society/policy/ecomm/library/communications_reports/annualreports/14th/index_en.htm

European Commission. (2010a). *15th progress report on the single european electronic communications market - 2009.* Retrieved March 5, 2012, from http://ec.europa.eu/information_society/policy/ecomm/library/communications_reports/annualreports/15th/index_en.htm

European Commission. (2010b). Broadband access in the EU: Situation at 1 July 2010. *Digital Agenda for Europe.* Retrieved March 5, 2012, from http://ec.europa.eu/information_society/newsroom/cf/item-detail-dae.cfm?item_id=6502&language=default

European Commission. (2012). *Europa digital agenda scoreboard - Pillar 4.* Retrieved August 11, 2012, from http://ec.europa.eu/information_society/digital-agenda/scoreboard/docs/download/broadband%20_country_charts_2012.xls

Eurostat. (2012). *Broadband and connectivity - Households.* Retrieved August 13, 2012, from http://appsso.eurostat.ec.europa.eu/nui/show.do?dataset=isoc_bde15b_h&lang=en

Fijnaut, C., & Paoli, L. (2004). *Organised crime in Europe: Concepts, patterns and control policies in the European Union and beyond.* Berlin, Germany: Springer.

Forlano, L., Powell, A., Shaffer, G., & Lennett, B. (2011). From the digital divide to digital excellence: Global best practices to aid development of municipal and community wireless networks in the United States. *New America Foundation.* Retrieved from http://www.newamerica.net/sites/newamerica.net/files/policydocs/NAFMunicipalandCommunityWirelessReport.pdf

Fryšták, V. (2007). *System for localization of users in a wireless network.* (MSc. Thesis). FAI UTB in Zlin.

Grange, B. (1999). *Internet par le câble en France: Le comparatif.* Retrieved March 7, 2012, from http://www.journaldunet.com/dossiers/cable/comparatif.shtml

Grygárek, P. (2012). *SPS course - Student projects results*. Retrieved August 29, 2012, from http://wh.cs.vsb.cz/sps/index.php/ SPSWiki:Port%C3%A1l#Realizovan.C3.A9_ studensk.C3.A9_projekty

Haleš, B. (2005). *A wireless network implementation*. (BSc. thesis). FAI UTB ve Zlíně.

Jaroš, J. (2005). *Building wifi network with internet access in a remote rural area*. (BSc. Thesis). FAI UTB ve Zlíně.

Jiří Peterka. (2001). *Historie naší liberalizace, díl VIII: Tarif internet 99*. Retrieved March 5, 2012, from http://www.earchiv.cz/b01/b1127001.php3

Johnson, S., Porta, R. L., de Silanes, F. L., & Shleifer, A. (2000). Tunnelling. *SSRN eLibrary*. Retrieved from http://papers.ssrn.com/sol3/papers.cfm?abstract_id=204868

Kapadia, V. V., Patel, S. N., & Jhaveri, R. H. (2010). Comparative study of hidden node problem and solution using different techniques and protocols. *arXiv:1003.4070*. Retrieved from http://arxiv.org/ abs/1003.4070

Kirstein, P. T. (2004). European international academic networking: A 20 year perspective. In *Selected Papers from the TERENA Networking Conference*. TERENA.

Kliment, M., Fibich, O., Sviták, J., Dulík, T., Ševčík, R., Daněk, P., & Rozehnal, M. (2007). *FreenetIS*. Retrieved March 9, 2012, from http:// freenetis.org

Klír, J. (2012). *Mapstats*. Retrieved May 31, 2012, from http://www.jklir.net/?p=mapstats

Krupa, J. (2012a). *CaLStats*. Retrieved May 31, 2012, from http://www.mobilnews.cz/honza/ calstats

Krupa, J. (2012b). *WeWiMo*. Retrieved May 31, 2012, from http://www.mobilnews.cz/honza/ wewimo

Kulhavý, K. (2010). *Ronja: Free FSO link project*. Retrieved July 8, 2012, from http://ronja. twibright.com/

Lin, S. C. (2002). Guerilla.net/Lucent/Maxrad *collinear omni*. Retrieved August 22, 2012, from http://www.lincomatic.com/wireless/collinear. html

Malinen, J. (2012). *Host AP linux driver for intersil prism2/2.5/3 wireless LAN cards and WPA supplicant*. Retrieved July 8, 2012, from http:// hostap.epitest.fi/

Marshall, T. (2001). *Biquad feed for primestar dish*. Retrieved August 22, 2012, from http:// www.trevormarshall.com/biquad.htm

Marshall, T. (2002). *Slotted waveguide 802.11b WLAN antennas*. Retrieved August 22, 2012, from http://www.trevormarshall.com/waveguides.htm

Martell, M. (2009). *Build A 9dB, 70cm collinear antenna*. Retrieved August 22, 2012, from http:// www.rason.org/Projects/collant/collant.htm

Matúšů, J. (2008). *A system for monitoring large area networks*. (MSc. Thesis). FAI UTB in Zlin.

Mikrotik Inc. (2003). *RouterBoard products*. Retrieved August 27, 2012, from http://router-board.com/

Mikrotik Inc. (2012). *The dude*. Retrieved April 12, 2012, from http://www.mikrotik.com/thedude.php

Miracle Group. (2001). *Company milestones*. Retrieved May 30, 2012, from http://miracle.cz/ czech/group_mezniky.htm

Moon, I., Songatikamas, T., & Wall, R. (2009). *Antenna project*. Retrieved from http://sjsulug. engr.sjsu.edu/rkwok/projects/Omni_and_Bi-quad_antenna_2009.pdf

Morton, C. (2010). *Building the wifi parabolic antenna by dxzone.com*. Retrieved August 22, 2012, from http://www.dxzone.com/cgi-bin/dir/ jump2.cgi?ID=17641

Mozer, Ł., Zapalski, M., Antoniuk, R., Drewicz, K., Machniak, A., Chiliński, T., et al. (2012). *LMS - LAN management system.* Retrieved April 12, 2012, from http://www.lms.org.pl/?lang=en

Myslík, V. (2008). *Crusader FSO link.* Retrieved July 8, 2012, from http://crusader.eu/

NET Service Solution. s.r.o. (2012). *ISPadmin: Systém pro správu sítí.* Retrieved April 12, 2012, from http://www.ispadmin.eu/

Novák, P. (2009). 1993 - Lidé a firmy. *Orcave Company Blog.* Retrieved July 8, 2012, from http://blog.orcave.com/index.php?itemid=48&catid=1

Orange & Green. (2012). *CIBS - Clever ISP business service.* Retrieved April 12, 2012, from http://www.cibs.cz/

PC Engines GmbH. (2007). *WRAP - Wireless router application platform.* Retrieved August 25, 2012, from http://www.pcengines.ch/wrap.htm

PC Engines GmbH. (2008). *ALIX system boards.* Retrieved August 25, 2012, from http://pcengines.ch/alix.htm

Pehe, J. (1994). Civil society at issue in the CR. *RFE/RL Research Report, 3*(32).

Peterka, J. (2001). *Historie naší liberalizace, díl III: Liberalizace Internetu v ČR.* Retrieved February 27, 2012, from http://www.earchiv.cz/b01/b1023001.php3

Peterka, J. (2002). *ČTÚ zakázal Telecomu jeho ADSL!* Retrieved March 7, 2012, from http://www.earchiv.cz/b02/b0702001.php3

Petr Anderle. (2005). *Máme na čem stavět: Útržky z historie občanské společnosti.* Retrieved March 8, 2012, from http://www.cs-magazin.com/index.php?a=a2005032099

Pithart, P., Suk, J., Přibáň, J., Čarnogurský, J., Marek, J., & Motej, O. (2006). *Transformation: The Czech experience.* Prague, The Czech Republic: People in Need.

Pot, M. (2005). *Home-brew Compact 6dBi collinear antenna.* Retrieved August 22, 2012, from http://martybugs.net/wireless/collinear.cgi

Procházka, V. (2012). *Vczela.* Retrieved from http://www.czela.net/pub/czela/info/

Rehm, G. (2007). *How to build a tin can waveguide antenna.* Retrieved August 22, 2012, from http://www.turnpoint.net/wireless/cantennahowto.html

Robison, B. (2012). *WaveLAN cards.* Retrieved from http://en.wikipedia.org/wiki/WaveLAN

Šlinz, P. (2012). *Issues of highly stressed wireless networks based on 802.11b/g standards.* (MSc. Thesis). Masarykova univerzita v Brně. Retrieved from http://is.muni.cz/th/208329/fi_m/

Smithfarm, L., Lemmio, J., & Voženílek, P. (2011). Eurotel company: The history. *Wikipedia.* Retrieved March 5, 2012, from http://en.wikipedia.org/wiki/Eurotel

Sova, M. (2009). *Plzeňské PilsFree trnem v oku komerčním providerům: Komu slouží dezinformace?* Retrieved July 8, 2012, from http://www.internetprovsechny.cz/plzenske-pilsfree-trnem-v-oku-komercnim-providerum-komu-slouzi-dezinformace/

Sova, M. (2010). Soumrak volného 10 GHz pásma aneb kdo má zájem na jeho zarušení? *Internet pro všechny.* Retrieved July 8, 2012, from http://www.internetprovsechny.cz/soumrak-volneho-10-ghz-pasma-aneb-kdo-ma-zajem-na-jeho-zaruseni/

Sporek, J. (2005). *Server, router and WiFi AP providing access to university network for students and guests.* (BSc. Thesis). FAI UTB ve Zlíně.

Super Micro Computer. (2009). *Supermicro solutions based on Intel® Atom™ processors.* Retrieved August 27, 2012, from http://www.supermicro.com/products/nfo/atom.cfm

Sviták, J. (2007). *Modeling, simulation and throughput analysis for the 802.11 protocols family.* (MSc. Thesis). FAI UTB in Zlin.

Turek, L. (2009). *Řízení toku v přístupových bodech bezdrátové sítě IEEE 802.11*. (Diplomová Práce). Praha: MFF UK. Retrieved from http://8an.cz/papers/RizeniTokuWiFi.pdf

Turek, L. (2012). *VisualOSPF*. Retrieved May 31, 2012, from http://intra.praha12.net/ospf/

Ubiquiti Networks, Inc. (2012). *AirControl beta*. Retrieved April 12, 2012, from http://www.ubnt.com/aircontrol

UnArt. O. S. (2009). *UnArt revue*. Retrieved from http://unart.cz/casopis.htm

Vágner, A. (2011). *Real performance of devices operating on 802.11n*. (MSc. Thesis). Brno University of Technology. Retrieved from https://www.vutbr.cz/en/studies/final-thesis?zp_id=40262

Vilímovský, M. (2005). *Community computer networks*. (BSc. Thesis). Vysoká škola ekonomická v Praze, Fakulta informatiky a statistiky, Katedra informačních technologií, Praha.

Vondráček, M. (2004). *CZFree forum - First fiber optics project in Czela.net*. Retrieved May 30, 2012, from http://czfree.net/forum/showthread.php?s=&threadid=10139&perpage=16&highlight=&pagenumber=1

Wikipedia. (2004). Capacitor plague. *Wikipedia, the free encyclopedia*. Retrieved from http://en.wikipedia.org/w/index.php?title=Capacitor_plague&oldid=508636355

Wikipedia. (2012). Linksys WRT54G series. *Wikipedia, the free encyclopedia*. Retrieved from http://en.wikipedia.org/w/index.php?title=Linksys_WRT54G_series&oldid=508160017

Wireless Gumph. (2002). *Easy homemade 2.4 Ghz omni antenna - Gumph*. Retrieved August 22, 2012, from http://wireless.gumph.org/articles/homemadeomni.html

Žáček, P. (2007). *Measurement and optimalization of routing performance*. FAI TBU in Zlin.

Zandl, P. (2001). Internet proti monopolu - Tři roky poté. *Lupa.cz*. Retrieved August 29, 2012, from http://www.lupa.cz/clanky/internet-proti-monopolu-tri-roky-pote/

ENDNOTES

[1.] In 2001, the minimum salary in the Czech Republic was 130 USD/month.

[2.] The average Czech electricity tariff 0.20 USD/kWh was used to calculate this number.

[3.] NCR was acquired by AT&T in 1991, then spun-off as Lucent in 1995, subsidized as Agere Systems in 2000 and acquired by Proxim Wireless Corporation in 2002.

Chapter 13
Lessons Learned from Grassroots Wireless Networks in Europe

Gwen Shaffer
California State University – Long Beach, USA

ABSTRACT

Grassroots groups in a number of European countries are building Community Wireless Networks (CWN) on small budgets. In underserved regions, CWNs are even surfacing as the principal Internet Service Providers (ISPs). These networks have identified and implemented innovative strategies for providing connectivity—encompassing aspects ranging from software development to infrastructure design and skills training. In other words, these grassroots Wi-Fi networks mobilize human, technical, and financial resources to create sustainable alternatives to telephone and cable companies. This chapter provides an understanding of both the strengths and weaknesses of these initiatives. The authors use data from action research and interviews with leaders and participants of six successful community Wi-Fi networks in Europe. The findings show that these ad hoc initiatives are forcing local incumbent ISPs to lower prices and alter terms of service agreements. In addition, these projects broaden the public sphere, create opportunities for civic engagement, and transfer knowledge among community members. The chapter suggests that community wireless networks should be fostered by governments and the European Union in order for them to function as true alternatives to conventional ISPs, particularly in the last mile. They conclude the chapter with key learned lessons and policy implications.

DOI: 10.4018/978-1-4666-2997-4.ch013

BACKGROUND

Grassroots groups in a number of European countries are building large-scale public telecommunication infrastructures on extremely small budgets. The majority of these projects rely on a model of peer-to-peer networking. Instead of information passing from "one to many," it may travel from "many to many" (Castells, 2007; Rafaeli & LaRose, 1993, p. 277). Because a great deal of excess capacity exists in corporate-owned broadband networks and personal networks, this type of bandwidth redistribution and sharing is costless to the giver. Mesh networks, which are created by users themselves, grow *virally*. The design includes at least one access point with a direct connection to the Internet—via fiber, cable, or satellite link—and nodes that hop from one device to the next. As the popularity of these networks expands and their social objectives are advertised, new people join the network and share their nodes. As a result, signals have shorter distances to hop, higher bandwidth is available, and more redundancy is built into the system, ultimately strengthening the network (Rowell, 2007).

In underserved regions of Europe, wireless community networks are even surfacing as the principal Internet Service Providers (ISPs). In other words, the rapid growth of these projects suggests that community, neutral and free networks can function as true alternatives to conventional operators, particularly in the last mile. This research finds that these projects are creating opportunities for civic engagement and public participation for their members. Additionally, the data collected for this study finds out that grassroots Wi-Fi networks in Europe are helping to mobilize human, technical, and financial resources—simultaneously providing affordable broadband connectivity and advancing technology. It also examines the practical and theoretical implications of these initiatives. In particular, the research examines how both ISPs and government entities are responding to CWNs that use mesh technology, and whether their reactions signify a shift in the political economy of telecommunications. This study is based on action research supported by data collected through qualitative interviews with both leaders and participants of six established mesh networks in Europe.

THEORETICAL FRAMEWORK

The resource mobilization theory applies economic and organizational concepts to contemporary social movement theory (Meyer, 2005), and considers social movements as augmenting mainstream politics rather than as offering an alternative to them. This approach offers an ideal framework for understanding how peer-to-peer broadband networks emerged and how participants sustain them. The resource mobilization approach emerged as a sub-discipline of social movement theory during the early 1970s, a historic period that bore witness to large-scale protests and high-profile political actions. The Civil Rights and anti-Vietnam War movements, along with various groups struggling against colonialism in Asia, Latin America, and Africa (Little, 2008), forced sociologists around the world to adjust the lens through which they studied social movements by explaining the rational, purposive facets of activism (Waterman, 1981). Subsequently, communication scholars began using these concepts to ground their own research. While this approach is not universally accepted, a critical point made by resource mobilization theory is that average citizens would lack the know-how to participate in political action and, thus, must rely on professional advocacy organizations. Therefore, core group members develop a strategy to catalyze the sentiments expressed by those who feel alienated (McCarthy & Zald, 1987). They attract financial and human resources, seize media attention, foster relationships with people in power, and develop an organizational structure (Kendall, 2006; Hannigan, 1985). Resource mobilization theory assumes that a social movement will fail to produce change without adequate resources and alliances (Gamson, 1975; Tilly, 1978; McAdam, 1982).

The activists behind these CWNs possess access to the media and relationships with policy-makers. People initiating and joining non-profit signal sharing networks, the core of this grassroots movement, typically have technology and IT knowledge (Cho, 2008; Abdelaal & Ali, 2012). In fact, they are in a position to mobilize resources only because they have acquired these resources. This research finds that those who join signal-sharing communities are linked through common interests, including the desire for ubiquitous connectivity, digital inclusion, and forcing change in the telecommunications industry. Finally, many broadband subscribers who choose to share their wireless signals do so for ideological reasons, with the intention of making a political statement (Lawrence, Bina, Culjak, & El-Kiki, 2007). Still, the question arises: Why do people join social movements when they can benefit from the work of others who are willing to bear the costs of achieving a common good? What are the incentives to contribute rather than free-ride (Olson, 1965)? These questions must be posed because the costs of defending an interest are obvious. These costs (e.g. time, money, efforts, and even safety) are often sacrificed for the common good. As with CWNs, successful new social movements result in collective benefits, and participants often get involved in hopes of obtaining some personal resource (McCarthy & Zald, 1987).

It would be remiss to not acknowledge arguments articulated by critics of resource mobilization theory. In general, these theorists contend that this sub-discipline of social movement theory deemphasizes both the complexities that define grievances and the role that ideology plays in social movements. Scholars point out that informal actors and networks, not just those who are socially integrated, participate in social movements (Fox, Piven, & Cloward, 1991). Additionally, because

the approach pays scant attention to historical contexts—especially to the structural inequality inherent in power relationships—its usefulness for explaining collective action is limited (Buechler, 2000; Canel, 1997; Kendall, 2008). New social movement theorists, most notably Touraine (1985), criticize resource mobilization proponents for defining actors by their strategies, as opposed to by key social relationships.

CONTEXT

The number of citizens using the Internet varies considerably among the countries studied as part of this research. As of 2010, the Internet penetration rate was about 75% in Austria; 71% in Czech Republic; 47% in Greece; 66% in Spain; and 83% in Germany (Council of Europe/ERI Carts, 2012; Internet World Stats, 2012). Denmark is among the most connected nations in the world, with an Internet penetration rate nearing 90% (Internet World Stats, 2012). The European Commission is attempting to bolster broadband access and services for all EU citizens by proposing to spend nearly €9.2 billion from 2014 to 2020 on related projects throughout its member countries. The funding is part of the proposed Connecting Europe Facility initiative, and at least €7 billion would be available for investment in high-speed broadband infrastructure (European Commission, 2011).

Even without this massive government effort, the cost for high-speed broadband is dropping throughout Europe (International Telecommunication Union, 2011). Technological advancement is a key factor driving this decline. Specifically, as next generation technology and government policy have matured, it is cheaper for new entrants to offer higher speed broadband (Analysys Mason, 2011). Of course, as more ISPs compete, Internet fees drop. However, an inability to afford an ISP subscription is not the main reason informants cited for participating in a CWN. In fact, participants are part of a movement that kicked off in 2004, during the first National Summit for Community Wireless Networks, held in the United States. That movement now includes "tens of thousands of community and municipal broadband initiatives" across the globe (International Summit for Community Wireless Networks, 2010). In recent years, grassroots broadband initiatives have moved away from building Wi-Fi hotspots that blanket neighborhoods and, rather, are experimenting with technology that allows residents to securely share existing wireless connections. Projects using this type of mesh architecture are the focus of this study.

RESEARCH METHOD

Qualitative researchers develop particular ideas, perspectives, or hunches relating to the topic under investigation (Cormack, 1991) and allow the theory to emerge from the findings. This type of inquiry is meant to enhance understanding of the world from the perspective of the subject—not to impose the researcher's outlook on study participants. With this in mind, I primarily collected data through qualitative semi-structured interviews. Specifically, I conducted face-to-face interviews with leaders and participants of European community mesh initiatives between February 27, 2009 and March 18, 2009. In addition to interviews, data collection involved observation of access points and demonstrations of hardware and software programs used to manage the networks. Broadband activists were asked about the role volunteers play in sustaining their projects, as well as about their relationships with the political establishment and with ISPs. Questions also probed how policy reforms could facilitate their work.

I selected the six CWNs based on their demonstrated sustainability, number of participants, innovation and commitment to digital inclusion. The following is a brief description of the studied networks and sources of data:

- Freifunk in Berlin is an open access mesh network run entirely by volunteers since 2003. The informant for Freifunk is one of the original co-founders.
- Funkfeuer is a free network in Vienna, with a Wi-Fi signal covering about one-third of the city. The Funkfeuer informant co-founded this project.
- The Athens Wireless Metropolitan Network was created by volunteers in 2002. The informants for this project included five participants.
- Guifi.net began in a rural, underserved community in Central Catalonia, Spain. The informants for this project included the network's co-founder and two other active participants.
- Czfree.net consists of many community ISPs across the Czech Republic, connected to one another via peering agreements. I collected that data through interviews with members of KlFree.net in Kladno; KHnet.info in Kutná Hora; and Spojovaci.net in Prague.
- Djurslands.net serves about 8,000 households, institutions and firms in rural Denmark. The founder of Djurslands.net was the informant for this research.

DESCRIPTION OF THE CASES

I have studied a mix of urban and rural wireless networks. The following is a brief description of the investigated cases:

Urban Networks

Freifunk Network

Following reunification of East and West Germany in 1989, the national telephone company ripped out old copper lines before determining that laying fiber would be prohibitively expensive. As former

East German neighborhoods gentrified, the tech-savvy residents moving in became frustrated at the lack of high-speed Internet access. "I moved into this area and, as a computer specialist, I couldn't stand living in a place with no broadband," reported a co-founder of the Freifunk initiative in Berlin. In 2002, this informant placed antennas on the roof of his building, connecting 35 residents to the network. His friends, then, set up more wireless nodes and began sharing bandwidth in Berlin, with the goal of creating a highly decentralized network with no ownership. They formalized Freifunk, meaning "free radio," in 2003.

As of March 2009, this mesh network included about 1,000 nodes—blanketing 10% of the city in free Wi-Fi. To host an access point for Freifunk, participants can rent or purchase a router for about $50, then "reflash" it by downloading firmware from the group's website. Any wireless device may be used to connect to the network, and the traffic is not centrally managed. Network participants are not asked to sign a terms of service agreement either. "There is social understanding that accompanies a shared network but no written policy," the Freifunk co-founder said. Participants who subscribe to our network donate a percentage of their bandwidth to the network. "It is important that people have the freedom to decide how much and how often they want to share," the informant reported. Coverage and speed of the network varies throughout Berlin. "If you live next to someone with a fiber ISP connection, you are lucky," he added. To bolster coverage, Freifunk members installed dedicated links—antennas that extend wireless signals 200 to 300 meters—on church steeples. The Berlin initiative has inspired smaller "Freifunk" networks in the German cities of Leipzig and Weimar.

Athens Wireless

Another urban initiative is the Athens Wireless Metropolitan Network (AWMN), which has began under parallel circumstances. In 2002, high-speed Internet access remained unavailable in various sections of the city. Although the incumbent phone company did offer DSL service, it was "slow" and "expensive," according to informants. Frustrated, a group of about 10 friends from a technical Web forum connected their computers. As a result, a large signal sharing initiative has started and it claimed 3,000 participants at the time of data collection. While online gaming remains a popular aspect of AWMN, its members have created dozens of services and applications that reflect their personal interests. About one-third of AWMN participants have installed $1,300 mesh "backbone nodes" on their rooftops, informants reported. These antennas communicate with one another and serve as the primary infrastructure for the network, providing average connection speeds of 130 megabits per second (mgbs). Another 2,000 "clients" have installed the network's routing software, enabling them to connect to backbone nodes but not to extend Wi-Fi signals. AWMN is concentrated in Athens, but strategically located access points—typically on the sides of mountains—and university-owned backhauls link the network to emerging Wi-Fi projects on the islands of Evia, Aigina, and Salamina.

Czfree.net

Political realities also played a direct role in creating a CWN in the Czech Republic. After 41 years of communist rule, the peaceful "Velvet Revolution" allowed Czechoslovakia to revert to a liberal democracy in November 1989. When the country split into Slovakia and the Czech Republic three years later, the Czech government swiftly privatized industries such as banks and manufacturing. Under corporate ownership, the phone company instituted a rate structure beyond the means of the typical Czech family. In the mid-1990s, Internet subscriptions cost as much as $110 per month and required customers to sign contracts committing them to up to five years of service, informants reported. When wireless routers became available in 1998, students in Prague purchased dial up service and began sharing bandwidth. Commercial DSL service made it possible for projects to formalize. They created websites and actively recruited members, and many began writing their own routing protocols and building antennas.

Today, dozens of community initiatives throughout the country belong to the umbrella organization Czfree.net. Although Czfree.net is loosely organized, most participating networks have agreed to peer—or seamlessly transmit data over their infrastructures—creating a de facto nationwide network with two key benefits. First, interconnectivity greatly improves the flow of data files. Second, individual networks gain leverage when negotiating bandwidth prices with ISPs. KlFree.net in Kladno is among the largest community Wi-Fi networks in the Czech Republic, with about 5,000 access points. It has evolved from wireless signal sharing to 75% of participants directly connected to fiber. Spojovaci.net is a smaller initiative with 200 meshnodes and 5,000 participants. "We started out using Pringles cans because a real antenna was too expensive," a key leader reported. Today, members invest about $200 each to buy open source antennas and Wi-Fi cards. Between 2003 and 2009, KHnet.info grew to include 120 mesh nodes and one direct gateway to the Internet. At the time of data collection, more than 2,000 households paid about $20 per month to subscribe to the network. Czfree.net initiatives support an array of applications: Web hosting, email, anti-virus software, game servers, voice, and video.

The evolution of each of the community Wi-Fi networks discussed in this section is unique. Even so, they all serve as examples of self-interest group that expanded to encompass the public interest. By decreasing the fee of Internet access, these viral networks enable more people to access information and express ideas online. In step with the political model of resource mobilization theory is the concept that social actors realize the flaws inherent in the existing power structure and make

up their minds to resolve the problem themselves (McAdam, 1982). The decision by early adopters of mesh technology to share bandwidth can be explained no other way. These actors did not wait for government intervention or a shift in corporate policy. Instead, they created broadband networks that fit both their technology needs and their budgets.

Finally, the ingenuity exhibited by the founders of these networks makes the entrepreneurial model of resource mobilization theory is particularly applicable to European Wi-Fi initiatives. The theory purports that key organizations act as "carriers" (McCarthy & Zald, 1987, p. 12) of social movements, bringing together various groups invested in a mutual cause—affordable broadband, in this case. The theory argues that social actors embrace new technologies as they become available. Some of the networks originated with dial-up bandwidth sharing, then migrated to DSL bandwidth sharing. Today, several grassroots Wi-Fi communities are laying fiber. Another tenet of resource mobilization theory asserts that established institutions participate in social movements, even when doing so is secondary to their mission (McCarthy & Zald, 1987). This reflects a key aspect of European mesh networks, which count schools, hospitals, libraries, and non-governmental agencies among their members. For instance, one network in the Czech Republic partners with a town government to provide free broadband in schools, and allows medical providers and social service agencies to use bandwidth free of charge. Non-profit institutions and schools receive free connections through another Czfree.net network. This participation is significant, according to McCarthy and Zald (1987), because resources must also flow *toward* wireless community networks in order for them to persuade others to join the cause.

In opposition to a proposed Internet data retention law, online privacy activists set up a single Wi-Fi hotspot near Vienna's Museum Quarter in 2003. They intended only to make a political statement, but the effort led the activists to consider long-term uses for wireless nodes, which were new on the market at the time. They contacted a wireless ISP that had gone bankrupt and the company agreed to give away its Wi-Fi transmitters on the condition that the devices not be used for a commercial network. Suddenly, the activists owned 10 strategically located access points around the city. In late 2003, the co-founders of Funkfeuer hosted a public meeting where they recruited volunteers to help create a mesh network in Vienna. As of March 2009, Funkfeuer had 400 users, including 240 node hosts that place mesh radios and antennas on their rooftops (an upfront investment of about $130). The network owns a 5-gigahertz fiber optic link to the Vienna Internet Exchange, allowing members to share bandwidth for free.

The trajectory of these events vividly illustrates how community networks are redefining the political economy of telecommunications. When the commercial ISP lost money, Funkfeuer converted the company into a project driven by ideology, rather than revenue. Under this new structure, profits materialized in the form of technological innovation and a strengthened sense of community—as opposed to dividends paid to shareholders. Funkfeuer shifted the typical perception of "success" away from its monetary connotation and toward an explanation that privileges public good. This newly defined political economy has gained traction in cities throughout Europe. As Silverstone (2004) noted, it is possible for regulations to also tilt away from corporate priorities and toward critical social and cultural practices.

Rural Networks

In addition to these CWNs in the urban areas of Berlin, Vienna, Athens, and Prague, this study includes analysis of two rural European wireless community initiatives. Both emerged out of necessity, after incumbent ISPs declined to deploy reliable broadband in these sparsely populated regions.

Djurslands.net

In the late 1990s, residents of Denmark's Djursland Peninsula repeatedly asked Tele Denmark to provide residential DSL service. "We realized rural people would fall behind if we didn't do something about it," the project founder said. When the incumbent phone company declined, the informant negotiated with 35 smaller Danish ISPs. "One after another, they said building the infrastructure for rural people was too expensive," he reported. In 2000, the price of access points had fallen to about $10,000, making it financially feasible to connect the region wirelessly. The Djurslands.net informant purchased discounted radios in bulk. Meanwhile, the village of Glesborg built a 50-meter high tower—selling it to the Internet activists for a symbolic single kronen (less than $1). Omni antennas on the tower communicated with devices on the roof of a sports hall 1.5 km away—creating the network's initial link. Djurslands.net officially launched in May 2003, and has evolved to include about 10 fiber gateways to the Internet; hundreds of strategically located access points in villages throughout the peninsula; and wireless connections to the 8,000 households that subscribe (as of mid-2009).

Guifi.net

Although Spanish incumbent carrier Telefónica offered DSL in some parts of rural Catalonia, Spain—about 120 kilometers outside of Barcelona—the service was expensive and unreliable, according to a Guifi.net informant. In 2004, this technology activist conceived the idea to create a CWN by attracting entire village governments, as opposed to individuals. Five years later, 23 town councils subscribed to an ISP, in turn sharing bandwidth with residents via a wireless backhaul. Governments installed 100 antennas on street lamps and roofs throughout their villages, and each of these access points has the capacity to support 30 Internet connections. With an average population of 2,000 to 3,000 residents, it costs just a few thousand dollars to deploy nodes throughout an entire village. In order to connect to the Guifi. net signal, individual residents and businesses purchase rooftop antennas.

The origins of these two rural networks, Djurslands.net and Guifi.net, are examples of how the political economy of the telecommunications industry deters incumbent ISPs from entering certain markets, regardless of demand for connectivity. The privatization of the European telecommunications industry has led to the deployment of infrastructure and services targeting customers with the most potential to generate revenue, "even if that means greater attention to linking metropolitan centers in global networks, rather than extending networks into rural and generally underserved regions" (Mosco, 1996, p. 202). As Internet infrastructure and computers permeate society and become vital aspects of economic performance, low-income rural regions risk lagging further behind (Parker, 2000). Economic development in rural areas depends significantly on IT access for businesses, non-profits and government (Lentz & Oden, 2001).

While free market principles influenced Spanish telecommunications policymaking during the early years of the Internet, the government also enacted regulations aimed at ensuring that Telefónica remained the dominant carrier (Souvirón & María, 1999). Along these same lines, Denmark's telecommunications industry transitioned from state control to private ownership in the late 1990s. Tele Denmark was forced to relinquish its monopoly in 1998, but remained by far the largest operator (Paldam & Christoffersen, 2004). Grimes (2003) notes that since the liberalization of EU telecommunications policies, universal service obligations have been abandoned and it is difficult for rural communities to compete. The political economic conditions in Spain and Denmark help explain why the founders of Djurslands.net and Guifi.net opted to create their own wireless community networks, as opposed to waiting for

incumbent ISPs to connect their sparsely populated regions. These social actors took steps to prevent being cut off from the information society. For instance, farming, fishing, and manufacturing traditionally comprised the core of Djursland's economy. Today, industry has all but vanished from the region, while former agricultural and fishing communities rely heavily on tourism. Since the late 90s, the Grenaa ferry to Sealand ceased operating, the *Daily News Djursland* folded, and the Grenaa Hospital closed along with many shops. While 15 broadband ISPs operate in the region, only Djurslands.net reaches farms and the smallest villages. Djurslands.net can be credited with ensuring the Peninsula did not become geographically and culturally isolated.

The Role of Volunteers

The success of each initiative studied in this chapter depends heavily upon contributions from volunteers. Freifunk participants in Berlin contribute a wide range of time and skills. The most passive form of involvement might mean simply installing an antenna on one's roof to support the network's backhaul. At the other end of the spectrum, 5% to 10% of members sustain the network by hosting meetings, answering technical questions in online forums, and developing software, the informant said. A unique role played by Guifi.net volunteers involves presenting information about the initiative to residents of neighboring villages. "We call this 'the wheel' because it turns on and, if we do it right, it creates momentum," one study informant reported. In fact, public presentations describing the necessary equipment, time commitments and costs are Guifi.net's most important recruiting tool. Once a village commits to participating, Guifi.net members deploy the nodes within a few months to ensure local residents remain "motivated" and "optimistic" about the network. Individual node owners are ultimately responsible for maintaining antennas on their own rooftops, but volunteers respond to questions posted to online forums and, sometimes, physically assist with repairs. Not only are Guifi.net participants resisting the dominant corporate culture, they are creating a new culture. Just as personal computers in the peer-to-peer networking movement each contribute a little bit of power to a much larger Web of machines, Guifi.net members each share their own base of knowledge with neighbors to extend and sustain the network. As Guifi.net deploys nodes throughout entire villages, the influence exerted by incumbent carrier Telefónica lessens—an example of "the edge becoming the core" (Hagel & Seeley Brown, 2005).

While a majority of Djurslands.net subscribers are "passively involved" in the network, a minority are engaged at every level. Volunteers develop hardware, repair equipment, and regularly attend management meetings hosted by one of 10 community boards throughout the sprawling service area. Djurslands.net also hosts a well-attended annual forum where attendees elect new board members, review financial information, and vote on principles for future management. "This is why it is a community network, not just a physical infrastructure," the Djurslands.net founder reported. Similarly, a "core" group of about 30 Funkfeuer subscribers participate in on-line forums, attend weekly meetings and deploy nodes. In contrast to some European networks, whose members are self-identified "techies," a significant contingent of Funkfeuer's active members are not professionally involved in IT. "We have lawyers, a heart surgeon, construction workers, and even a recovering heroin addict," the informant said.

Because each of the approximately 20 Czfree.net initiatives is unique, it is impossible to generalize about the role played by volunteers. The most formalized projects have paid staff and charge their members for highly reliable broadband connectivity; more loosely organized initiatives rely exclusively on unpaid labor and experimental technology. All three Czfree.net networks in this study depend on volunteers to help deploy and repair nodes. An informant for KlFree.net said he devotes more time and "mental energy" to the

project than to his professional job. "I spend several hours every evening dealing with administrative and technical aspects of the initiative," a Spojovaci.net leader reported. Additionally, informants for these networks agreed that word-of-mouth endorsements are their primary means of recruitment. "The best method is *jednapanípovídala*, which means 'one woman said.' It means that new members usually get information from their friends, relatives, or neighbors who are involved," the KHnet.info activist explained.

These projects depend on volunteer efforts to constantly broaden the coverage area—a kind of viral marketing effort that creates innovations and expands the public sphere with each neighbor brought into the fold. The commitments made by volunteers with these six European wireless community networks underscore the social movement principles used to frame this research.

Most members of these wireless communities care more about civic engagement than free Internet access. For those deeply involved in the peer-to-peer movement, these initiatives nurture other networks—social, technical, and economic (British Columbia Wireless Network Society, 2006). Others say any drawbacks related to sharing their bandwidth and skills are offset by the convenience of borrowing other people's free bandwidth when necessary (Efstathiou, Frangoudis, & Polyzos, 2005), as well as by the opportunity to experiment with open source technology. However, McChesney (2009) puts forth a more radical theory to explain why social movements, such as peer-to-peer networking, are gaining momentum. McChesney (2009) describes a "severe social disequilibrium" (p. 44) in which the existing system has broken down and reformers are organizing to fill the gap.

Used Technology

Participants in European Wi-Fi initiatives tend to be deeply immersed in the open source software movement, working to improve existing routing protocols and mesh equipment. As many as 60

Freifunk participants with an interest in developing firmware and other technology-related projects are known to drop by the "Hackers Lab" held each Wednesday evening. Freifunk members have optimized mesh routing firmware, including the B.A.T.M.A.N. protocol used today by community wireless initiatives around the world—as well as in the OpenMesh router sold commercially. Guifi.net volunteers developed a program that uses a proxy system to track traffic, an informant said. At the time of data collection, Funkfeuer developers were testing "a more user-friendly firmware," as well as building 5-gigahertz ring around the city. This static network was designed to extend the main uplink to all of Vienna and "allow people to connect more directly and with fewer hops," the Funkfeuer informant said. Innovation is also thriving in the Czech Republic. Among the most creative efforts involves a large community network in a city outside of Prague. Rather than spending potentially millions of dollars to dig up streets and lay fiber cables, these network members threaded fiber through dormant steam radiators.

The Athens network functions as a laboratory for members. It is a venue for them to develop antennas and satellite dish feeders used on backbone nodes, as well as to refine routing protocols and network management software. However, the dozens of applications created exclusively for the Athens wireless community are what makes AWMN distinct. As the names of some of these services imply, they mirror sites found on the public Internet: the auction site Wbay; search engines Woogle and Wahoo; and wTube, to name a few. Some members use the network to float ideas for online services they intend to eventually introduce commercially. For instance, a movie application led to negotiations with an ISP to create a video-on-demand service within the network. "[AWMN] is more important to us than the outside network because it gives us an opportunity to experiment," an informant pointed out.

Management and Organizational Concerns

Decentralization is a key aspect of peer-to-peer architecture—from distributed storage and processing, to information sharing. No server exists to coordinate the activities of the system, and no database stores global information about traffic transmissions (Pourebrahimi, Bertels, & Vassiliadis, 2005). This is analogous to how informants described *management* of their community networks, from decentralized ownership to consensus-based decision-making. Both these technical and social arrangements reflect an open source approach that privileges collaboration, sharing, and a lack of restrictions on use of infrastructure. This philosophy is in direct conflict with traditional social shaping of technology, which reflects "broad and long-term institutional power" (McDowell, Steinberg, & Tomasello, 2008, p. 37). Corporate broadband services are designed and deployed to fulfill the goals of influential organizations and interest groups—likely governments and telecommunications companies. By reinforcing the status quo, the traditional social shaping of technology has strengthened the need for alternatives and, ironically, encouraged community Wi-Fi efforts.

The six community broadband projects examined in this chapter have varied relationships with government officials and commercial ISPs—from non-existent interactions to true partnerships. These relationships have political economic implications, as well as consequences for the broadband reform movement as a whole. Freifunk represents one end of the spectrum, as this network has no official association with governmental or non-profit organizations. However, Freifunk's popularity convinced incumbent carriers to amend their terms of service agreements and allow DSL bandwidth sharing among subscribers. In addition, at the time of data collection, the group was talking to Berlin officials about "peering" with an open wireless network being deployed near tourist destinations. Such a move would dramatically expand Freifunk coverage, the network co-founder said. Some Czfree.net participants are convinced that their projects impact incumbent carriers. For instance, the cost of a typical DSL subscription has fallen to about $58 per month, "and continues to drop," one informant said. At the opposite end of the spectrum, Guifi.net's model depends heavily upon Catalonia officials signing on to the network and installing mesh antennas throughout their villages. The return on their investments comes in the form of free Internet connections for village employees, as well as new economic development opportunities. The Catholic Church also allows members of Guifi.net to install antennas on steeples—typically, the highest points in town. However, Church officials are uninvolved in network leadership "because you can't mix bishops with anarchists," the informant said. Djurslands. net's positive relationship with both local ISPs and elected leaders has resulted in two key benefits. First, the network leases municipally owned fiber, which it relies on for connecting directly to the Internet. Second, villages throughout Djurslands have built towers that host antennas. Djurslands.net also participates in two rural innovation projects sponsored by the European Union. Significantly, because of Funkfeuer's former status as a commercial ISP in Vienna, the network is a voting member of Internet Service Providers Austria. Despite membership in the association, the co-founder insisted Funkfeuer poses no threat to Telecom Austria. "It takes determination to build a node. Telecom Austria realizes most people are content to pay €25 each month for Internet," he said.

A typical AWMN participant is more interested in experimentation than connectivity. In fact, the 3,000 members of this Athens network are exclusively "technical guys," an active member said. "Installing the routing software is complicated so all the people connected to our network have IT knowledge," he said. AWMN members characterize themselves as activists. "We don't just complain about technology, we do something

about it," another network leader said. In contrast to Freifunk, Guifi.net, and Funkfeuer—all of which are philosophically committed to a flat governance structure—a legally recognized association runs AWMN. However, just about 200 people pay the $65 annual fee required to join the association. This means just 10% of AWMN members participate in officer elections and set policy for the network. It is association members who develop new software protocols, install strategic nodes, and host workshops and "Antenna Fests." Despite these contributions, some network participants vocally oppose the existence of a governing body. "The association tries to control everything. Individual members should be able to do what we want," insisted an informant who develops routing protocols. The Czfree.net broadband initiative is also run by a board of directors. Spojovaci. net's management decisions are made by elected officers, who host monthly meetings open to all members. Participants in another project, KHnet. info, elect seven new commissioners every three years, and the general membership keeps up with new developments by checking the website and blog.

While two networks examined for this study, AWMN and Guifi.net, are legally incorporated, community Wi-Fi initiatives typically function as "a movement of equals" (Neumann, 2007) that take advantage of decentralized wireless infrastructure to recruit new participants and ensure sustainability. Bauwens (2005b) coined the phrase "equipotency" to explain the concept that all participants in a project *begin* as equals. It is only through the subsequent "practice of cooperation" that levels of leadership are dictated. The equipotency characterizing a peer-to-peer network does not reject the idea of management; but it does rebuff the notion that hierarchy is predetermined by characteristics other than expertise, initiative, and ability. It is a spirit of collaboration—devoid of competition or greed—that defines the community ad hoc networking movement.

BENEFITS OF COMMUNITY WIRELESS NETWORKS

The examined case studies show that CWNs have the following benefits:

Knowledge Transfer

Reflecting the global nature of the Wi-Fi signal sharing movement, as well as its public good ethos, members of European initiatives have traveled to underserved communities across the globe to share their knowledge of building grassroots ISPs. Specifically, participants in Freifunk and Guifi.net met with broadband activists in India and Africa, while members of Djurslands.net spent time helping deploy wireless connectivity in Lithuania and Kakistan. "We just want teach people how to create open networks and improve the tools for deploying them," the Guifi.net founder reported (convictions such as this helped Guifi.net win Spain's 2008 National Telecommunications Award, accompanied by about $20,000. With this money, Guifi.net established a foundation to help develop open, free networks worldwide).

Civic Engagement

The strength of democracy may be measured against how effectively members of a society participate in the political process (Rheingold, 1993). In terms of this study, participation in the open source and broadband reform movements also symbolize forms of civic engagement. A healthy public sphere not only promotes a broad spectrum of ideas, but also ensures access to these ideas (Napoli, 1999). While the members of European Wi-Fi communities are not political in the traditional sense—they may not attend protests or campaign for candidates—they are clearly challenging the status quo. They do this in multiple virtual spaces, such as online forums and blogs, as well as during "hacker nights" and community summits. With each new innovation developed by peer-to-peer network participants, underserved members of society acquire broadband connectivity. More significantly, they obtain access to information and ideas that were, previously, within the grasp of only a select few.

This understanding of civic engagement goes beyond deliberative discourse to "developing the combination of knowledge, skills, values, and motivation to make that difference...through both political and non-political processes" (Ehrlich, 2000, p. vi). By defying customer service agreements, participants in the signal sharing movement introduce resistance to corporate policy as yet another permutation of civic engagement. Not only are these participants resisting the dominant corporate culture, they are creating a new culture. Just as networked personal computers in the peer-to-peer movement each contribute a little bit of power to a much larger Web of machines, participants each share their own base of knowledge with neighbors to extend and sustain the network.

Digital Inclusion

The goal of achieving digital inclusion varies from one network to another. Informants representing Guifi.net, Djurslands.net and Czfree.net—networks that partially or exclusively cover rural areas—were resolute that residents and businesses would suffer if they did not provide broadband access. Despite the goal of closing the digital divide, none of these grassroots initiatives purchase or subsidize equipment for low-income members. "It protects us as an association if we don't legally own the nodes...An ISP approached us but Funkfeuer doesn't own the access points so we can't sell them," the Funkfeuer leader pointed out. The Guifi.net informant agreed that "subsidizing equipment is a bad idea," but for a different reason. "It is important for users to understand that when joining the network, they are providers too. You can't be opportunistic if this is going to work," he said. Freifunk's mesh architecture also means that each node host owns an equal portion of the network. "The network is a concept, it is not an entity," this study participant reported. This structure facilitates non-hierarchal management of both Freifunk and Funkfeuer. For instance, Funkfeuer has an "official" president, but decisions are typically made by consensus, the informant said. However, this commitment to openness and equality can create new challenges. Funkfeuer's leadership struggles with how to implement stronger security measures capable of preventing spam and viruses, without infringing upon user privacy. Ultimately, software may not provide the solution. "The trick is to involve everybody in the network. If they helped build it, they will want to protect it," he said.

While digital inclusion plays an undeniable role in the mesh networking movement, pragmatism appears central to the creation of some CzFree.net initiatives. "The entire point is to get Internet for as cheaply as possible," said an informant with one regional network, Kladno.net. For other wireless community projects in the Czech Republic, expanding Internet access remains a key tenet. KLFree.net partners with a town government to provide free broadband in schools, and allows medical providers and social service agencies to use bandwidth free of charge. Non-profit institutions and schools get free connections through another Czfree.net participant called Khnet.info, which has also created several hotspots accessible to non-members. While Djurslands.net does not offer discounts for low-income families or institutions, in 2008 the network deployed 30 hotspots in Ebeltoft and Grenaa, the "big" town on the Jutland peninsula. Free access points are planned for other villages, as well. Historically, digital inclusion has been a peripheral concern for AWMN leaders. However, the demand for ubiquitous connectivity has spurred the group to deploy hotspots throughout Athens, enabling anyone with a mobile device to freely connect to the Internet from these locations. By creating free hotspots, Khnet.info, Djurslands.net and AWMN recognize that different consumers possess differ-

ent telecommunications needs. As a result, "the value proposition" for broadband infrastructures "also needs to incorporate consumer choice and flexibility" (Jayakar, 2009, p. 194). In other words, as the use of mobile devices becomes commonplace and demands for ubiquitous connectivity increase, affordability alone is not the key to increasing broadband penetration and closing the digital divide. The reality, however, is that providing free hotspots does not result in short-term profits. Therefore, the private sector is not interested in "the positive externalities" (Tapia, 2009, p. 225) of making free Wi-Fi widely available. In this sense, then, the signal sharing movement in Europe also functions as a form of resistance against capitalism, which commodifies Internet access.

A Guifi.net informant reported that the ability for businesses to run Internet applications ranging from VoIP to surveillance cameras has boosted the local economy. Local hog and cattle farmers also rely on Guifi.net to accomplish routine tasks, such as transmitting test results to veterinarians, he added. "If we don't have Guifi.net, I'm not even able to live here," said the informant, who frequently works from home. Beyond economic development, open access principles play an important role in sustaining Guifi.net. "We are trying to extend Internet neutrality to the edge by providing an alternative to the ISPs," the network founder reported. Similarly, Djurslands.net helped rescue the regional economy, as traditional sources of income like farming and fishing vanished. "People would have had to leave the area in order to compete, and only the poor would be left behind," the Djurslands.net leader said. He credits the initiative with creating 100 new jobs in each village, pointing to a printing press that subscribes to the network as an example. His goal is to connect half of Djursland's 82,000 residents. Network members pay a $325 initiation fee, which goes toward the cost of household equipment and toward a fund for future network maintenance. Additionally, participants pay $15 monthly. This revenue is just enough to cover network expenses, the Djurslands.net informant said.

With an emphasis on digital inclusion, all six of these Wi-Fi initiatives are helping create a "networked public sphere" (Benkler, 2006) that enables individuals to work from home, to become politically engaged, to participate in community decisions, and to glimpse the world beyond their own backyards. Theoretically, a networked community offers new modes of exhibiting strength, as it is able to both disperse and to concentrate power (van Dijk, 1999). It may be argued that incumbent ISPs *concentrate* power by dictating which services will be offered, in which regions and at which price-points. Wireless community projects offer an antidote to the capitalistic nature of commercial telecommunications companies by *dispersing* power. Specifically, ad hoc initiatives enable users themselves to control where nodes are deployed and which technologies to use. By bypassing entrenched institutions and corporations, ad hoc networks also play a critical role in expanding the public sphere. In contemporary society, entrée into the polity and civic engagement hinge upon access to a reliable and affordable Internet infrastructure. Networks also unify participants around "a sense of shared values and mutual interdependence that comes from social interaction" (Schement, 2009, p. 7). What these grassroots initiatives recognize is that communication technologies enhance not only personal quality of life—by enabling people to keep in touch with friends on Facebook or to look for a new job—but also facilitate the kind of social cohesion that strengthens entire societies.

LESSONS LEARNED

The data collected for this study suggests a number of steps that could be taken by both CWNs and governmental regulators to help create and sustain grassroots broadband networks.

Lessons for Public Sectors

The EU should allocate additional unlicensed spectrum for wireless devices such as mesh routers. Currently, WiFi devices transmit in the 2.4 GHz frequency—the same "junk band" used by microwave ovens and cordless telephones. Currently, the EU is "painfully slow" (Rodriguez del Corral, 2011) when it comes to allocating wireless spectrum for new entrants. Furthermore, existing allocation policies enable incumbents to invest in additional spectrum for the sole purpose of blocking new entrants (Rodriguez del Corral, 2011). However, EU regulators could make available additional unlicensed frequencies, similar to a beach that is free and open to the general public. This move would allow mesh routers to transmit stronger signals, creating more robust networks. It would also reduce costs associated with WiFi transmissions and lead to opportunities for new wireless services and products (Peha, 2009). Media reform groups such as Free Press pushed for additional open spectrum in the United States, and European activists are following their lead. "Once we have that property, we can build totally scalable networks with multiple fiber uplinks," the Funkfeuer co-founder noted. Additional spectrum for wireless devices has the potential to expand the public sphere by creating new virtual spaces for civic participation. Unlicensed bandwidth also has implications for the political economy, as it would encourage development of open source mesh software capable of delivering reliable, inexpensive networking. This is in contrast to the development of proprietary software programs that operate exclusively in spectrum owned by specific operators.

Invest research and development dollars in technology that strengthens privacy and security of Internet networks. Additional EU funding for research on network protection would help ease fears that sharing wireless bandwidth results in increased vulnerability to viruses, hacker attacks, phishing, or other online threats. With greater security assurances, more Europeans are likely to participate in the CWN movement. As resource mobilization theory suggests, the behavior of government will influence how social movements are designed and deployed. If this policy recommendation were adopted, it could also lead to increased participation in the public sphere by those who currently fear joining a non-commercial broadband project.

Provide micro-grants and other subsidies for community initiatives. CWN participants reported that grants of even a few thousand dollars could cover the costs associated with critical needs— such as hosting servers, marketing their initiatives, purchasing bandwidth and conducting research. These basic functions are necessary for growing ad hoc communities, yet European networks are not necessarily able to fully execute them. By contrast, some of the most successful initiatives are benefitting from governmental support. For instance, Djurslands.net participates in two rural innovation projects sponsored by the EU, and Guifi.net partners with the public sector at both the local and Catalan levels.

Lessons for CWNs

Focus equally on innovation and broadband connectivity. One promising aspect of mesh architecture is that "out-of-the-box" technology requires minimal computer networking knowledge. A set of pre-configured routers may be plugged into the wall and, instantly, a mesh network emerges. For this reason, wireless community activists should continue directing resources toward this robust but simple means of broadband connectivity. At the same time, data collected for this study illustrates that the signal sharing movement is driven, in large part, by "techies" who appreciate the opportunity to develop routing protocols, create new applications, and build mesh hardware. As a result, activities such as "hack nights" and "play days" should remain prominent aspects of

peer-to-peer initiatives. As resource mobilization theory points out, individuals are motivated to join social movements because of the emotional and intellectual connection they feel with a broader community when working toward a mutual goal. This perception of a shared status and positive feelings for other members of the group help sustain these projects.

Deploy networks in low-income and rural communities. In responses to qualitative interviews, proponents of peer-to-peer signal sharing reported a genuine desire to help close the digital divide. Nevertheless, these initiatives tend to emerge in gentrified residential neighborhoods and near tourist areas. Unless grassroots networks spread to disenfranchised communities—where residents will benefit most from free broadband connectivity—the movement's potential to serve the public good will fall short. As repeatedly noted in this study, those who lack communication technology will miss out on opportunities to fully participate in the information society. Furthermore, in order to create ubiquitous connectivity, CWNs must be deployed across diverse communities. Otherwise, they could grow "like a spot of oil," as the Guifi. net informant characterized his initiative. Specifically, access points could be densely concentrated in areas that lack connectivity to one another.

LIMITATIONS OF RESEARCH

This research is limited by the fact that it examines just six of the scores of community mesh networks thriving throughout Europe. However, key aspects of these projects overlap. Therefore, many of the findings in this chapter are generalizable to other European initiatives.

As previously noted, participation in a wireless signal sharing community has the potential to increase opportunities for civic engagement. Therefore, the study also could have examined the peer-to-peer networking phenomenon through a social capital framework. As Granovetter (1985) observed, individual goals and social influence are understood best when viewed in the context of long-standing social patterns. Another notes that "social capital is produced by the intentional activities of individuals who are connected to one another by ongoing networks of social relationships" (La Due Lake & Huckfeldt, 1998, p. 569).

Yet Sennett's (2006) understanding of social capital is perhaps most applicable to the CWN movement. He focuses on the judgments actors form of their own participation. According to Sennett (2006), "social capital is low when people decide their engagements are of poor quality, high when people believe their associations are of good quality" (pp. 63-64). When viewed this way, social capital is key for sustaining ad hoc communities. If members do not feel loyalty and trust toward the network, they will not make an effort to ensure remains survivable. On the other hand, if members feel a strong sense of loyalty and trust to their network, they are willing to make significant sacrifices to guarantee its sustainability. Another characteristic of social capital involves institutional knowledge (Sennett, 2006). When people understand how a system works, they are more competent at manipulating that system. In the case of CWNs, participants who understand the bureaucracy will have a better chance of persuading politicians, ISPs and other stakeholders to support their goals. However, networks must hold on to volunteers long enough to benefit from this institutional knowledge.

CONCLUSION

The European projects examined in this chapter are truly grassroots efforts, but they also recognize the value of forming partnerships with non-governmental organizations, ISPs and the public sector. While the Wi-Fi initiatives in rural regions are primarily focused on closing the digital divide, many urban signal-sharing networks in Europe center on the development of open source software and hardware. Informants from all six European networks opposed distributing free mesh routers to participants. On ideological grounds, they believe participants should jointly own the network infrastructure, leading them to feel a deeper stake in maintaining the infrastructure. On more practical grounds, distributed ownership protects wireless community projects from legal liability, as well as from buy-out attempts initiated by commercial ISPs. According to informants, European networks are beginning to create free hotspots in response to demands for ubiquitous connectivity. Such moves also expand the public sphere by providing access to non-network members, as well.

European networks depend nearly exclusively on volunteers to help deploy nodes, troubleshoot and recruit new members. AWMN, Djurslands. net and various initiatives in Prague all rely on volunteer board members to set policy and make management decisions. Because European network members are also heavily involved in the open source technology movement, these networks function as laboratories for members to develop hardware and software tools. Ad hoc networks in Europe have also fostered relationships with local officials and institutions. In most cases, these relationships are reciprocal—with governments and social service agencies benefiting from free Internet access, and networks greatly expanding their coverage areas. Future research studies might examine the long-term sustainability of CWN projects using mesh technology in Europe, as well as their impact on local telecommunications policies.

REFERENCES

Abdelaal, A., & Ali, H. (2012). Measuring social capital in the domain of community wireless networks. In *Proceedings of the 2012 45th Hawaii International Conference on System Sciences*. Maui, HI: IEEE Press.

Analysys Mason. (2011). *Ultrafast broadband prices fall to new low in Europe and the U.S.* Retrieved July 16, 2012, from http://www.analysysmason.com/About-Us/News/Insight/Ultra-fast-broadband-prices-fall-to-a-new-low-in-Europe-and-the-USA/

Bauwens, M. (2005b). Peer-to-peer and human evolution. *Integral Visioning*. Retrieved April 16, 2009, from http://integralvisioning.org/article.php?story=p2ptheory1#_Toc107024680

Becker, J. (2002). Theses on information. In Mansall, R., Samarajiva, R., & Mahan, A. (Eds.), *Networking Knowledge for Information Societies: Institutions & Intervention* (pp. 216–219). Delft, The Netherlands: Delft University Press.

Benkler, Y. (2006). *The wealth of networks: How social production transforms markets and freedom.* New Haven, CT: Yale University Press.

Canel, E. (1997). New social movement theory and resource mobilization theory: The need for integration. In Kaufman, M., & Dilla Alfonso, H. (Eds.), *Community Power and Grassroots Democracy* (pp. 189–221). London, UK: Zed Books.

Castells, M. (2007). *Communication technology, media and power.* Retrieved March 23, 2007, from http://web.mit.edu/newsoffice/2007/media-technology-0314.html

Cherney, L. (1999). *Conversation and community: Discourse in a social MUD.* Chicago, IL: Chicago University Press.

Cho, H. (2008). Towards peer-place community and civic bandwidth: A case study in community wireless networking. *Journal of Community Informatics, 4*(1).

Cormack, D. (1991). *The research process.* Oxford, UK: Black Scientific.

Council of Europe/ERICarts. (2012). *Internet penetration rate in Europe.* Retrieved July 16, 2012, from http://www.culturalpolicies.net/web/compendium.php

Czfree.net. (2009). Homepage. *Wireless Broadband Community.* Retrieved February 23, 2009, from http://czfree.net/home/index.php

Djursland International Institute of Rural Wireless Broadband. (2009). *Homepage.* Retrieved February 23, 2009, from http://diirwb.net/

Efstathiou, E., Frangoudis, P., & Polyzos, G. (2005). *Stimulating participation in wireless community networks.* Retrieved April 21, 2009, from http://mm.aueb.gr/technicalreports/2005-MMLAB-TR-01.pdf

Ehrlich, T. (2000). *Civic responsibility and higher education.* New York, NY: Oryx Press.

European Commission. (2011). *Digital agenda: Commission proposes over €9 billion for broadband investment.* Retrieved July 16, 2012, from http://ec.europa.eu/information_society/newsroom/cf/itemdetail.cfm?item_id=7430

Fox Piven, F., & Cloward, R. (1991). Collective protest: A critique of resource mobilization theory. *International Journal of Politics Culture and Society, 4*(4), 435–458. doi:10.1007/BF01390151

Freifunk. (2009). *Was ist Freifunk? Homepage.* Retrieved February 23, 2009, http://start.freifunk.net/

Funkfeuer. (2009). *Don't log onto the net—Be the net!* Retrieved February 23, 2009, from http://www.funkfeuer.at/

Gamson, W. (1975). *The strategy of social protest.* Homewood, IL: Dorsey Press.

Granovetter, M. (1985). Economic action and social structure: The problem of embeddedness. *American Journal of Sociology, 91*, 481–510. doi:10.1086/228311

Grimes, S. (2003). The digital economy challenge facing peripheral rural areas. *Progress in Human Geography, 27*(2), 174–193. doi:10.1191/0309132503ph421oa

Guifi.net. (2009). *Homepage*. Retrieved February 23, 2009, from, http://guifi.net/

Hannigan, J. (1985). Alain Touraine, Manuel Castells and social movement theory: A critical appraisal. *The Sociological Quarterly*, *26*(4), 435–454. doi:10.1111/j.1533-8525.1985.tb00237.x

Hegel, G. (1820). *Philosophy of right*. (H. Reeve, Trans.). Retrieved April 19, 2007, from http://www.marxists.org/reference/archive/hegel/index.htm

International Telecommunication Union. (2011). *ICT services getting more affordable worldwide*. Retrieved July 16, 2012, from http://www.itu.int/net/pressoffice/press_releases/2011/15.aspx

Internet World Stats. (2012). *Internet and Facebook usage in Europe*. Retrieved July 16, 2012, from http://www.internetworldstats.com/stats4.htm

Jayakar, K. (2009). Universal service. In Schejter, A. (Ed.), *And Communications for All* (pp. 181–202). Lanham, MD: Rowman & Littlefield.

Kendall, D. (2008). *Sociology in our times* (7th ed.). Belmont, CA: Wadsworth.

La Due Lake, R., & Huckfeldt, R. (1998). Social capital, social networks and political participation. *Political Psychology*, *93*, 567–584. doi:10.1111/0162-895X.00118

Lentz, R., & Oden, M. (2001). Digital divide or digital opportunity in the Mississippi Delta region of the U.S. *Telecommunications Policy*, *25*(5), 291–313. doi:10.1016/S0308-5961(01)00006-4

Little, D. (2008). Collective behavior and resource mobilization theory. *Understanding Society*. Retrieved February 29, 2008, from http://understandingsociety.blogspot.com/2008/01/collective-behavior-and-resource.html

McAdam, D. (1982). *Political process and the development of black insurgency, 1930-1970*. Chicago, IL: The University of Chicago Press.

McCarthy, J., & Zald, M. (1987). *Social movements in an organizational society: Collected essays*. New Brunswick, NJ: Transaction Publishers.

McChesney, R. (2009). Public scholarship and the communications policy agenda. In Schejter, A. (Ed.), *And Communications for All* (pp. 41–46). Lanham, MD: Rowman and Littlefield.

Meyer, D. (2005). Social movements and public policy: Eggs, chicken, and theory. In Meyer, D., Jenness, V., & Ingram, H. (Eds.), *Routing the Opposition: Social Movements, Public Policy, and Democracy* (pp. 1–26). Minneapolis, MN: University of Minnesota Press.

Mosco, V. (1996). *The political economy of communication*. London, UK: Sage.

Napoli, P. (1999). The marketplace of ideas metaphor in communications regulation. *The Journal of Communication*, *49*(4), 151–169. doi:10.1111/j.1460-2466.1999.tb02822.x

Neumann, J. (2007). Community wi-fi in UK and Germany: A roundup. *Nettime Discussion Board*. Retrieved February 14, 2008, from http://www.nettime.org/Lists-Archives/nettime-l-0710/msg00034.html

Olson, M. (1965). *The logic of collective action: Public good and the theory of groups*. Cambridge, MA: Harvard University Press.

Paldam, M., & Christoffersen, H. (2004). Privatization in Denmark, 1980-2002. *Social Science Research Network*. Retrieved April 13, 2009, from http://papers.ssrn.com/sol3/papers.cfm?abstract_id=514242

Parker, E. (2000). Closing the digital divide in rural America. *Telecommunications Policy*, *24*(4), 281–290. doi:10.1016/S0308-5961(00)00018-5

Peha, J. (2009). A spectrum policy agenda. In Schejter, A. (Ed.), *And Communications for All: A Policy Agenda for a New Administration* (pp. 137–152). Lanham, MD: Lexington Books.

Poster, M. (1995). *CyberDemocracy: Internet and the public sphere*. Retrieved April 16, 2009, from http://www.hnet.uci.edu/mposter/writings/democ.html

Pourebrahimi, B., Bertels, K., & Vassiliadis, S. (2005). A survey of peer-to-peer networks. In *Proceedings of the 16th Annual Workshop on Circuits, Systems and Signal Processing*. Veldhoven, The Netherlands: IEEE.

Rafaeli, S., & LaRose, R. (1993). Electronic bulletin boards and 'public goods' explanation of collaborative mass media. *Communication Research*, *20*(2), 277–297. doi:10.1177/009365093020002005

Rheingold, H. (1993). *The virtual community*. Reading, MA: Addison-Wesley Publishing Co.

Rodriguez del Corral, A. (2011). Financing high speed broadband—What do you think? *European Broadband Portal*. Retrieved July 16, 2012, from http://broadbandeurope.wordpress.com/2011/05/01/financing-high-speed-broadband-what-do-you-think/

Rowell, L. (2007). Can mesh networks bring low-cost, wireless broadband to the masses? *netWorker*, *11*(4), 26–33. doi:10.1145/1327512.1327514

Schement, J. (2009). Broadband, internet and universal service. In Schejter, A. (Ed.), *And Communications for All: A Policy Agenda for a New Administration* (pp. 3–27). Lanham, MD: Lexington Books.

Sennett, R. (2006). *The culture of the new capitalism*. New Haven, CT: Yale University Press.

Silverstone, R. (2004). Regulation, media literacy and media civics. *Media Culture & Society*, *26*(3), 440–449. doi:10.1177/0163443704042557

Souvirón, M., & María, J. (1999). *The process of liberalization and the new telecommunications regulations*. Granada: Editorial Comares.

Tapia, A. (2009). Municipal broadband. In Schejter, A. (Ed.), *And Communications for All*. Lanham, MD: Lexington Books.

Tilly, C. (1978). *From mobilization to revolution*. Reading, MA: Addison-Wesley.

Touraine, A. (1985). An introduction to the study of social movements. *Social Research*, *52*(4), 749–787.

vanDijk, J. (1999). *The network society: Social aspects of new media* (Spoorenberg, L., Trans.). London, UK: Sage.

Waterman, H. (1981). Reasons and reason: Collective political activity in comparative and historical perspective. *World Politics*, *32*, 554–589. doi:10.2307/2010135

KEY WORDS AND DEFINITIONS

Community Network: A co-operative, non-commercial network that enables members to share not only bandwidth, but also skills and knowledge.

Digital Divide: That gap that separates those who have access to the best in information technology from those who do not.

Digital Inclusion: Realizing the goal of bringing the benefits of Internet connectivity, skills and hardware to everyone who wants them.

Mesh Network: A networking technology that enables signals to hop from node to node, choosing the most efficient path. Just one gateway node must have access to the Internet.

Political Economy of Telecommunication: The production, distribution, exchange and consumption of the Internet and other digital mass media—with a focus on how these values influence those in power and societal change.

Resource Mobilization Theory: A social movement framework that focuses on the critical role that resources such as money, knowledge, ties to the establishment, and media access play in helping a social movement succeed.

Telecommunications Policy: Rules developed by governments, in consultation with various stakeholders, regarding how telecommunications systems will operate.

Chapter 14

A Critical Analysis of the Limitations and Effects of the Brazilian National Broadband Plan

André Lemos
Federal University of Bahia, Brazil

Francisco Paulo Jamil Almeida Marques
Federal University of Ceará, Brazil

ABSTRACT

This chapter examines the limitations and the socio-political effects of the Brazilian National Broadband Plan (PNBL: is its Portuguese acronym). The discussion considers the main transformations witnessed in the telecommunications landscape in Brazil during the second half of the twentieth century. On the one hand, the end of state monopoly of telecommunications services and the provision of such services by the private sector called for greater investments in infrastructure. On the other hand, the Brazilian regulatory agencies have failed to lower prices, promote competition, and spread broadband access to remote and underserved areas. The PNBL was launched in order to deal with these difficulties. The plan, however, has at least three important problems: (1) the low-speed connection offered to users, (2) the unattractive prices, and (3) the lack of reflection on issues such as net neutrality. The text argues that only by taking such issues into consideration will the plan ensure innovation, economic growth, diversity, and freedom of access to information.

DOI: 10.4018/978-1-4666-2997-4.ch014

INTRODUCTION

There are a series of studies and reports that indicate that Brazil is one of the leading countries of the current digital culture. This is what can be gleaned from different articles, which are coverage targets of the national press (IDGNOW, 2011a, 2011b; Portal G1, 2011; UOL Notícias, 2011). However, expanding the digital infrastructure and access to online resources in such country faces a number of challenges. These challenges include the high cost of Internet access and the difficulty of serving remote areas.

In 2010, the Brazilian Government proposed the PNBL (Plano Nacional de Banda Larga, or National Broadband Plan) for the purpose of lowering the prices of internet access and offering connectivity solutions to the Brazilian society at large.

Indeed, there were previous attempts at the municipal, state, and federal levels to combat digital exclusion. These attempts included the creation of telecenters, equipping public schools with computers, and decreasing taxes on IT products (Costa, 2007; Lemos, 2007). One of the lessons learned was that only access to equipment was not sufficient. Therefore, providing affordable and reliable high-speed Internet access has become a priority in public policy at the state and federal levels. The PNBL is the main result of the country's recent efforts.

This chapter addresses the following questions: what are the distinctive characteristics of the PNBL? What are the main tension hot spots generated from the moment that telecommunications operators take on a more interventionist profile similar to that of public agencies? To what degree can the PNBL address the challenges facing users, companies, and governments?

While answering these questions, we focus on the social and economic effects of the PNBL. We will also identify its limitations, discuss its current status and provide suggestions for future improvements.

The text also presents the current Brazilian regulatory framework, completely modified following the telecommunications monopoly breakdown, which occurred in the late 1990s. This discussion allows us to understand the correlation of forces, which have surrounded users, companies, and governments for over a decade. For the purpose of contextualization, we present data regarding the use of telecommunications (principally of the Internet) in Brazil.

If, on one hand, Brazilians make up a consumer market in full expansion in the telecommunications sector, on the other hand, the difficulties to take advantage of digital communication services in the country are clear. Mobile telephony, for example, beats line creation records every month; but the majority of these subscriptions relate to prepaid service, given that many users do not have the financial conditions to deal with the expenses tied to (a) the operation costs and (b) the high tax burden witnessed in Brazil.

The chapter also discusses the tensions concerning the declared resistance of a few telecommunications companies to expand their Internet connection service to less profitable areas. Finally, the text points towards the problem of net neutrality and of digital inclusion—central issues to the understanding of the PNBL.

THE CONTEXT OF BRAZILIAN TELECOMMUNICATIONS

Since the 1917 Decree 3,296 (issued to affirm the "exclusive responsibility of the Federal Government regarding radiotelegraphic and radiotelephonic services in the Brazilian territory"), until the last decade of the 20th century, one can note a strong centralization of telecommunications services in the hands of the Brazilian State.

The development and greater use of radio broadcasting, started in the 1930s, generated a new list of concerns for the Government. Besides transmission itself, it became necessary to keep watch on the type of content offered by radio stations and later on, by television networks (even though the State was not the main responsible for its production). President Getúlio Vargas, while dictator (1937-1945), as well as elected president (1951-1954), considered it essential to monitor (and even censor) the elaboration of materials which could influence the formation of the government's public image (Jambeiro, et al., 2004). During most of the first half of the 20th century, the impact of television and radio on the formation of public opinion remained a concern of Brazilian political leaders.

Stage 1: The Monopoly of Telecommunications Services

Up until the 1960s, Brazil was still going through the implantation phase of its telecommunications system (Sundfeld, 2006, p. 2). The telecommunications operators operated locally, being that only the telegraph system and some of the television networks were under the Union's control. From the 1960s on, the situation changed substantially.

The strategy of the governments in that period consisted of setting up an institutional arrangement, which divided the responsibility for the operation of the telecommunications sector and the radio-broadcasting sector. The telecommunications sector was left to state-run companies, as a monopoly. The radio broadcasting sector had a strong influence from the private sector, albeit the existence of stations directly controlled by state institutions.

The government approved the Código Brasileiro de Telecomunicações (CBT) (Brazilian Telecommunications Code) that took effect in 1962. The CBT was an instrument, which regulated both telecommunications as well as radio broadcasting in Brazil since its inception for decades up to the 1990s. Sundfeld (2006, p. 2) referred to this as the "nationalization phase" of telephony services.

Coming in power in 1964, the military maintained strict observation of the communications sector. They had as their objective the implementation of a national security doctrine, which demanded a strategic command over, for example, telecommunications via satellite. At the same time, the Military Forces were concerned with the creation of a favorable public opinion towards themselves, exerting influence (or even control) over radio broadcasting and the press.

The dedication of the military became evident upon the editing of two measures: Decree-law 200, in 1967, which dealt with the creation of the Communications Ministry. This decree was directed towards establishing policies specifically for telecommunications and for radio broadcasting. In the same year, it also issued Decree-law 236, which revoked 41 provisions of the CBT (approved only five years earlier). In addition, the Decree added 14 more articles to the CBT, which provided, for example, stronger punishment in the case of the use of communications different from that previously established by the government. The CBT went through other occasional changes during the 1970s and 1980s. With the re-democratization of the country along with the enactment of a new Federal Constitution in 1988, this provision lost most of its strength.

The creation of Telebrás (the state-run Brazilian Telecommunications Company), in 1972, was one of the greatest changes in the telecommunications sector promoted by the military at that time. Telebrás was a holding company for 28 companies (Embratel + 27 state operators) whose objective was to unify telephony services and to allow the operation of the state monopoly.

Stage 2: The Breakdown of the State Monopoly

It must be emphasized that changes in the political plan were not the single aspect, which influenced the telecommunications sector in Brazil during the last few years. According to Ramos (2000), the technological landscape (the convergence among telecommunications, information systems, and the production of communications content), the internationalization (regarding both the supply and consumption of media materials), and the expansion of the range of actors involved in the regulatory debate added up to create the transformations whose understanding is fundamental in order to have an comprehensive overview of the sector in contemporary Brazil and of the current PNBL.

It can be stated that the main transformation in this area was about the breakdown of the telecommunications monopoly during the 1990s. This process integrated a set of strategies adopted by the Government in order to attract foreign investment. The goal was to stimulate the recuperation of the economic activity following the crisis of the previous decade. The flexibilization phase in the Brazilian communications sector (Sundfeld, 2006) initiated in 1995 when the constitutional amendment permitting the breakdown of the state monopoly was approved. This paved the way for a subsequent stage: privatization.

Stage 3: Privatization and Its Aftermaths

The auction of companies that made up the Telebrás system occurred in July 1998. In addition, the holding company was dismembered into 12 companies. Moreover, the Brazilian State had removed itself from the operation of the sector instead acted through a regulatory authority, Anatel (Agência Nacional de Telecomunicações – National Telecommunications Agency), with the role of defining rates, imposing fines and approving concessions. The fundamental legal guideline followed by the Agency, the General Telecommunications Law (Lei Geral de Telecomunicações [LGT, Lei nº 9.472/97 – Law 9,472/97]), had been approved in 1997.

The General Telecommunications Law revoked a large part of the already outdated Brazilian Telecommunications Code (Código Brasileiro de Telecomunicações), deepening the difference (in terms of regulation) between the telecommunications and radio broadcasting sectors.

There is no doubt that the privatization of telecommunications brought benefits to Brazilian society, fulfilling a repressed demand in the telephony sector and setting goals for the universalization of telecommunications services. One used to wait up to two years for the installation of new telephone terminals, even in urban areas. It can be estimated that in 1997 there were 17 million landlines; while in 2007, approximately 40 million terminals were already operating (Info Magazine, 2007). There was an increase in investments in the sector including the influx of foreign capital.

On the other hand, the privatization process (the decline in the role of the state to control companies previously under its rule) had initially foreseen greater competition in the sector that would bring about the tendency to lower existing prices in the market. However, this did not occur; neither in large cities, nor in the interior of the country. In many rural areas, there are services, which are not even provided on a regular basis. Today, it is surely easier and faster to acquire a phone line in certain locations, but the cost of the service is considerably higher. Excessive taxation also exists which hinders a broader adoption of these services by the population.

It can be said that the current period of "post-privatization" (Sundfeld, 2006, p. 8) witnessed the emergence of a new reality of confrontations leading to tensions between the government and private companies. A power play is highlighted, for example, in the pressure that the government has been exerting over the last years to improve Internet connection services.

The PNBL is one more test to see if the principles of universalization can take root in an environment where the interests of private enterprises (not always in sync with those of the Government) have almost always prevailed.

INTERNET USE IN BRAZIL

The privatization of telecommunications in Brazil took the country to a new level in terms of the development of digital communication. For instance, the number of cellular phone lines leaped from 800 thousand in 1994 to 7.4 million in 1998 (Brazil, 2001). In September 2011, Anatel registered more than 227.4 million mobile phone lines (Folha-Online, 2011a). It is important to note that the current population of the country is approximately 193 million people.

The number of users with Internet access also increased consistently. In July 2011, the Nielsen Online Survey stated that Brazil had reached the mark of 58.6 million people with Internet access (Folha-Online, 2011b). This survey took into account only users who have network access at work and in their homes. In January 2010, the study pointed 46.8 million users (Folha-Online, 2010a).

In spite of problems related to access and high costs, the Brazilian digital culture continues to grow. According to research by Ibope/NetRatings, in January 2007, Brazilian users spent, on average, 21 hours and 20 minutes per month surfing the Internet (Folha-Online, 2007). In February 2010, data showed that Brazilians stayed connected for 26.4 hours monthly (Portal IG, 2010).

Upon presenting the collected data in February 2011, in the article "Comparing Global Internet Connections" (Nielsenwire, 2011), The Nielsen Company consulting firm revealed that, among the nine countries then researched (Switzerland, United States, Germany, Australia, United Kingdom, France, Spain, Italy, and Brazil), Brazil was at the top in terms of amount of time spent, reaching an average of 30 to 31 hours per person per month.

In a paradoxical manner, the same document shows that Brazil leads (percentage-wise) in the number of users who access slow speed Internet. The Nielsen Company classified access speed into four groups:

- **Slow Browser Speed:** Up to 512 Kbps;
- **Medium Browser Speed:** From 512 Kbps to 2 Mbps;
- **Fast Browser Speed:** From 2 to 8 Mbps; and
- **Super-Fast Browser Speed:** Above 8 Mbps.

The range corresponding to medium browser speed is utilized by 48% of Internet users in Brazil. In all the other 8 countries mentioned above, majority of accesses are via fast connections (between 2 and 8 Mbps). In Brazil, 31% of Internet users browse with a speed below 512 Kbps and only 6% enjoy connections with a speed over 8 Mbps.

In March 2011, FIRJAN (Federação das Indústrias do Estado do Rio de Janeiro – Federation of Industries of the State of Rio de Janeiro) published a report indicating that enterprises in Brazil pay more for broadband service than in countries like Germany, Canada and the United States, without necessarily receiving any additional quality in the service provided. According to the report, the 1 Mbps connection in Brazil costs, on average, US$42.73 per month. The monthly amount in Germany is US$9.30; in Mexico and in Colombia the values are US$16.50 and US$16.70, respectively; in Canada, one pays US$28.60; while in the United States, the value can reach up to US$40.

In Brazil, for broadband DSL technology for businesses, the access cost is US$41.67 [R$70.85 – Real-Dollar exchange rate as at December 2011: 1 United States Dollar = 1.70 Brazilian Real.] on average for a package with a download speed of 1 Mbp, considered the minimum necessary to fulfill the simplest needs of micro and small businesses. This cost, however, is not the same for all states. In fact, it varies up to 650% across the country, having the lowest price of US$33.76 [R$57.40] in the states of Alagoas and Espírito Santo and the highest price of US$252.88 [R$429.90] in Amapá, where the maximum available speed is only 600 kbps (FIRJAN, 2011, p. 5).

As can be noted, the issue pointed out in the FIRJAN report refers not only to low speed and to high price but also points out the inequality of speed connections among different regions of Brazil. It is true that the number of homes with computers and Internet access have been increasing substantially in the last few years. According to the Brazilian Institute of Geography and Statistics (Brazil, 2010), in the year 2009, 27.4% of the surveyed households had Internet access; in 2008, this number was 23.8%. The same survey demonstrates that, in 2009, 67.9 million Brazilians over the age of 10 confirmed using the Internet. Four years prior, in 2005, only 31.9 million users were registered.

During the period 2005-2009, thus, the increase in number of Internet users in Brazil was 112.9%. Nevertheless, the same survey revealed that, in the Southeast region (the richest in the country), households which had computers (remembering that not all of these had Internet access) represented 43.7% of the sample. The index is much higher compared to the data verified in poorer regions, such as the North (where computers are present in 13.2% of households) and the Northeast (with 14.4% of the researched households having computers).

To summarize, it is important to highlight that the OECD provided the following diagnosis on connection costs in Brazil:

Some countries have large absolute amounts of infrastructure, which is one indication of national capacity for receiving ICT-enabled offshore services. For example, China has more PCs than Germany and more Internet subscribers than the United States. Brazil, India and Russia each have about as many PCs as Canada or Italy, and Brazil and India have slightly fewer Internet subscribers than Canada. However, apart from China, these countries' broadband subscriber numbers are much lower, and broadband costs are much higher than in most OECD countries in all of them (OECD, 2009, p. 32).

INTERNET AND TELECOMMUNICATIONS POLICIES

Judging from the studies above, Brazilians face connection difficulties when compared to users of other countries; and even among Brazilians, the discrepancy of connected households dependent on the region of the country is acute. This is the case, even when the population may be willing to pay more for the service, while the service is not offered by operators in those locations. This diagnosis generates a set of criticism vis-à-vis these companies and even the State itself. The latter should supervise and exert pressure on these operators with the intent to guarantee more ambitious goals to service universalization and increase in connection quality.

Anatel (the regulatory agency) has been activated to check market practices adopted by telephony companies, given the noted discrepancies, in the supply of connection speeds and prices (Folha-Online, 2010b). The friction between the government and the operators has increased and, in fact, is not recent. However, the discrepancies and accountabilities of such companies are not restricted to the area of Internet access only (Folha-Online, 2011c).

As noted above, the number of cellular phones in Brazil has already reached more than 227 million lines. However, only 18.36% (41.7 million) of these lines belong to ongoing contract plans. The rest (81.64%, which corresponds to 185.6 million) are pre-paid and belong, in general, to a low-income population, mainly due to the high rates charged in the country.

Over the last year, most of the companies (especially those providing mobile phone services) have been accused by IT equipment suppliers of decreasing investments in service betterments (Folha-Online, 2010c). As a consequence, a constantly growing number of customers complain about the bottlenecks in computer network traffic when trying to browse or simply sending text messages. The companies defend themselves alleging that the high prices are mostly due to the high tax burden built into the supply of services. Moreover, they also argue that the high costs relative to infrastructure investment justify the difference in prices among regions, since some remote areas of the country can only be reached by air or inland waterway transport. In reality, however, even in cities of Rio de Janeiro's metropolitan area, users are penalized via the payment of much higher rates than those found in the capital (Folha-Online, 2010b).

The Government, through the Communications Ministry, has made efforts to pressure companies to accept bolder goals that very often end up in the courtroom. One of the latest confrontations was tied to the telecommunications companies' resistance against the "Plano Nacional de Banda Larga" (PNBL) – National Broadband Plan. Since of its proposal, some of private operators threatened to contest the PNBL in Court (Folha-Online, 2010d). Following an intense negotiation process (and concessions on behalf of the Government), some of the companies adhered to the plan which took effect in mid-2011.

The characteristics of the PNBL are detailed below. The next section examines its potentials and limitations, in order to verify up to what point the project outlined by the Federal Government fulfills society's demands in the context of production and consumption of digital information characteristic of the 21st century.

AN OVERVIEW OF THE BRAZILIAN NATIONAL BROADBAND PLAN

The National Broadband Plan (PNBL) was officially launched in May 2010. Its objective was to increase Internet access from 12 million to 40 million households in all regions of the country. The management of the plan was under the coordination of the state-owned company Telebrás (Telecomunicações Brasileiras S.A), at an estimated cost of US$6.5 billion for the 2010-2014

period. Initially, 100 cities would be serviced, while it is expected by 2014, all Brazilian cities would have rapid Internet service available.

First Steps

The PNBL is seen as a plan for digital inclusion. The discussion regarding digital divide has been on the public agenda since 1999, when the Federal Government launched the "Programa Sociedade da Informação" (Information Society Program), by means of Decree 3,294 on December 15th, which culminated in the elaboration of the so-called "Livro Verde" (Green Book), in 2000 (Takahashi, 2000).

The "Livro Verde da Sociedade da Informação no Brasil" (Green Book of the Information Society in Brazil) is a set of goals initialed in 2000 with the objective of placing the country in the so-called "information society." The elaboration of the document began in 1996, under the auspices of the National Science and Technology Council (Conselho Nacional de Ciência e Tecnologia [CCT]), a department of the Ministry of Science and Technology that counted on extensive discussions amongst experts and society. The "Livro Verde" proposed a group of projects for the period of 2000-2003, with forecasted investments of US$ 1.7 billion. Specifically, its main lines of action were: market, jobs, and opportunities; universalization of services for citizenship; education in the information society; content and cultural identity; the government at everyone's reach; research and development; advanced infrastructure and new services.

Regrettably, he Brazilian Information Society Program lost its direction and was discontinued as a new government with a divergent political-ideological tendency assumed power in 2003 and decided to change the direction of the digital inclusion initiatives.

Afterwards, the project "Computadores para Todos" (Computers for All) (2005-2008) was launched. Its objective was to reduce the price of hardware and make access to computers easier. The "Computadores Para Todos" was the Lula Government's first initiative of digital inclusion. According to the project, the purchased computer should come with free operating system and software applications along with the hardware being able to access the Internet. To facilitate the purchase of the equipment, the Federal Government offered more advantageous financing. Currently, there are two approved lines of credit: the "Fundo de Amparo ao Trabalhador" (FAT) (Workers' Aid Fund), run by public banks; and another administered by the Banco Nacional de Desenvolvimento Econômico e Social (BNDES) (National Development Bank). According to information outlined in the project, *The Computer for All* should be sold to the consumers at a maximum price of US$ 700.00. However, the market currently offers prices, which are a lot more attractive than this.

Machines Are Not Enough: A New Plan to Increase Internet Access

According to the Federal Government's portal, there is no plan in the world like the Brazilian PNBL. There are similar initiatives, but not with the same kind of governmental weight behind it. The analysis which justifies the current PNBL refers to the fact that the Brazil is lagging behind when compared to other countries coined as "emerging markets." For example, data shows that among the countries with the largest percentages of the population who use high speed Internet, South Korea is in first place, 97%, followed by Switzerland, 90%, Norway, 84%, and Holland, 83%. The United States is in 15th place, while Brazil is in 53rd, with 19% (Superinteressante, 2010). Figure 1 and 2 show the broadband policy in some countries in terms of speed, investment and rights acquired by citizens.

In addition, data from Teleco (Inteligência em Telecomunicações – Intelligence in Telecommunications) shows that the broadband market in Brazil today is highly concentrated in specific areas, divided amongst the operators: Oi (30.5%), Net (25.4%), Telefônica (22.8%), GVT (8.8%), CTBC (1.7%), and others (10.8%) (Estadão, 2011). To sum up, the country counts on slow and expensive services, concentrated in few regions and states (Rio and São Paulo).

Therefore, decentralizing access, reducing costs, expanding the types of connections, and improving the Internet services in Brazil are the challenges of the PNBL, proclaimed by former President Luis Inácio Lula da Silva, and implemented by the current government, run by Dilma Rousseff.

Notwithstanding, the PNBL underwent many modifications since it was created (mainly due to pressures of private companies). One of the most anxious moments of the negotiation was the inflexible position of President Rousseff regarding the minimum speed to be offered in 2014, which went from 600 Kbps to 1 Mbps, for the price of R$35,00 (approximately US$20.00) per month.

Framing the Main Tensions and Problems of the PNBL

Current policy recommends that the government needs to be aggressive and guarantee a truly national infrastructure in order to foment something that is already in the Brazilian digital culture. However, the future does not seem very promising.

First of all, it should be highlighted that the PNBL is a plan of digital inclusion which is going through the expansion of network infrastructure and its capacity of data transmission. Here, one can find another problem typical of the Brazilian case, concerning the difference between the speed contracted by the user and that which effectively is delivered by the companies. Currently, the operators guarantee the provision of, at least, 10% of the contracted speed, which does not seem reasonable, since the user is paying the full rate. Another focal point of tension found is the lack of clear criteria for acceptable connection parameters, even when the users are willing to pay (a lot) for the service.

Figure 1. Policies guarantee access for all (adapted from Brazil, 2011)

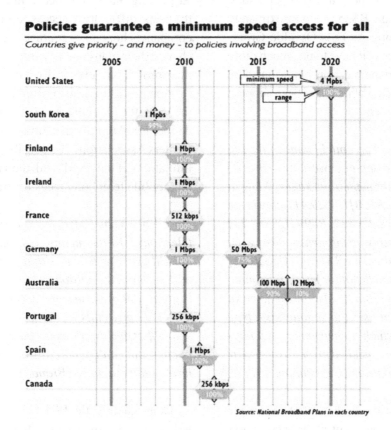

Figure 2. Countries' investments in broadband access (adapted from Brazil, 2011)

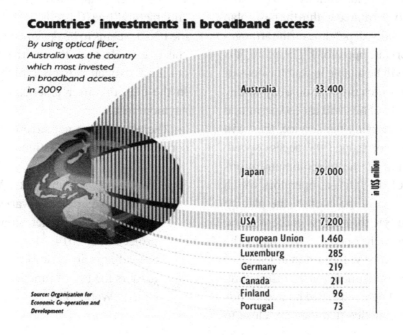

However, experts argue that the final cost shall be much higher than R$35.00 (approximately US$20.00) per month. As Luís Osvaldo Grossmann explains, users will have to pay much more to guarantee a 1 MBps connection, being that there are different approaches which have been adopted by the service providers.

... Oi, Telefônica and Sercomtel [some of the big telecommunication companies working in Brazil] include activation fees in their packages. While at Sercomtel this means R$50 [US$29.41] more in the first month, at Oi the amount is R$99 [US$58.23] divided into 10 monthly installments causing the R$35 [US$20.58] to become R$44.90 [US$26.41] during this period. On the other hand, Telefônica charges R$150 [US$88.23], also in the form of an activation fee, but uses this amount as a type of loyalty scheme which the customer only pays should the service be cancelled prior to a one-year completion of the contract (Grossmann, 2011).

Secondly, the price is also more expensive because a speed of 1 Mbps provides a computer network traffic (download) of about 300Mbp for fixed access and 150 Mbp for mobile access (according to the Communications Ministry, the download capacity permitted within the basic plan will increase gradually, depending on the arrangement with each private operator.). Therefore, it is likely that users will be unable to watch videos or update programs, which they use on a day-to-day basis as the quantity of bits will exceed permitted quotas. To continue using the service, they will have to pay more, or reduce the access speed. Thus, the lobbying of the telecommunications enterprises and of the Internet operators which dominate the market in the country are placing very concrete limits to the success of the venture.

The advantage to the operators in limiting the download speed can be witnessed in at least two aspects: 1) lower investments in infrastructure, once the amount of data requested by users tends to decrease as they notice that they are close to

surpassing the established limit; 2) by knowing the users' difficulty to access certain contents, the companies hope to encourage their customers to change their plans and pay more. In other words, their profits will eventually increase.

According to Mário Brandão, president of the ABCID (Associação Brasileira de Centros de Inclusão Digital – Brazilian Digital Inclusion Centers Association), it is possible to compare the limitations of the PNBL with the same difficulties that mobile telephony faces in Brazil.

Do all of us, in Brazil, have access to cellular telephony? To the means, yes; but to the usage, no! I'll explain: Brazil with a population of 180 million has 200 million cellular phones, which is great. But of these same 200 million, 2/3 have not made any calls in the last six months. This shows that the lower income population owns cell phones, but can't make calls because they do not have credits to do so (Brandão, 2011).

Consequently, the PNBL, if thus established, will be a project which aims to include, as a matter of fact, excludes the already excluded, giving them a false sense of inclusion. The inclusion will be of an "interacted" citizen and not of an "interacting" one (Castells, 1996). In other words, the PNBL seems to be designed for a citizen who checks e-mails, surfs the Web a bit, and who does not access social networks, blogs, or YouTube. In the end, this user practically does not exist anymore today.

Therefore, three negative points can be discerned in a deeper evaluation of the PNBL:

1. Taking into account that 1 Mbps is already very low to perform the varied tasks required in our daily lives, let alone access speed projected for 2014. Moreover, the proposed upload capacity is slower (128 Kbps), which hinders the use of functionalities of the so-called "Web 2.0";

2. The telecommunications companies will be paid according to the volume of content accessed by users, changing the logic currently in place on the Brazilian network. Adopting this posture may violate the basic principle of net neutrality, since the flow of data will be treated in a discriminatory fashion. Moreover, the operators will unreasonably limit the amount of data which the user can download per month (300 Mb);

3. It can be estimated that, because of download and upload limits, the final costs to users will be much greater than the US$20.58 initially foreseen in the PNBL. In some cases, users need to purchase a modem in order to have wireless access.

These are indications that the PNBL may fail in its lofty goals. Marcelo Branco, coordinator of the digital campaign of President Dilma Rousseff during the 2010 elections, asked for greater participation from the Brazilian States in order to improve the negotiations with the telecom companies. The reason being, the state and municipal governments spend the most on telecommunications services and Internet access. According to Branco:

The state government has invested much money in Telecommunications services. (...) If a company has a gigantic account with a supplier, it has the bargaining power to obtain a few advantages in a negotiation. Surely, the largest customer of Telecom in Rio Grande do Sul is the state government. Thus, what is the role of the Tarso administration [Tarso Genro, Governor of the State of Rio Grande do Sul] in this discussion? (Branco, 2011).

As mentioned above, the objective of the PNBL is to provide good quality access with a minimally acceptable speed in order to universalize the service and try to reduce the regional disparities. Notwithstanding, should the current terms be maintained, the goal is unlikely to be reached. In order to reach its objectives, the PNBL should be projected within a strategic view, which will guarantee individual freedom, privacy, anonymity and effective participation and use of the Internet.

In the proposed configuration, there will only be "interacted" citizens, that is, with limited autonomy to produce and consume content. It is known that there is a direct relation between the increase in GDP and the access speed and universalization of Internet services. According to data from the World Bank, for each 10% increase in broadband services, there is an increment of 1.4% in the GDP (Superinteressante, 2010). As projected in the Government Plan, access will be expensive, with few usage options for citizens and with the grave precedent of being able to break the paradigm of net neutrality. The following section further explores this issue.

CHALLENGES AND SUGGESTIONS FOR FUTURE IMPROVEMENTS

Net Neutrality and the PNBL

Net Neutrality is the Internet's guiding principle. It preserves our right to communicate freely online. This is the definition of an open Internet. Net Neutrality means that Internet service providers may not discriminate between different kinds of online content and apps. It guarantees a level playing field for all websites and Internet technologies. Net Neutrality is the reason the Internet has driven online economic innovation, democratic participation, and free speech. It protects our right to use any equipment, content, application, or service without interference from the network provider. With Net Neutrality, the network's only job is to move data—not choose which data to privilege with higher-quality service and which to relegate to a slower lane (Free Press, 2012).

To achieve "Net Neutrality," it is necessary to guarantee not only bandwidth, but also price, inclusion and regional balance. To insert another 40 million households into the digital world is an audacious bet and it is a fact that the Brazilian government will be unable to execute a project of this magnitude without the participation of the telecom companies.

Nevertheless, the partnership should be established with a view of guaranteeing a good price, the speed effectively contracted and net neutrality. In its current state, the PNBL ends up becoming a threat to the founding principle of freedom on the Internet and its neutrality. If such principles were to be violated, the network would transform itself into a type of cable TV for computers, where some channels (packages) are accessed through specific payment. It is, therefore, fundamental to guarantee that the reduction of inequality in Internet access through the PNBL will not violate net neutrality. According to Marcelo Branco:

Until today, within the operating logic of the Internet, the creator of the content on the net is the one who can charge for its access and not the operators. Upon the agreement signed with the Communications Ministry, the Telecom companies, besides being paid for the bandwidth they provide, for the transmission speed, will begin to limit the amount of content that the user can download during the month (Branco, 2011).

Moreover, Branco states that:

Net neutrality is our main battle at this moment. In the United States, the Federal Communications Commission (FCC) ... also undergoes strong pressures to violate Internet neutrality. Private telecommunications operators are lobbying to violate neutrality. There is a concern in Brazil that Anatel will begin to regulate the Internet. The managing committee of Brazilian Internet is a role model for other countries and for international governance of the net. The 'Comitê Gestor da Internet no Brasil' (Management Committee of Internet in Brazil) a much more democratic agency, is who administers the Internet in Brazil. It is a reference for other countries in the democratic management of the Internet and a model for

worldwide governance of the net. The managing committee relies on the participation of private entrepreneurs, of civil society, of the government and universities. It would be inadmissible for an agency like Anatel, which represents only one of the sectors, to begin to regulate Internet in Brazil (Branco, 2011).

The dimension of the debate on net neutrality is spatial and political. It can be understood as a topological equality of the network nodes, which maintain all information at an equal degree of accessibility, independent of type (data, text, sound, image) or application. Thus, it can be said that the principle of neutrality is based on network architecture, which is built as a neutral space (politically open), independent of the type of content and application produced, distributed or consumed. In this manner, the space created by net neutrality states that the access "locations" are the same in terms of accessibility.

In order to guarantee neutrality, the access providers (which offer the highways for travel through this "space") cannot institute browsing zones, block or limit access (by means of charges or speed reduction to browse sites, etc.). Access to applications cannot be different from the browsing of other locations of the net.

Using cities as an analogy, it is as if the government permitted travel to some places, but not to others; it is as if, in order to go from one place to another, citizens should pay for their choices (if walking on a street or avenue, for example). In some cases, this even happens, however, under very well established legal milestones, such as tolls for automobiles. However, even here, the service is charged (for the maintained highway).

Thus, the Brazilian government (as well as all governments) should guarantee that the space be neutral and public. A service provider cannot charge more for the user to access YouTube at a certain speed, for example, under allegations that the videos use up more broadband capacity of the net. This would bring the democratic, transversal, and conversational dynamics, which characterize digital communication to an end. This is the main guarantee that the Internet will continue to be a "space" without restrictions for technical innovations and for cultural, social, political, and economic development.

Internet, Legislation, and Net Neutrality in Other Countries

Jomar Silva (2010), in an article published on the "Trezentos" blog, states that the Comitê Gestor da Internet no Brasil (CGI.br) (Management Committee of Internet in Brazil) approved and published a resolution in 2009 which defines the "Princípios para a Governança e uso da Internet no Brasil" (Principles for Governance and Internet use in Brazil). Specifically regarding neutrality, Silva highlights item 6 of the document where one can read: "Filtering network traffic or traffic privileges should respect only technical and ethical criteria—political, commercial, religious, and cultural motives or any other form of discrimination or favoritism not being admissible" (Comitê Gestor, 2009). The guideline touches upon the issue, but it still seems too broad and imprecise.

Currently, Brazil is drawing up a bill of principles—unique in the world—dedicated to establishing citizens' rights and obligations with respect to Internet use in the country. This bill, elaborated through extensive consultation of the public, as an initiative of the Ministry of Justice, is the "Marco Civil da Internet Brasileira" (Brazilian Internet Civil Rights Framework), a project to be voted on still in 2012. The Civil Rights Framework states that net neutrality should be guaranteed as one of the principles of Internet use in the country. Such a provision displeases telecommunication companies which insist on managing the data traffic through, for example, the differentiation of type of content searched by Internet users. Article 9 of the Civil Rights Framework states that "the party responsible for the transmission, switching, or routing has the duty to treat any data bundle in an egalitarian manner,

without distinction with regards to content, origin and destination, service, terminal or application, thus prohibiting any discrimination or degradation of traffic not resulting from necessary technical requisites to the appropriate providing of services as per regulations." The proposal also states that "in providing Internet connection, paid or free, it is prohibited to monitor, filter, analyze or inspect data bundle content."

There are gaps, however, with respect to the violation of net neutrality, given that the Civil Rights Framework provides that this violation may occur should the "need for adequate service" be clear. However, as can be noted, the general principle of protection to the user is there. Recently, SindiTelebrasil (union of the telecom services operators) and Anatel are trying to make the gap left in the Civil Rights Framework more flexible. Now, with the PNBL, and with the backing of the Brazilian Government, the issue becomes even more acute. The future will be decided in the next few years. As affirmed by Magrani (2011):

The digital civil rights framework (currently under discussion at the Brazilian Congress) will establish a general rule regarding neutrality preventing any type of differential treatment of information, except for the hypothesis in which the discrimination or degradation should come about from technical requirements necessary for the provision of adequate service.(...) The regulation provides a general non-discrimination rule and goes beyond, making explicit the forbidding of content and application blocking.

Various countries have been adopting laws to guarantee net neutrality. Recently, the United States government sent rules—although timid in nature, but already advancement—to the FCC that aim to guarantee this neutrality in terms of transparency of service providers' practices, forbidding of blocking any content or applications, and discrimination of traffic content. According to Bruno Magrani:

Governments and regulators around the world, awoken by the ever more frequent examples of affront to net neutrality, have initiated a discussion and implementation process of the first norms regarding net neutrality. Following Chilean leadership, which in 2010 passed the first law regarding net neutrality in the world, Colombia recently adopted norms in its national development plan to stop information discrimination practices. Within the scope of the European Union, Netherlands was the pioneer in the adoption of specific norms. A comparative framework of the main net neutrality regulations and proposals for regulation in some Latin American countries can be seen below. Colors indicate the influence of one law over the laws of other countries; and so we are able to note that the Chilean law has a very strong influence over the Argentinean bill and the Colombian neutrality law. Brazil and Mexico, in turn, have discussed approaches different from those of Chile (Magrani, 2011).

Is the PNBL a Plausible Solution to the Digital Inclusion Problem?

Discussing digital inclusion is a thorny subject that demands the elaboration of policies, which encompass the access to new communication and information technologies as elements of social inclusion in its broadest sense.

Digital inclusion is not achieved simply when computers or Internet access is given, but rather, when the individual is placed in a broader process of full exercise of his citizenship. Digital inclusion should, consequently, be pondered thoroughly, from the enrichment of four basic capitals: social, cultural, intellectual, and technical capital (Lévy, 1998; Lemos, 2007). Social capital is that which values the dimensions of identity and community, social ties and political action (Bourdieu, 1980). Cultural capital is what refers to the history and symbolic assets of a social group, its past, its conquests, its art (Bourdieu, 1984). On the other hand, technical capital refers to the potential of action and communication. It is what permits a social group or an individual to be able to play its role in the world and communicate in a free and autonomous manner. Intellectual capital refers to a person's education, to the individual intellectual growth through learning; evidencing the exchange of knowledge and the accumulation of experiences.

Information and Communications and Technology (ICT) should be used as a means to expand and accumulate these kinds of capital. This could be achieved through guaranteeing connection quality, low prices, and net neutrality. This is the only way by which the PNBL will be able to potentialize the development of the four aforementioned types of capital.

Many who use these technologies in a compulsory manner are, as would say the Spanish sociologist Manuel Castells (2002), "interacted" and not "interacting"; that is, they utilize the devices and the electronic networks in a very rudimentary way, not achieving the ideal conditions in order to take advantage of all the cultural, social and economic benefits which digital media can offer.

With the current PNBL, the lack of guarantees of minimum speed, the limit on the amount of data to be downloaded, the uncertainty related to the variable price and the ghost of the violation of net neutrality can lead to an inclusion of only the "interacted."

Therefore, the greater challenge of citizen inclusion in the universe of digital culture is to enable individuals to produce their own content and distribute it freely, maintain ownership of one's personal data, guarantee topological net neutrality, privacy, and anonymity. Besides technical capital, digital inclusion projects should cause the growth of social, intellectual, and cultural capital. Inclusion presumes autonomy, freedom, and criticism.

Before finishing this section, it is important to highlight that the regional differences perceived in Brazil are another factor to be carefully considered for the improvement of the PNBL. In fact, even large cities in the poorest states face difficulties regarding broadband access to the Internet at competitive prices.

According to research published by the Center for Research on Information Technology and Communication (CETIC), only about 20% of households in the Brazilian North and Northeast have Internet access. The data reveal a noticeable gap when one considers the richest regions in Brazil: the Midwest has 39% of its households connected; in the South, the percentage reaches 45%, while in the Southeast, the number is about 49% (CETIC, 2012).

As expected, this difficulty is greater in rural areas, albeit progress was witnessed over the past years. The same survey released by CETIC revealed that only 10% of households in rural areas have Internet connections. This problem was reported in the document "Current results of the Broadband Access Program for Brazilian Public Schools in Urban Areas," published by the International Telecommunications Union (ITU):

According to the Brazilian experience, public policies directed towards different regions and with a nationwide scope must address local issues and take into consideration regional differences in order to succeed. The importance of adapting a policy to each state comes from the fact that Brazil consists of a vast territory and thus, different realities arise for different localities (ITU, 2011).

The Brazilian scenario becomes even less encouraging upon realizing that the Government gives up on more ambitious telecommunications universalization goals or that it does not utilize the reserved funds to expand the infrastructure due to resource budget restrictions. According to an article published on FolhaOnline, for example, "The telecommunications sectorial funds raised R$48.5 billion [US$28.52 billion] in ten years, but only R$2.6 billion [US$1.52 billion], that is, 5.4% of the total resources, were used for their original purpose: the expansion of telecommunications

services, including rural areas" (Folha-Online, 2011d).

In other words, the government policies need to be concerned not only with the setting in motion of all the processes involved in the battle against digital exclusion (less taxes on computer equipment and contracting of services), but they must also take into consideration the issues related to privacy and security.

The Brazilian National Broadband Plan in International Perspective

The ultimate goal of a broadband plan is to provide a fast and stable Internet access at low cost to the whole country. For example, a central customer service can be installed in any small town, even in remote areas, if a good Internet connection is available. In other words, having access to broadband is essential to increase productivity and competitiveness. That is why governments all over the world are taking into account the importance of digital communication and proposing different strategies to provide Internet access to their citizens. Figure 3 summarizes the national broadband plan of selected countries.

Besides the policies stated above, the main international experiences in the area make clear a second set of objectives of national broadband plans. In OECD countries, the proposals include: encouraging competition among service providers; a continuous investment in infrastructure; conducting programs directed to train users; adopting network neutrality and protecting privacy of users.

These bold goals have led countries like South Korea to be one of the leaders in innovation, research and development (Ovum Consulting, 2009; Wonki Min, 2010). However, we must realize that countries with large land areas face challenges different from those found in countries like Holland, Denmark (also ranked in the top of the list). Although the report "Broadband Growth and Policies in OECD Countries" (published by

Figure 3. National broadband plans around the world (adapted from Brazil, 2011)

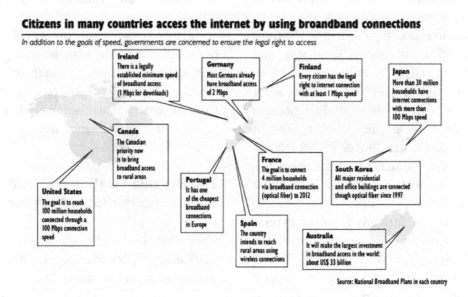

OECD in 2008) point that geographical variables are not related to an extensive coverage of DSL or cable networks (Canada has the highest penetration rate among the G7 countries), the case of developing countries with huge territorial extensions is different.

In India, for example, the national plan considers rural areas as the priority of investments in telecommunications. Initially, the intention of the government relied on to install towers in order to provide mobile telephony services. In a second stage, the intention is to offer broadband services (Kumar, 2008).

The Brazilian reality seems a bit more encouraging, since mobile services reach more than 90% of the country's 5.564 municipalities. However, one must keep in mind that there are areas where rural electrification is still being consolidated. Cable networks are not available even in some neighborhoods of large cities. This is evidence that wireless technology can offer a unique contribution. In addition, other challenges faced by Brazil regarding universal access are the establishment of an appropriate legal framework and a plan directed to lowering taxes.

Finally, one must remember that the provision of Internet access is only the means, not the end. Providing Internet access to remote areas is an opportunity not only to create jobs, but also to make it possible for citizens to enjoy public services and follow the performance of their representatives. In this context, the willingness of political actors is essential.

CONCLUSION

The main objective of the Brazilian National Broadband Plan (PNBL) is to increase Internet access from 12 million to 40 million households in all regions of the country. The estimated cost of the Plan is about 6.5 US billion dollar for the 2010-2014 period.

After presenting the main characteristics of the PNBL, the chapter pointed out to what degree the project outlined by the Federal Government can fulfill the demands of Brazilian society. We also discussed the tensions and the interests of the stakeholders involved. Moreover, the chapter presented the PNBL's main challenges, the lessons learned and gave suggestions for future improvements

As previously stated, it can be said that the last decade of the 20th century deeply transformed the telecommunications sector in Brazil. At least two great events changed the reality of the national market: (1) the end of the telecommunications monopoly and (2) the sale of state-owned companies. Within this new reality, telephony and Internet access end up being offered, fundamentally, by private enterprises, including multinationals. The operations are monitored by the regulatory authority, Anatel.

If privatization served as a justification to fulfill the needs for free trade and alignment with the global economy, it is plausible to affirm that the role of the authorities of the Brazilian State has still not had the desired effect in terms of universalization and the affordability of telecommunications services. Due to pressures from companies and political parties, there is great resistance to an effective control or regulation in the Brazilian communications sector.

In the telecommunications sector, the clash between government and companies takes place, largely, in the attempt to universalize services. The private sector is interested only in investing in the more profitable areas, leaving out remote regions. In this manner, there is no doubt regarding the expansion potential of telecommunications services in Brazil for purpose of including connection-hungry customers. However, the market reality does not appear favorable to competition, since there are cities in the Brazil that have only one mobile phone operator.

It is also fundamental to perfect the goals of connection supply in rural areas. The government has on its horizon of concerns Internet access in small cities, but as far as the areas farther away from city centers are concerned, there is little mobilization. Naturally, it is not worth it to install cable and fiber optics in all the corners of the country. That is exactly why the supply of connections by means of wireless networks is something to be emphasized in the current Brazilian framework.

One can consider that the PNBL will be an excellent project, but needs some adjustments without delay. Otherwise, it will fail even before being implemented either because of inefficiency of the public power or because of the strength of lobbies. The Plan could be a good inductor of competition for high quality services and lower costs. The government should adopt policies against lower speeds, higher prices, and violation of net neutrality. In other words, high quality, neutral operation, and fast Internet access should be guaranteed. Only in this manner it will be an adequate instrument to guarantee innovation, economic growth, informational plurality, political commitment, freedom and cultural growth.

REFERENCES

Bourdieu, P. (1980). Le capital social: Notes provisoires. *Actes de la Recherche en Sciences Sociales, 31,* 2–3.

Bourdieu, P. (1984). *Questions de sociologie.* Paris, France: Les Éditions de Minuit.

Branco, M. (2011). Interview on net neutrality on Brazil. Retrieved November 09, 2011, from http://goo.gl/5jQpP

Brandão, M. (2011). Interview on the Brazilian national broadband plan. Retrieved November 09, 2011, from http://goo.gl/WS14Z

Brazil, I. B. G. E. (2010). *PNAD 2009:* Income and number of registered workers rise; unemployment increases. Retrieved November 09, 2011, from http://goo.gl/Mu2eX

Brazil National Telecommunications Agency. (2001). Annual report. Retrieved November 09, 2011, from http://goo.gl/aA7oF

Brazil The Federal Senate. (2011). Em discussão magazine. Retrieved from http://www.senado.gov.br/noticias/Jornal/emdiscussao/banda-larga/banda-larga-no-mundo.aspx

Castells, M. (2002). *A sociedade em rede*. São Paulo, Brazil: Paz & Terra.

CETIC. (2012). Report: Access to information and communication technologies (ICT) in Brazil. Retrieved May 31, 2012, from http://cetic.br/usuarios/tic/2011-total-brasil/rel-geral-04.htm

Comitê Gestor da Internet no Brasil. (2009). *Resolução* CGI.br/RES/2009/003/P - *Princípios para a governança e uso da internet no Brasil*. Retrieved from http://www.cgi.br/regulamentacao/resolucao2009-003.htm

Costa, L. (2007). *Comunicação, novas tecnologias e inclusão digital – Uma análise dos projetos realizados na Bahia. (Dissertação de Mestrado)*. Salvador, Brazil: Universidade Federal da Bahia.

Estadão. (2011). A survey on the quality of broadband access in Brazil. Retrieved November 09, 2011, from http://goo.gl/IF5AQ

Firjan. (2011). Relatório: Quanto custa o acesso à Internet para as empresas no Brasil? Retrieved from http://www.firjan.org.br/lumis/portal/file/fileDownload.jsp?fileId=2C908CEC2F01DD5A012F225AE55B053C

Folha-Online. (2007). Brazilian users increase the time spent on the internet by more than 3 hours in a year. Retrieved November 09, 2011, from http://goo.gl/IAkmk

Folha-Online. (2010a). Active internet users reach 36.9 million in Brazil in January 2010. Retrieved November 09, 2011, from http://goo.gl/7rCW9

Folha-Online. (2010b). Anatel studies a more strict control over broadband prices. Retrieved November 09, 2011, from http://goo.gl/CM2U9

Folha-Online. (2010c). Telecommunication companies reduce investments by 40%. Retrieved November 09, 2011, from http://www.folha.com.br/me821929

Folha-Online. (2010d). Telecommunication companies will contest the Brazilian national broadband plan. Retrieved November 09, 2011, from http://goo.gl/nCugX

Folha-Online. (2011a). Brazil nears 230 million mobile phone lines. Retrieved November 09, 2011, from http://folha.com/no992891

Folha-Online. (2011b). Brazilian internet users are 58.6 million. Retrieved November 09, 2011, from http://folha.com/te940381

Folha-Online. (2011c). Anatel approves phones subscriptions at a US$8 cost. Retrieved November 09, 2011, from http://folha.com/me924732

Folha-Online. (2011d). Government spent only 5.4% of the amount reserved to improve telecommunications. Retrieved November 09, 2011, from http://folha.com/me945226

Free Press. (2012). *What is net neutrality?* Retrieved from http://www.savetheinternet.com/frequently-asked-questions

Grossmann, L. (2011). Popular broadband access: Cost is higher than negotiated between telecommunication companies and government. Retrieved November 09, 2011, from http://goo.gl/eE45L

IDGNOW. (2011a). Brazilians are the ones who most use social networks in the workplace. Retrieved November 09, 2011, from http://migre.me/66Ogy

IDGNOW. (2011b). Study reveals that 55% of Brazilians access pornography sites. Retrieved November 09, 2011, from http://migre.me/66Ofd

Info Magazine. (2007). Brazil has over 40 million landline subscribers. Retrieved November 09, 2011, from http://goo.gl/WzXfg

ITU. (2012). Current results of the broadband access program for Brazilian public schools in urban areas. Retrieved May 31, 2012, from http://www.itu.int/md/D10-SG01-C-0087/en

Jambeiro, O., Mota, A., Amaral, C., Santos, S., Ferreira, S. A., & Simoes, C. ... Costa, E. (2004). Tempos de vargas - O rádio e o controle da informação. Salvador, Brazil: Edufba.

Kumar, A. (2008). Regulatory initiatives to promote: Universal access to broadband. In *Proceedings of the ITU Workshop on Regulatory Policies on Universal Access to Broadband Services.* Retrieved from http://www.itu.int/ITU-D/treg/Events/Seminars/2008/2_1_Kumar.ppt

Lemos, A. (Ed.). (2007). *Cidade digital: Portais, inclusão e redes no Brasil.* Salvador, Brazil: Edufba.

Lévy, P. (1998). *A inteligência coletiva: Por uma antropologia do ciberespaço.* São Paulo, Brazil: Loyola.

Magrani, B. (2011). New developments on the regulation of network neutrality. Retrieved November 09, 2011 from http://goo.gl/HgY8v

Nielsenwire. (2011). Swiss lead in speed: Comparing global internet connections. Retrieved November 09, 2011, from http://shar.es/q7ur4

Notícias, U. O. L. (2011). Brazilian children are the youngest children to access social networks. Retrieved November 09, 2011, from http://migre.me/66Oo8

OECD. (2008). *Broadband growth and policies in OECD countries.* Paris, France: OECD.

OECD. (2009). *Broadband and the economy.* Seoul, South Korea: OECD.

Ovum Consulting. (2009). *Broadband policy development in the Republic of Korea.* Retrieved from http://unpan1.un.org/intradoc/groups/public/documents/un-dpadm/unpan041139.pdf

Portal, I. G. (2010). *Brazilian users are the ones who spend most time on the internet.* Retrieved November 09, 2011, from http://goo.gl/RJ8Ou

Portal G1. (2011). Brazilians are the ones who most watch videos and TV on the internet. Retrieved November 09, 2011, from http://migre.me/66O8V

Ramos, M. C. O. (2000). Às margens da estrada do futuro: Comunicações, políticas e tecnologias. Sao Paolo, Brasília: UnB.

Silva, J. (2010). What is net neutrality and why you should worry about it. Retrieved November 09, 2011, from http://www.trezentos.blog.br/?p=6465

Sundfeld, C. A. (2006). A regulação das telecomunicações: Papel atual e tendências futuras. Revista Eletrônica de Direito Administrativo Econômico, 8. Retrieved from http://www.direitodoestado.com/revista/REDAE-8-NOVEMBRO-2006-CARLOS%20ARI%20SUNDFELD.pdf

Superinteressante. (2010). Broadband incites economic growth. Retrieved November 09, 2011, from http://goo.gl/e7ADR

Takahashi, T. (Ed.). (2000). *Green book of the information society in Brazil.* Sao Paolo, Brasília: MCT. Retrieved from http://www.mct.gov.br/index.php/content/view/18878.html

Wonki Min. (2010). *Broadband policy in Korea.* Retrieved from http://siteresources.worldbank.org/BELARUSEXTN/Resources/1.77koreabb.pdf

ADDITIONAL READING

Galloway, A. (2004). *Protocol: How control exists after decentralization.* Cambridge, MA: MIT.

Goldsmith, J., & Wu, T. (2006). *Who controls the internet? Illusions of a borderless world.* Oxford, UK: Oxford University Press.

Lenard, T., & May, R. E. (Eds.). (2006). *Net neutrality or net neutering: Should broadband internet services be regulated?* New York, NY: Springer. doi:10.1007/0-387-33928-0

Lessig, L. (2000). *Code and other laws of cyberspace*. New York, NY: Basic Books.

Morales, D. (2011). *The national broadband plan: Analysis and strategy for connecting America*. New York, NY: Nova Science Pub Inc.

KEY TERMS AND DEFINITIONS

Broadband Access: High-speed Internet access. Year after year, users demand faster alternatives to access digital contents.

Digital Divide: Refers to inequalities in use and access to Information and Communication Technologies (ICTs). Includes, for example, differences in computing skills and in making use of the information available on the Internet.

Information and Communication Technologies (ICT): The entire set of digital technologies that allows the processing, storage and transmission of data through the use of devices (telephones, computers, etc.) and applications (software).

Net Neutrality: Principle according to which the Internet should be a platform free from the intervention of Internet service providers regarding the content that users access. Telecommunication companies are not supposed to charge subscribers according to the content sent over their networks.

Policy Making: The process of discussing and developing policies.

Regulation: All the tools that the States have to guarantee that services are being properly delivered to citizens or consumers.

Universal Access: Refers to allowing everyone to have access to telecommunications services at a reasonable cost.

Compilation of References

2020 Communications Trust. (2009). *2020 leader in communications ICT in New Zealand and the Pacific.* Retrieved March 12, 2009, from http://www.2020.org.nz/

Abdelaal, A., & Ali, H. (2007). Typology for community wireless network business models. In *Proceedings of the Thirteenth Americas Conference on Information Systems.* Keystone, CO: IEEE.

Abdelaal, A., & Ali, H. (2008). A graph theoretic approach for analysis and design of community wireless networks. In *Proceedings of Americas Conference on Information Systems (AMCIS).* Toronto, Canada: AMCIS. Retrieved October 15, 2011 from http://aisel.aisnet.org/amcis2008/310/

Abdelaal, A., & Ali, H. (2009). Analyzing community contributions to the development of community wireless networks. In *Proceedings of the European Conference for Information Systems.* IEEE.

Abdelaal, A., & Ali, H. (2009). *Community wireless networks: Collective projects for digital inclusion.* Paper presented at the International Symposium on Technology and Society (ISTAS 2009). Tempe, AZ.

Abdelaal, A., & Ali, H. (2012). Human capital in the domain of community wireless networks. In *Proceedings of the 45th Hawaii International Conference on System Sciences,* (pp. 3338-3346). IEEE Press.

Abdelaal, A., & Ali, H. (2012). Measuring social capital in the domain of community wireless networks. In *Proceedings of the 2012 45th Hawaii International Conference on System Sciences.* Maui, HI: IEEE Press.

Abdelaal, A., & Ali, H. H. (2009). Community wireless networks: Emerging wireless commons for digital inclusion. In *Proceedings of the IEEE International Symposium on Technology and Society (ISTAS 2009),* (pp. 1-9). IEEE Press.

Abdelaal, A., Ali, H., & Khazanchi, D. (2009). The role of social capital in the creation of community wireless networks. In *Proceedings of the 42nd Hawai'i International Conference on Systems Sciences.* Waikoloa, HI: IEEE Press.

Ackerman, L. S. (1982). Transition management: An in-depth look at managing complex change. *Organizational Dynamics, 82*(1), 46–66. doi:10.1016/0090-2616(82)90042-0

Adeyeye, M., & Gardner-Stephen, P. (2011). The village telco project: A reliable and practical wireless mesh telephony infrastructure. *EURASIP Journal on Wireless Communications and Networking, 78.* doi:10.1186/1687-1499-2011-78

Agrawal, R., Gupta, A., & Gupta, M. C. (2011). Financing of PPP infrastructure projects in India: Constraints and recommendations. *The IUP Journal of Infrastructure, 9*(1), 54–57.

Aitkin, H. (2009). Bridging the mountainous divide: A case for ICTs for mountain women. *Mountain Research and Development, 22*(3), 225–229. doi:10.1659/0276-4741(2002)022[0225:BTMDAC]2.0.CO;2

Akyildiz, I. F., Wang, X., & Wang, W. (2005). Wireless mesh networks: A survey. *Computer Networks, 47*(4), 445–487. doi:10.1016/j.comnet.2004.12.001

Albanesius, C. (2009). *Philadelphia repurchases city wi-fi network for $2M*. Retrieved December 21, 2009, from http://www.pcmag.com/article2/0,2817,2357395,00.asp#fbid=Y0Rf0aYjKVe

Albert, S., Flournoy, D., & LeBrasseur, R. (2009). *Networked communities: Strategies for digital collaboration*. Hershey, PA: IGI Global. doi:10.4018/978-1-59904-771-3

Albert, S., & Lebrasseur, R. (2010). Citizen engagement in the networked community. *The International Journal of Knowledge, Culture, and Change Management, 10*(4), 55–67.

Alfeus, o. (2012). *Systém komplex*. Retrieved April 12, 2012, from http://www.alfeus.sk/index.php/komplex/

American Public Power Association. (2004). State barriers to community broadband services. *APPA Fact Sheet*. Retrieved from http://www.baller.com/pdfs/Barriers_End_2004.pdf

Analysys Mason. (2011). *Ultrafast broadband prices fall to new low in Europe and the U.S.* Retrieved July 16, 2012, from http://www.analysysmason.com/About-Us/News/Insight/Ultra-fast-broadband-prices-fall-to-a-new-low-in-Europe-and-the-USA/

Auray, N., Charbit, C., Charbit, C., & Fernandez, V. (2003). *Wifi: An emerging information society infrastructure*. Retrieved from http://ses.telecom-paristech.fr/auray/Auray%20Beauvallet%20Charbit%20Fernandez.pdf

Azad Indian Foundation. (2012). *Website*. Retrieved on January 14, 2012 from http://azadindia.org/social-issues/rural-healthcare-in-india.html

Badger, E. (2011). *How the telecom lobby is killing municipal broadband*. Retrieved from http://www.theatlanticcities.com/technology/2011/11/telecom-lobby-killing-municipal-broadband/420

Bai, V. T., Prashant, S., & Jhunjhunwala, A. (2011). Rural health care delivery- Experiences with ReMeDi telemedicine solution in southern Tamilnadu. In *Proceedings of the International eHealth Telemedicine and Health ICT Forum for Educational, Networking and Business*. IEEE.

Baller, J. (2006a, September-October). Quoted in state broadband battles. *Public Power Magazine*. Retrieved from http://www.appanet.org

Baller, J. (2006). *Proposed barriers to state entry*. Retrieved from http://www.baller.com/pdfs/Baller_Proposed_State_Barriers.pdf

Baller, J. (2011). *State restrictions on community broadband services or other public communications initiatives*. Retrieved from http://baller.com/pdfs/BallerHerbstState-Barriers(7-1-12).pdf.

Ball-Rokeach, S. J., Gibbs, J., Hoyt, E. G., Jung, J. Y., Kim, Y. C., Matei, S., et al. (2002). *Metamorphosis project white paper #1 - The challenge of belonging in the 21st century: The case of Los Angeles*. Retrieved 2 September, 2002, from http://www-scf.usc.edu/~matei/stat/globalization.html

Bauwens, M. (2005). Peer-to-peer and human evolution. *Integral Visioning*. Retrieved April 16, 2009, from http://integralvisioning.org/article.php?story=p2ptheory1#_Toc107024680

BBC. (2001). *Village in the clouds embraces computers*. Retrieved July 22, 2011, from http://news.bbc.co.uk/2/hi/science/nature/1606580.stm

BBC. (2004). *Wi-fi lifeline for Nepal farmers*. Retrieved July 28, 2011, from http://news.bbc.co.uk/2/hi/technology/3744075.stm

BBC. (2010). *Poorer pupils to be given free laptops*. Retrieved July 25, 2011, from http://news.bbc.co.uk/2/hi/8449485.stm

BBC. (2011). *£98 PCs target UK digital divide*. Retrieved July 22, 2011, from http://www.bbc.co.uk/news/technology-12205412

Becker, J. (2002). Theses on information. In Mansall, R., Samarajiva, R., & Mahan, A. (Eds.), *Networking Knowledge for Information Societies: Institutions & Intervention* (pp. 216–219). Delft, The Netherlands: Delft University Press.

Benjamin, S., Lichtman, D. G., Shelanski, H., & Weiser, P. (2006). *Telecommunications law and policy*. Durham, NC: Carolina Academic Press.

Benjamin, S., Lichtman, D. G., Shelanski, H., & Weiser, P. (2011). *Telecommunications law and policy: 2011 supplement*. Durham, NC: Carolina Academic Press.

Benkler, Y. (2006). *The wealth of networks: How social production transforms markets and freedom*. New Haven, CT: Yale University Press.

Best, M. L., & Maclay, C. M. (2002). *Community internet access in rural areas: Solving the economic sustainability puzzle*. Retrieved from http://unpan1.un.org/intradoc/groups/public/documents/apcity/unpan008658.pdf

Biczók, G., Toka, L., Vidacs, A., & Trin, T. A. (2009). *On incentives in global wireless communities*. Paper presented at the International Conference on Emerging Networking Experiments and Technologies Archive. New York, NY.

Bimber, B. (1998). The internet and political transformation: Populism, community and accelerated pluralism. *Polity, 31*(1), 133–161. doi:10.2307/3235370

Bina, M. (2007). *Wireless community networks: A case of modern collective action*. (PhD Thesis). Athens University of Economics and Business. Athens, Greece.

Bina, M., & Giaglis, G. M. (2006). Unwired collective action: Motivations of wireless community participants. In *Proceedings of the 5th International Conference on Mobile Business (ICM)*. ICMB.

Bina, M., & Giaglis, G. M. (2006). *A motivation and effort model for members of wireless communities*. Paper presented at the European Conference on Information Systems. Göteborg, Sweden.

Bina, M., & Giaglis, G. M. (2005). Emerging issues in researching community-based WLANS. *Journal of Computer Information Systems, 46*(1), 9–16.

Blakely, R. (2008, March 12). State bank of India takes no-frills ATMs to masses. *Times (London, England)*.

Blumer, H. (1969). *Symbolic interactionism: Perspective and method*. Englewood Cliffs, NJ: Prentice Hall.

Botein, M. (2006). Regulation of municipal wi-fi. *New York Law School Law Review. New York Law School, 51*, 975–988.

Bourdieu, P. (1980). Le capital social: Notes provisoires. *Actes de la Recherche en Sciences Sociales, 31*, 2–3.

Bourdieu, P. (1984). *Questions de sociologie*. Paris, France: Les Éditions de Minuit.

Bourdieu, P. (1999). Structures, habitus, practices. In Elliott, A. (Ed.), *The Blackwell Reader in Contemporary Social Theory* (pp. 108–118). Malden, MA: Blackwell.

Bourgeois, L. J., & Brodwin, D. R. (1984). Strategic implementation: Five approaches to an elusive phenomenon. *Strategic Management Journal, 5*, 241–264. doi:10.1002/smj.4250050305

Branco, M. (2011). Interview on net neutrality on Brazil. Retrieved November 09, 2011, from http://goo.gl/5jQpP

Brandão, M. (2011). Interview on the Brazilian national broadband plan. Retrieved November 09, 2011, from http://goo.gl/WS14Z

Brazil National Telecommunications Agency. (2001). Annual report. Retrieved November 09, 2011, from http://goo.gl/aA7oF

Brazil The Federal Senate. (2011). Em discussão magazine. Retrieved from http://www.senado.gov.br/noticias/Jornal/emdiscussao/banda-larga/banda-larga-no-mundo.aspx

Brazil, I. B. G. E. (2010). *PNAD 2009:* Income and number of registered workers rise; unemployment increases. Retrieved November 09, 2011, from http://goo.gl/Mu2eX

Breitbart, J. (2008). The Philadelphia story: Learning from a municipal wireless pioneer. *New America Foundation*. Retrieved October 11, 2011, from http://www.newamerica.net/publications/policy/philadelphia_story

Brinkerhoff, J. M. (2008). What does a goat have to do with development diasporas, IT, and the case of thamel.com. *Information Technologies and International Development, 4*(4), 9–14. doi:10.1162/itid.2008.00023

Broadband Commission of ITU and UNESCO. (2012). *The state f broadband 2012: Achieving digital inclusion for all*. Retrieved from http://www.broadbandcommission.org

Broadband Technology Opportunities Program. (2011). The virtual village: Digital El Paso's pathway to success. *Broadband USA*. Retrieved October 15, 2011, from http://www2.ntia.doc.gov/grantee/city-of-el-paso

Buckley, S. (2012). *CenturyLink gets $35m in FCC CAF funding for broadband expansion*. Retrieved from http://www.fiercetelecom.com/story/centurylink-gets-35m-fcc-caf-funding-broadband-expansion/2012-07-25

Bula, J. (2007). *Measuring the RF parameters of WiFi devices.* (BSc. thesis). FAI UTB ve Zlíně.

Burt, R. S. (1999). *The social capital of opinion leaders.* New York, NY: American Academy of Political and Social Science.

Camponovo, G. (2011). *A review of motivations in virtual communities.* Paper presented at the European Conference on Information Management and Evaluation (ECIME). Geneva, Switzerland.

Camponovo, G., & Picco-Schwendener, A. (2010). *A model for investigating motivations of hybrid wireless community participants.* Paper presented at the 2010 Ninth International Conference on Mobile Business / 2010 Ninth Global Mobility Roundtable. New York, NY.

Camponovo, G., & Picco-Schwendener, A. (2011). *Motivations of hybrid wireless community participants: A qualitative analysis of Swiss Fon members.* Paper presented at Tenth International Conference on Mobile Business 2011. Athens, Greece.

Camponovo, G., Heitmann, M., Stanoevska-Slabeva, K., & Pigneur, Y. (2003). *Exploring the WISP industry Swiss case study.* Paper presented at the 16th Bled Electronic Commerce Conference eTransformation. Bled, Slovenia.

Canel, E. (1997). New social movement theory and resource mobilization theory: The need for integration. In Kaufman, M., & Dilla Alfonso, H. (Eds.), *Community Power and Grassroots Democracy* (pp. 189–221). London, UK: Zed Books.

Carlson, S. C. (1999). A historical, economic, and legal analysis of municipal ownership of the information highway. *Rutgers Computer & Technology Law Journal, 25*(1), 3–60.

Carroll, J., & Rosson, M. (2007). Participatory design in community informatics. *Design Studies, 28*, 243–261. doi:10.1016/j.destud.2007.02.007

Casswell, S. (2001). Community capacity building and social policy – What can be achieved? *Social Policy Journal of New Zealand, 17*, 22–35.

Castells, M. (2007). *Communication technology, media and power.* Retrieved March 23, 2007, from http://web.mit.edu/newsoffice/2007/media-technology-0314.html

Castells, M. (2002). *A sociedade em rede.* São Paulo, Brazil: Paz & Terra.

Castells, M. (2006). *The network society: From knowledge to policy.* Washington, DC: Johns Hopkins Center for Transatlantic Relations.

Castells, M., Fernandez-Ardevol, M., Qiu, J. L., & Sey, A. (2006). *Mobile communication and society: A global perspective.* Cambridge, MA: MIT Press. doi:10.1111/j.1944-8287.2008.tb00398.x

Cellular Authorities of India. (2004). *GSM mobile statistics.* New Delhi, India: Cellular Authorities of India.

Census of India. (2011). *Government of India, ministry of home affairs, census India.* Retrieved on January 14, 2012, from http://censusindia.gov.in/

Census of India. (2011). *Provisional population totals: Rural urban distribution.* Retrieved on January 14, 2012 from http://www.censusindia.gov.in/2011-prov-results/paper2/data_files/india/paper2_at_a_glance.pdf

Cequel III Communications v. Local Agency Formation Commission of Nevada, 149 Cal.App.4th 310 (2007).

CESNET. (2012). *Czech internet celebrates its 20th anniversary today.* Retrieved February 27, 2012, from http://www.ces.net/doc/press/2012/pr120213.html

CETIC. (2012). Report: Access to information and communication technologies (ICT) in Brazil. Retrieved May 31, 2012, from http://cetic.br/usuarios/tic/2011-total-brasil/rel-geral-04.htm

Chapleau Portal. (2007). *The Chapleau project highlighted in Nortel e-seminar.* Retrieved December 26, 2011, from http://www.chapleau.ca/portal/en/connectingchapleau?paf gear_id=1000025&itemId=3100038&returnUrl=%2Fportal%2Fen%2Fconnectingchapleau%3Bjsessionid%3DJIYQRVKO2BQBNTRPH3XHLRQ

Chapleau Portal. (2011). *Nortel learn IT.* Retrieved December 26, 2011, from http://www.chapleau.ca/portal/en/connectingchapleau/nortellearnit

Cherney, L. (1999). *Conversation and community: Discourse in a social MUD.* Chicago, IL: Chicago University Press.

Cho, H. (2008). Towards peer-place community and civic bandwidth: A case study in community wireless networking. *Journal of Community Informatics*, *4*(1).

Christensen, A. D. (2006). 'Wi-Fi'ight them when you can join them? How the Philadelphia compromise may have saved municipally-owned telecommunications service. *Federal Communications Bar Journal*, *58*(3), 683–704.

CIA. (2012). *Nepal - CIA worldfactbook*. Retrieved 12 July, 2012, from http://www.theodora.com/wfbcurrent/nepal/nepal_introduction.html

Cisco Systems, Inc. (2006). Municipalities adopt successful business models for outdoor wireless networks. White Paper (C11-325079-00). *Cisco*. Retrieved October 20, 2011, from http://www.cisco.com/en/US/prod/collateral/wireless/ps5679/ps6548/prod_white_paper0900aecd80564fa3.pdf

Cisco Systems, Inc. (2006). Evolution of municipal wireless networks. White Paper (C11-378713-00). *IDG Connect*. Retrieved on October 26, 2011, from http://www.idgconnect.com/view_abstract/2254/evolution-municipal-wireless-networks

Cisco. (2007). *Municipalities adopt successful business models for outdoor wireless networks*. Retrieved from http://www.cisco.com/en/US/netsol/ns621/networking_solutions_white_paper0900aecd80564fa3.shtml

City of Abilene v. FCC, 164 F.3d 49 (D.C. Cir. 1999).

City of Bristol v. Earley, 145 F.Supp.2d 741 (W.D. Va. 2001).

City of El Paso. (2006). *The digital El Paso project*. Retrieved November 10, 2011, from http://www.elpasotexas.gov/it/_documents/3T%20Quality%20of%20Place%20Proposal_Final.pdf

City of El Paso. (2011). *El Paso Lyceum*. Retrieved November 1, 2011, from http://www.elpasotexas.gov/mayor/elpaso_lyceum.asp

Clara, S. (2011). *Website*. Retrieved from http://www.ci.santa-clara.ca.us

Clarke, P., & Healy, K. (2003). Investigating aspects of public private partnerships in Ireland. *The Irish Journal of Management*, *24*(2), 20–30.

Clary, E. G., Snyder, M., Ridge, R., Copeland, J., Stukas, A., & Haugen, J. (1998). Understanding and assessing the motivations of volunteers: A functional approach. *Journal of Personality and Social Psychology*, *74*(6), 1516–1530. doi:10.1037/0022-3514.74.6.1516

Clinton, J., Hattie, J., & Dixon, R. (2007). *Evaluation of the flaxmere project: When families learn the language of school*. Auckland, New Zealand: Ministry of Education.

Coleman, R., Behunin, J., & Harvey, M. (2012). *A performance audit of the Utah telecommunication open infrastructure agency*. Retrieved from http://le.utah.gov/audit/12_08rpt.pdf

Coleman, J. (1988). Social capital in the creation of human capital. *American Journal of Sociology*, *94*, S95–S120. doi:10.1086/228943

Collins, S. (2012). Auckland: A city divided by income. *New Zealand Herald*. Retrieved from http://www.nzherald.co.nz/nz/news/article.cfm?c_id=1&objectid=10783692

Collins, D., Morduch, J., Rutherford, S., & Ruthven, O. (2009). *Portfolio of the poor*. Princeton, NJ: Princeton University Press.

Columbia Telecommunications Corporation. (2009). *Enhanced communications in San Francisco: Phase II fiber optics feasibility report*. Kensington, MD: Columbia Telecommunications Corporation.

Comitê Gestor da Internet no Brasil. (2009). *ResoluçãoCGI.br/RES/2009/003/P- Princípios para a governança e uso da internet no Brasil*. Retrieved from http://www.cgi.br/regulamentacao/resolucao2009-003.htm

Community Broadband Networks. (2012). *Community broadband network map*. Retrieved from http://muninetworks.org/communitymap

Community Wireless Resource Center. (2009). *Status report on wireless Africa initiative in Uganda. Technical Report*. Kampala, Uganda: Makerere University.

Community Wireless Resource Center. (2011). *Examining the impact of public access to information and communications technologies (ICTs): Selected IDRC-funded telecentres and the associated wireless networks in rural Uganda. Technical Report*. Kampala, Uganda: Makerere University.

Computer Access New Zealand. (2009). *Computer access trust NZ: Refurbishing office computers for schools and the community*. Retrieved January 31, 2012, from http://www.canz.org.nz/

Computers in Homes. (2011). *Website*. Retrieved April 16, 2011, from http://www.computersinhomes.org.nz/

Connect, C. C. (2012). *About*. Retrieved from http://www.connectcc.com/aboutconnect.html

Connecting America. (2010). *The national broadband plan*. Retrieved from http://download.broadband.gov/plan/national-broadband-plan.pdf

Connecting America. (2012). The national broadband plan. *Download the Plan*. Retrieved March 26, 2012 from http://www.broadband.gov/download-plan

Cooper, M. (2012). *Efficiency gains and consumer benefits of unlicensed access to the public airwaves: The dramatic success of combining market principles and shared access*. Retrieved from http://www.consumerfed.org/pdfs/EFFICIENCYGAINS-1-31.pdf

Cormack, D. (1991). *The research process*. Oxford, UK: Black Scientific.

Costa, L. (2007). *Comunicação, novas tecnologias e inclusão digital – Uma análise dos projetos realizados na Bahia. (Dissertação de Mestrado)*. Salvador, Brazil: Universidade Federal da Bahia.

Council of Europe/ERICarts. (2012). *Internet penetration rate in Europe*. Retrieved July 16, 2012, from http://www.culturalpolicies.net/web/compendium.php

Crawford, S. P. (2008). *Information/telecommunications services*. Retrieved from http://scrawford.net/blog/informationcommunications-services/1181

Crawford, S. P. (2011). The communications crisis in America. *Harvard Law & Policy Review, 5*(2), 245–263.

Crow, B., Miller, T., & Powell, A. (2006). *A case study of ISF 'free' hotspot owners and users*. Paper presented to the Canadian Communications Association. Ottawa, Canada.

ČTÚ. (1997). Gerální povolení GP - 01/1994. *Telekomunikační věstník, 1*. Retrieved from http://www.ctu.cz/cs/download/vseobecna-opravneni/archiv/tv_mimoradna_castka_01_1997.pdf

ČTÚ. (2000). Gerální licence GL - 12/R/2000. *Telekomunikační věstník, 9*, 15–16.

Cullen, R. (2001). Addressing the digital divide. *Online Information Review, 25*(5), 311–320. doi:10.1108/14684520110410517

Cupertino. (2008). *Wesbsite*. Retrieved from http://www.cupertino.org

Czfree.net. (2009). Homepage. *Wireless Broadband Community*. Retrieved February 23, 2009, from http://czfree.net/home/index.php

Damsgaard, J., Parikh, M. A., & Rao, B. (2006). Wireless commons: Perils in the common good. *Communications of the ACM, 49*(2). doi:10.1145/1113034.1113037

Das, D. (2005). *How do we measure if closing the digital divide addresses barriers to social inclusion?* Wellington, New Zealand: Victoria University of Wellington.

Deci, E. L., & Ryan, R. M. (1985). *Intrinsic motivation and self-determination in human behavior*. New York, NY: Plenum.

Demirag, I., & Khadaroo, I. (2011). Accountability and value for money: A theoretical framework for the relationship in public-private partnerships. *Journal of Management & Governance, 15*(2), 271–296. doi:10.1007/s10997-009-9109-6

Department of Internal Affairs. (2010). 3,000 New Zealand families to learn new digital skills. *Computers in Homes*. Retrieved from http://www.computersinhomes.org.nz/

Department of Internal Affairs. Department of Labour, Department of Prime Minister & Cabinet, Ministry of Agriculture & Forestry, Ministry of Economic Development, Ministry of Education, et al. (2002). *Connecting communities: A strategy for government support of community access to information and communications technology*. Retrieved from http://www.dol.govt.nz/PDFs/cegBooklet2000.pdf

Devera, M. (2002). *Hierachical token bucket theory*. Retrieved August 30, 2012, from http://luxik.cdi.cz/~devik/qos/htb/manual/theory.htm

Díaz Andrade, A., & Urquhart, C. (2009). The value of extended networks: Social capital in an ICT intervention in rural Peru. *Information Technology for Development, 15*(2), 108–132. doi:10.1002/itdj.20116

Dingwall, C. (2006). Municipal broadband: Challenges and perspectives. *Federal Communications Bar Journal, 59*(1), 67–103.

Djursland International Institute of Rural Wireless Broadband. (2009). *Homepage*. Retrieved February 23, 2009, from http://diirwb.net/

Dobričić, D. (2007). Efficient feed for offset parabolic antennas for 2.4 GHz. *antenneX, 128*. Retrieved from http://www.qsl.net/yu1aw/ANT_VHF/fid24ghz.pdf

Dulík, T. (2012). *Methods for interference mitigation in wireless networks*. (Doctoral Thesis). FAI TBU in Zlin. Retrieved from http://zamestnanci.fai.utb.cz/~dulik/dissertation/

Dunne, M. (2007). Let my people go (online): The power of the FCC to preempt state laws that prohibit municipal broadband. *Columbia Law Review, 107*(5), 1126–1164.

Dutta, S., & Mia, I. (2011). *The global information technology report 2010–2011: Transformations 2.0*. Washington, DC: World Economic Forum-INSEAD.

EarthLink/Google. (2006). *Presentation*. Paper presented at San Francisco TechConnect. San Francisco, CA.

Efstathiou, E., Frangoudis, P., & Polyzos, G. (2005). *Stimulating participation in wireless community networks*. Retrieved April 21, 2009, from http://mm.aueb.gr/technicalreports/2005-MMLAB-TR-01.pdf

Ehrlich, T. (2000). *Civic responsibility and higher education*. New York, NY: Oryx Press.

El Paso Region Creative Cities Leadership Project. (2011). *City of El Paso*. Retrieved October 31, 2011, from http://www.elpasotexas.gov/mcad/cclp.asp

E-Networking Research and Development. (2009). *Final report of APT ICT J3 project in Nepal*. Paper presented to Asia Pacific Telecommunity. Bangkok, Thailand.

Esmond, J., & Dunlop, P. (2004). *Developing the volunteer motivation inventory to assess the underlying motivational drives of volunteers in Western Australia*. Perth, Australia: CLAN WA.

Estadão. (2011). A survey on the quality of broadband access in Brazil. Retrieved November 09, 2011, from http://goo.gl/IF5AQ

Eubanks, V. E. (2007). Trapped in the digital divide: The distributive paradigm in community informatics. *Journal of Community Informatics, 3*(2).

European Commission. (2009). *14th report on the implementation of the telecommunications regulatory package - 2008*. Retrieved March 5, 2012, from http://ec.europa.eu/information_society/policy/ecomm/library/communications_reports/annualreports/14th/index_en.htm

European Commission. (2010). *15th progress report on the single european electronic communications market - 2009*. Retrieved March 5, 2012, from http://ec.europa.eu/information_society/policy/ecomm/library/communications_reports/annualreports/15th/index_en.htm

European Commission. (2010). Broadband access in the EU: Situation at 1 July 2010. *Digital Agenda for Europe*. Retrieved March 5, 2012, from http://ec.europa.eu/information_society/newsroom/cf/item-detail-dae.cfm?item_id=6502&language=default

European Commission. (2011). *Digital agenda: Commission proposes over €9 billion for broadband investment*. Retrieved July 16, 2012, from http://ec.europa.eu/information_society/newsroom/cf/itemdetail.cfm?item_id=7430

European Commission. (2012). *Europa digital agenda scoreboard - Pillar 4*. Retrieved August 11, 2012, from http://ec.europa.eu/information_society/digital-agenda/scoreboard/docs/download/broadband%20_country_charts_2012.xls

Eurostat. (2012). *Broadband and connectivity - Households*. Retrieved August 13, 2012, from http://appsso.eurostat.ec.europa.eu/nui/show.do?dataset=isoc_bde15b_h&lang=en

FCC Press Release. (2010). *FCC frees up vacant TV airwaves for "super wi-fi" technologies*. Retrieved from http://hraunfoss.fcc.gov/edocs_public/attachmatch/DOC-301650A1.pdf

FCC v. Pacifica Foundation, 438 U.S. 726 (1978).

FCC. (2012). *Connect America fund (CAF) phase I*. Retrieved from http://www.fcc.gov/maps/connect-america-fund-caf-phase-i

Ferlander, S. (2003). *The internet, social capital and local community*. Stirling, UK: University of Stirling.

Fernback, J. (2005). Information technology, networks, and community voices – Social inclusion for urban regeneration. *Information Communication and Society, 8*(4), 482–502. doi:10.1080/13691180500418402

Fijnaut, C., & Paoli, L. (2004). *Organised crime in Europe: Concepts, patterns and control policies in the European Union and beyond*. Berlin, Germany: Springer.

Firjan. (2011). Relatório: Quanto custa o acesso à Internet para as empresas no Brasil? Retrieved from http://www.firjan.org.br/lumis/portal/file/fileDownload.jsp?fileId=2C908CEC2F01DD5A012F225AE55B053C

Fisher, K. (2010). ACT4PPP – A transnational initiative to promote public-private cooperation in urban development. *European Public Private Partnership Law Review, 5*(2), 106–111.

Flamm, K., & Anindya, C. (2005). *An analysis of the determinants of broadband access*. Paper presented at the Telecommunications Policy Research Conference. Washington, DC.

Florida, R. (2002). *The rise of the creative class: And how it's transforming work, leisure, community and everyday life*. New York, NY: Basic Books.

Folha-Online. (2007). Brazilian users increase the time spent on the internet by more than 3 hours in a year. Retrieved November 09, 2011, from http://goo.gl/IAkmk

Folha-Online. (2010). Active internet users reach 36.9 million in Brazil in January 2010. Retrieved November 09, 2011, from http://goo.gl/7rCW9

Folha-Online. (2010). Anatel studies a more strict control over broadband prices. Retrieved November 09, 2011, from http://goo.gl/CM2U9

Folha-Online. (2010c). Telecommunication companies reduce investments by 40%. Retrieved November 09, 2011, from http://www.folha.com.br/me821929

Folha-Online. (2010d). Telecommunication companies will contest the Brazilian national broadband plan. Retrieved November 09, 2011, from http://goo.gl/nCugX

Folha-Online. (2011). Brazil nears 230 million mobile phone lines. Retrieved November 09, 2011, from http://folha.com/no992891

Folha-Online. (2011). Brazilian internet users are 58.6 million. Retrieved November 09, 2011, from http://folha.com/te940381

Folha-Online. (2011c). Anatel approves phones subscriptions at a US$8 cost. Retrieved November 09, 2011, from http://folha.com/me924732

Folha-Online. (2011d). Government spent only 5.4% of the amount reserved to improve telecommunications. Retrieved November 09, 2011, from http://folha.com/me945226

Ford, G. (2005, February). Does municipal supply of communications crowd-out private communications investment? An empirical study. *Applied Economic Studies*.

Forlano, L. (2008). Anytime? Anywhere? Reframing debates around community and municipal wireless networking. *Community Informatics*. Retrieved November 3, 2011 from http://ci-journal.net/index.php/ciej/article/view/438

Forlano, L., Powell, A., Shaffer, G., & Lennett, B. (2011). From the digital divide to digital excellence: Global best practices to aid development of municipal and community wireless networks in the United States. *New America Foundation*. Retrieved from http://www.newamerica.net/sites/newamerica.net/files/policydocs/NAFMunicipalandCommunityWirelessReport.pdf

Forrest, R., & Kearns, A. (2001). Social cohesion, social capital and the neighbourhood. *Urban Studies (Edinburgh, Scotland), 38*(12), 2125–2143. doi:10.1080/00420980120087081

Fox Piven, F., & Cloward, R. (1991). Collective protest: A critique of resource mobilization theory. *International Journal of Politics Culture and Society, 4*(4), 435–458. doi:10.1007/BF01390151

Francisco, S. (2005). Request for information and comment (RFI/C). *San Francisco techconnect community wireless broadband initiative*. Retrieved from http://www.sfgov.org/site/uploadedfiles/dtis/tech_connect/BroadbandFinalRFIC.doc

Francisco, S. (2005). *RFP 2005-19: Request for proposals: TechConnect community wireless broadband network*. Retrieved from http://www.sfgov.org/site/uploadedfiles/dtis/tech_connect/TechConnectRFP_2005-19_12-22-05Rev1-17-06.pdf

Francisco, S. (2011). *San Francisco TechConnect*. Retrieved from http://www.sfgov3.org/index.aspx?page=1432

Free Press. (2012). *What is net neutrality?* Retrieved from http://www.savetheinternet.com/frequently-asked-questions

Freifunk. (2009). *Was ist Freifunk? Homepage*. Retrieved February 23, 2009, http://start.freifunk.net/

Friedkin, N. E. (2004). Social cohesion. *Annual Review of Sociology*, *30*, 409–425. doi:10.1146/annurev.soc.30.012703.110625

Frischmann, B. M. (2012). *Infrastructure: The social value of shared resources*. Oxford, UK: Oxford University Press. doi:10.1093/acprof:oso/9780199895656.001.0001

Frohlich, D. M., Bhat, R., & Jones, M. (2009). *Democracy, design, and development in community content creation: Lessons from the StoryBank project*. Los Angeles, CA: University of Southern California.

Fryšták, V. (2007). *System for localization of users in a wireless network*. (MSc. Thesis). FAI UTB in Zlin.

Fukuyama, F. (1995). Social capital and the global economy. *Foreign Affairs*, *74*(5), 89–103. doi:10.2307/20047302

Funkfeuer. (2009). *Don't log onto the net—Be the net!* Retrieved February 23, 2009, from http://www.funkfeuer.at/

Gagné, M., & Deci, E. L. (2005). Self-determination theory and work motivation. *Journal of Organizational Behavior*, *26*(4), 331–362. doi:10.1002/job.322

Gamson, W. (1975). *The strategy of social protest*. Homewood, IL: Dorsey Press.

Ganapati, S., & Schoepp, C. F. (2009). The wireless city. In Reddick, C. G. (Ed.), *Handbook of Research on Strategies for Local E-Government Adoption and Implementation* (pp. 554–568). Hershey, PA: IGI Global. doi:10.4018/978-1-60566-282-4.ch029

Gillett, S. (2012). *FCC launches connect America fund*. Retrieved from http://www.fcc.gov/blog/fcc-launches-connect-america-fund

Goel, V. (2007, November 4). Vindu's view from the valley. *San Jose Mercury News*.

Google. (2012). *Free wi-fi access for mountain view*. Retrieved from http://Wi-Fi.google.com/

Gordon, D. (2005). Indicators of poverty and hunger. *Expert Group Meeting on Youth Development Indicators*. Retrieved on January 14, 2012 from http://www.un.org/esa/socdev/unyin/documents/ydiDavidGordon_poverty.pdf

Gorney, D. (2010). Minneapolis unplugged. *The Atlantic*. Retrieved March 27, 2012 from http://www.theatlantic.com/personal/archive/2010/06/minneapolis-unplugged/57676/

Government of India. (2004). *Annual Report*. New Delhi, India: Government of India.

Grange, B. (1999). *Internet par le câble en France: Le comparatif*. Retrieved March 7, 2012, from http://www.journaldunet.com/dossiers/cable/comparatif.shtml

Granovetter, M. (1985). Economic action and social structure: The problem of embeddedness. *American Journal of Sociology*, *91*, 481–510. doi:10.1086/228311

Grayson, T., Crawford, S., & Baller, J. (2012). *The present and future of alternative fiber networks*. Unpublished.

Greeley, B., & Fitzgerald, A. (2011). *Pssst ... Wanna buy a law?* Retrieved from http://www.businessweek.com/magazine/pssst-wanna-buy-a-law-12012011.html

Gregory v. Ashcroft, 501 U.S. 452 (1991).

Grimes, S. (2003). The digital economy challenge facing peripheral rural areas. *Progress in Human Geography*, *27*(2), 174–193. doi:10.1191/0309132503ph421oa

Grossmann, L. (2011). Popular broadband access: Cost is higher than negotiated between telecommunication companies and government. Retrieved November 09, 2011, from http://goo.gl/eE45L

Grygárek, P. (2012). *SPS course - Student projects results*. Retrieved August 29, 2012, from http://wh.cs.vsb.cz/sps/index.php/SPSWiki:Port%C3%A1l#Realizovan.C3.A9_studensk.C3.A9_projekty

Guifi.net. (2009). *Homepage*. Retrieved February 23, 2009, from, http://guifi.net/

Haleš, B. (2005). *A wireless network implementation*. (BSc. thesis). FAI UTB ve Zlíně.

Hall, P. M. (2003). Interactionism, social organization and social processes: Looking back and moving ahead. *Symbolic Interaction*, *26*(1), 33–55. doi:10.1525/si.2003.26.1.33

Hamel, J.-Y. (2010). *ICT4D and the human development and capability approach: The potentials of information and communication technology*. Geneva, Switzerland: UNDP.

Hampton, K. (2002). Place-based and IT mediated "community". *Planning Theory & Practice, 3*(2), 228–231. doi:10.1080/14649350220150099

Hampton, K. (2007). Neighborhoods in the network society: The e-neighbors study. *Information Communication and Society, 10*(5), 714–748. doi:10.1080/13691180701658061

Hannigan, J. (1985). Alain Touraine, Manuel Castells and social movement theory: A critical appraisal. *The Sociological Quarterly, 26*(4), 435–454. doi:10.1111/j.1533-8525.1985.tb00237.x

Harsany, J. (2003). *Taking wifi to new heights*. Retrieved July 25, 2011, from http://www.pcmag.com/article2/0,2817,1365140,00.asp

Harsany, J. (2004). *Wireless networks open up Nepal*. Retrieved July 25, 2011, from http://abcnews.go.com/Technology/ZDM/story?id=99622&page=1

Hays, R. A., & Kogl, A. M. (2007). Neighbourhood attachment, social capital building, and political participation: A case study of low- and moderate-income residents of Waterloo, Iowa. *Journal of Urban Affairs, 29*(2), 181–205. doi:10.1111/j.1467-9906.2007.00333.x

Heeks, R. (2008). ICT4D 2. 0: The next phase of applying ICT for international development. *Computer, 41*(6), 26–33. doi:10.1109/MC.2008.192

Heeks, R., & Kanashiro, L. (2009). Telecentres in mountain regions - A Peruvian case study of the impact of information and communication technologies on remoteness and exclusion. *Journal of Mountain Science, 6*(4), 320–330. doi:10.1007/s11629-009-1070-y

Hegel, G. (1820). *Philosophy of right*. (H. Reeve, Trans.). Retrieved April 19, 2007, from http://www.marxists.org/reference/archive/hegel/index.htm

Herselman, M., & Britton, G. K. (2002). Analyzing the role of ICT in bridging the digital divide amongst learners. *South African Journal of Education, 22*(4), 270–274.

Hiner, J. (2007). *US cities are jumping off the municipal wireless bandwagon*. Retrieved from http://blogs.techrepublic.com.com/hiner/?p=545

Holson, L., & Helft, M. (2008, August 14). T-Mobile to offer first phone with Google software. *New York Times*.

Horrigan, J. (2009). *Broadband adoption and use in America*. Working Paper. Washington, DC: Federal Communications Commission.

Housing New Zealand Corporation. (2006). *Community renewal*. Retrieved August 29, 2012, from http://www.hnzc.co.nz/about-us/research-and-policy/housing-research-and-evaluation/summaries-of-reports/community-renewal-march-2006/

Hudson, H. E. (2007, January 30). Why San Francisco should approve the EarthLink/Google wireless contract now. *San Francisco Examiner*.

Hudson, H. E. (2005). *Comments on request for information and comment (RFI/C): San Francisco TechConnect community wireless broadband initiative*. San Francisco, CA: Government of San Francisco.

Hudson, H. E. (2010). Municipal wireless broadband: Lessons from San Francisco and Silicon Valley. *Telematics and Informatics, 27*(1), 1–9. doi:10.1016/j.tele.2009.01.002

ICF – Fredericton. (2008). *Top 7 intelligent community award*. Retrieved August 25, 2008, from http://www.intelligentcommunity.org/index.php?src=gendocs&ref=Top7_2008_Frederict

ICF. (2012). *Intelligent community forum: Smart 21*. Retrieved December 26, 2011, from https://www.intelligentcommunity.org/index.php?src=gendocs&ref=Smart21&category=Events&link=Smart21

IDGNOW. (2011). Brazilians are the ones who most use social networks in the workplace. Retrieved November 09, 2011, from http://migre.me/66Ogy

IDGNOW. (2011). Study reveals that 55% of Brazilians access pornography sites. Retrieved November 09, 2011, from http://migre.me/66Ofd

IL & FS. (2006). *The common service entre scheme*. Retrieved on January 14, 2012 from http://www.ilfsindia.com/downloads/bus_concept/CSC_ILFS_website.pdf

In Re Missouri Municipal League, 299 F.3d 949 (8th Cir. 2002).

Indian Knowledge@Wharton. (2010). *MNCs in rural India: At a turning point*. Retrieved on January 14, 2012 from http://knowledge.wharton.upenn.edu/india/article.cfm?articleid=4472

Info Magazine. (2007). Brazil has over 40 million landline subscribers. Retrieved November 09, 2011, from http://goo.gl/WzXfg

Institute of Community Cohesion. (2009). *The nature of community cohesion*. Retrieved May 20, 2009, from http://www.cohesioninstitute.org.uk/Resources/Toolkits/Health/TheNatureOfCommunityCohesion

Intel. (2005). *Website.* Retrieved on January 14, 2012 from ftp://download.intel.com/museum/Moores_Law/Printed_Materials/Moores_Law_2pg.pdf

International Telecommunication Union. (2005). *WSIS outcome documents*. Retrieved from http://www.itu.int

International Telecommunication Union. (2009). *Information society statistical profiles 2009 – Africa*. Retrieved from http://www.itu.int

International Telecommunication Union. (2011). *ICT services getting more affordable worldwide*. Retrieved July 16, 2012, from http://www.itu.int/net/pressoffice/press_releases/2011/15.aspx

International Telecommunication Union. (2011). *Measuring the information society*. Retrieved from http://www.itu.int

International Telecommunications Union. (2009). [*ICT facts and figures*. Retrieved from http://www.itu.int]. *WORLD (Oakland, Calif.), 2009*

Internet World Stats. (2012). *Internet and Facebook usage in Europe*. Retrieved July 16, 2012, from http://www.internetworldstats.com/stats4.htm

Internet World Usage Statistics. (2009). *Website.* Retrieved February 28, 2009, from http://www.internetworldstats.com/stats.htm

Ishmael, J., Bury, S., Pezaros, D., & Race, N. (2008). Deploying rural community wireless mesh networks. *IEEE Internet Computing*, *12*(4), 22–29. doi:10.1109/MIC.2008.76

ITU. (2000). *The internet from the top of the world: Nepal case study.* Retrieved from http://www.itu.int/ITU-D/ict/cs/nepal/material/nepal.pdf

ITU. (2012). Current results of the broadband access program for Brazilian public schools in urban areas. Retrieved May 31, 2012, from http://www.itu.int/md/D10-SG01-C-0087/en

ITU-D 19. (2010). *Recommendation ITU-D 19: Telecommunication for rural and remote areas*. Retrieved on September 19, 2012, from http://www.itu.int/dms_pubrec/itu-d/rec/d/D-REC-D.19-201003-I!!PDF-E.pdf

ITU-D Focus Group 7. (2001). *New technologies for rural applications*. Retrieved September 19, 2012, from http://www.itu.int/ITU-D/fg7/pdf/FG_7-e.pdf

ITU-D SG2 Rapporteur. (2006). *Analysis of case studies on successful practices in telecommunications for rural and remote areas*. Retrieved September 19, 2012, from http://www.itu.int/pub/D-STG-SG02.10.1-2006/en

ITU-D SG2 Rapporteur. (2009). *For telecommunications for rural and remote areas: Analysis of case studies on successful practices in telecommunications for rural and remote areas.* Retrieved from http://www.itu.int

ITU-D SG2. (2004). *Rapporteur for telecommunications for rural and remote areas: Analysis of replies to the questionnaire on rural communications*. Retrieved from http://www.itu.int

ITU-D SG2. (2010). *Recommendation D-19 telecommunication for rural and remote areas*. Retrieved from http://www.itu.int

Jambeiro, O., Mota, A., Amaral, C., Santos, S., Ferreira, S. A., & Simoes, C. … Costa, E. (2004). Tempos de vargas - O rádio e o controle da informação. Salvador, Brazil: Edufba.

Jaroš, J. (2005). *Building wifi network with internet access in a remote rural area*. (BSc. Thesis). FAI UTB ve Zlíně.

Jayakar, K. (2009). Universal service. In Schejter, A. (Ed.), *And Communications for All* (pp. 181–202). Lanham, MD: Rowman & Littlefield.

Jensen, R. (2007). The digital provide: Information (technology), market performance, and welfare in the south Indian fisheries sector. *The Quarterly Journal of Economics*, *122*(3), 879–924. doi:10.1162/qjec.122.3.879

Jhunjhunwala, A. (2006). Rural connectivity towards enhancing health care. In *Proceedings of the Pugwash Conference on HIV/AIDS*. MSSRF.

Jhunjhunwala, A., & Ramachandran, A. (2004). n-Logue: Building a sustainable rural services organization, case study on communication for rural and remote areas. *International Telecommunication Union*.

Jhunjhunwala, A., Narayanan, B., Prashant, S., & Arjun, N. N. (2011). *Innovative services & mobile access to drive broadband in rural India- Learning from recent successes and failures.* Unpublished.

Jhunjhunwala, A., Prashant, S., & Kittusami, S. P. (2011). *Can information and communication technology initiatives make a difference in healthcare? Case study: Rural India.* Unpublished.

Jhunjhunwala, A. (2004). *Applications and services through e-kiosks towards enabling rural India.* New Delhi, India: Asian Development Bank.

Jhunjhunwala, A. (2011). *Can ICT make a difference? Case study: Rural India.* Copenhagen, Denmark: Infrastructures for Health Care.

Jhunjhunwala, A., Anandan, V., Prashant, S., & Sachdev, U. (2011). Experiences on using voice based data entry system with mobile phone in rural India. In *Proceedings of Infrastructures for Health Care* (pp. 59–67). Copenhagen, Denmark: Infrastructures for Health Care.

Jhunjhunwala, A., Jalihal, D., & Giridhar, K. (2000). Wireless in local loop, some fundamentals. *Journal of IETE, 46*(6).

Jhunjhunwala, A., Kumar, R., Prashant, S., & Singh, S. (2009). *Technologies, services and new approaches to universal access and rural telecoms.* International Telecommunication Union.

Jhunjhunwala, A., Ramachandran, A., & Bandhopadyay, A. (2004). n-Logue: The story of a rural service provider. *The Journal of Community Informatics, 1*(1), 30–38.

Jhunjhunwala, A., & Ramamurthi, B. (1995). Wireless in local loop: Some key issue. *Journal of IETE, 12*(5&6), 309–314.

Jhunjhunwala, A., Ramamurthi, B., & Gonsalves, T. A. (1998). The role of technology in telecom expansion in India. *IEEE Communications Magazine, 36*(11). doi:10.1109/35.733480

Jiří Peterka. (2001). *Historie naší liberalizace, díl VIII: Tarif internet 99.* Retrieved March 5, 2012, from http://www.earchiv.cz/b01/b1127001.php3

Johnson, S., Porta, R. L., de Silanes, F. L., & Shleifer, A. (2000). Tunnelling. *SSRN eLibrary.* Retrieved from http://papers.ssrn.com/sol3/papers.cfm?abstract_id=204868

Joint Venture. (2012). *Census data listed in the SAM-CAT RFP.* Retrieved from http://www.jointventure.org/programs-initiatives/smartvalley/projects/wirelesssv/documents

Jung, J., Qiu, J., & Kim, Y. (2001). Internet connectedness and inequality. *Communication Research, 28*(4), 507–535. doi:10.1177/009365001028004006

Kanjlal, B., Mandal, S., Samanta, T., Mandal, A., & Singh, S. (2007). *A parallel healthcare market: Rural medical practitioners in west Bengal.* India: Institute of Health Management Research.

Kapadia, V. V., Patel, S. N., & Jhaveri, R. H. (2010). Comparative study of hidden node problem and solution using different techniques and protocols. *arXiv:1003.4070.* Retrieved from http://arxiv.org/abs/1003.4070

Karisny, L. (2010). Year-end review: Economic recovery through municipal wireless networks. *MuniWireless.* Retrieved November 3, 2011 from http://www.muniwireless.com/2010/01/01/year-end-review-economic-recovery-through-municipal-wireless-networks/

Kavanaugh, A., & Patterson, S. (2001). The impact of community computer networks on social capital and community involvement. *The American Behavioral Scientist, 45*(3), 496–509. doi:10.1177/00027640121957312

Kawasumi, Y. (2002, March). Challenges for rural communications development. *ITU News.*

Kawasumi, Y. (2004). Deployment of WiFi for rural communities in Japan and ITU's initiative for pilot projects. In *Proceedings of the 6ᵗʰ International Workshop on Enterprise Networking and Computing in Healthcare Industry, HEALTHCOM,* (pp. 200-207). HEALTHCOM.

Kawasumi, Y. (2005). *Maitland+20-Fixing the missing link, focus on rural connectivity.* London, UK: The Anima Center.

Kearns, A., & Forrest, R. (2000). Social cohesion and multilevel urban governance. *Urban Studies (Edinburgh, Scotland), 37*(5-6), 995–1017. doi:10.1080/00420980050011208

Kendall, D. (2008). *Sociology in our times* (7th ed.). Belmont, CA: Wadsworth.

Kim, Y.-C., Jung, J.-Y., Cohen, E. L., & Ball-Rokeach, S. J. (2004). Internet connectedness before and after September 11 2001. *New Media & Society*, *6*(5), 20. doi:10.1177/146144804047083

Kirstein, P. T. (2004). European international academic networking: A 20 year perspective. In *Selected Papers from the TERENA Networking Conference*. TERENA.

Kliment, M., Fibich, O., Sviták, J., Dulík, T., Ševčík, R., Daněk, P., & Rozehnal, M. (2007). *FreenetIS*. Retrieved March 9, 2012, from http://freenetis.org

Klír, J. (2012). *Mapstats*. Retrieved May 31, 2012, from http://www.jklir.net/?p=mapstats

Knology, Inc. v. Insight Communications Company, 393 F.3d 656 (6th Cir. 2004).

Krupa, J. (2012). *CaLStats*. Retrieved May 31, 2012, from http://www.mobilnews.cz/honza/calstats

Krupa, J. (2012). *WeWiMo*. Retrieved May 31, 2012, from http://www.mobilnews.cz/honza/wewimo

Kulhavý, K. (2010). *Ronja: Free FSO link project*. Retrieved July 8, 2012, from http://ronja.twibright.com/

Kumar, A. (2008). Regulatory initiatives to promote: Universal access to broadband. In *Proceedings of the ITU Workshop on Regulatory Policies on Universal Access to Broadband Services*. Retrieved from http://www.itu.int/ITU-D/treg/Events/Seminars/2008/2_1_Kumar.ppt

Kumar, R., & Best, M. (2006). Social impact and diffusion of telecenter use: A study from the sustainable access in rural India project. *The Journal of Community Informatics*, *2*(3), 1–21.

Kwak, Y. H., Chih, Y., & Ibbs, C. W. (2009). Towards a comprehensive understanding of public private partnerships for infrastructure development. *California Management Review*, *51*(2), 51–78. doi:10.2307/41166480

La Due Lake, R., & Huckfeldt, R. (1998). Social capital, social networks and political participation. *Political Psychology*, *93*, 567–584. doi:10.1111/0162-895X.00118

Lakshmipathy, N., Meinrath, S., & Breitbart, J. (2007, December 11). The Philadelphia story: Learning from a municipal wireless pioneer. *New America Foundation*.

Laroche, M., Merette, M., & Ruggeri, G. C. (1999). On the concept and dimensions of human capital in knowledge- based economy context. *Canadian Public Policy*, *25*(1), 87–100. doi:10.2307/3551403

Lavallee, A. (2008). A second look at citywide wi-fi. *The Wall Street Journal*. Retrieved March 27, 2012 from http://online.wsj.com/article/SB122840941903779747.html

Lawrence, E., Bina, M., Culjak, G., & El-Kiki, T. (2007). *Wireless community networks: Public assets for 21st century society*. Paper presented at the International Conference on Information Technology. Las Vegas, NV.

Lawson, S. (2012). San Jose tries again with free downtown wi-fi. *NetworkWorld*. Retrieved March 27, 2012 from http://www.networkworld.com/news/2012/031212-san-jose-tries-again-with-257199.html?page=1

Lehr, W., Sirbu, M., & Gillett, S. (2006). Wireless is changing the policy calculus for municipal broadband. *Government Information Quarterly*, *23*, 435–453. doi:10.1016/j.giq.2006.08.001

Lemos, A. (Ed.). (2007). *Cidade digital: Portais, inclusão e redes no Brasil*. Salvador, Brazil: Edufba.

Lentz, R., & Oden, M. (2001). Digital divide or digital opportunity in the Mississippi Delta region of the U.S. *Telecommunications Policy*, *25*(5), 291–313. doi:10.1016/S0308-5961(01)00006-4

Levett, C. (2004). *Himalayan village joins wireless road*. Retrieved July 23, 2011, from http://www.smh.com.au/news/World/Himalayan-village-joins-wireless-world/2004/12/26/1103996439623.html?oneclick=true

Lévy, P. (1998). *A inteligência coletiva: Por uma antropologia do ciberespaço*. São Paulo, Brazil: Loyola.

Lewin, K. (1948). *Resolving social conflicts: Selected papers on group dynamics*. New York, NY: Harper & Row.

Lewin, K., & Grabbe, P. (1945). Conduct, knowledge and acceptance of new values. *The Journal of Social Issues*, *2*.

Lin, S. C. (2002). Guerilla.net/Lucent/Maxrad *collinear omni*. Retrieved August 22, 2012, from http://www.lincomatic.com/wireless/collinear.html

Little, D. (2008). Collective behavior and resource mobilization theory. *Understanding Society*. Retrieved February 29, 2008, from http://understandingsociety.blogspot.com/2008/01/collective-behavior-and-resource.html

Loader, B. D., & Keeble, L. (2004). *Challenging the digital divide? A literature review of community informatics initiatives*. York, UK: The Joseph Rowntree Foundation/YPS, for the Community Informatics Research and Applications unit at the University of Teesside.

Magrani, B. (2011). New developments on the regulation of network neutrality. Retrieved November 09, 2011 from http://goo.gl/HgY8v

Mahanta, S. (2012). *Why are telecom companies blocking rural America from getting high-speed internet?* Retrieved from http://www.tnr.com/article/politics/102699/rural-broadband-internet-wifi-access

Maitland Commission. (1985). *The missing link*. Retrieved on 9/19/2012 from http://www.itu.int/osg/spu/sfo/missinglink/The_Missing_Ling_A4-E.pdf

Malik, O. (2006). *Google launches wi-fi network in Mountain View*. Retrieved from http://gigaom.com/2006/08/15/google-launches-wi-fi-network-in-mountain-view/

Malinen, J. (2012). *Host AP linux driver for intersil prism2/2.5/3 wireless LAN cards and WPA supplicant*. Retrieved July 8, 2012, from http://hostap.epitest.fi/

Mandviwalla, M., Jain, A., Fesenmaier, J., Smith, J., Weinberg, P., & Meyers, G. (2008). Municipal broadband wireless networks: Realizing the vision of anytime, anywhere connectivity. *Communications of the ACM, 51*(2), 72–80. doi:10.1145/1314215.1314228

Marcus Cable Associates v. City of Bristol, 237 F.Supp.2d 675 (W.D. Va. 2002)

Marshall, T. (2001). *Biquad feed for primestar dish*. Retrieved August 22, 2012, from http://www.trevormarshall.com/biquad.htm

Marshall, T. (2002). *Slotted waveguide 802.11b WLAN antennas*. Retrieved August 22, 2012, from http://www.trevormarshall.com/waveguides.htm

Martell, M. (2009). *Build A 9dB, 70cm collinear antenna*. Retrieved August 22, 2012, from http://www.rason.org/Projects/collant/collant.htm

Matúšů, J. (2008). *A system for monitoring large area networks*. (MSc. Thesis). FAI UTB in Zlin.

McAdam, D. (1982). *Political process and the development of black insurgency, 1930-1970*. Chicago, IL: The University of Chicago Press.

McCarthy, J., & Zald, M. (1987). *Social movements in an organizational society: Collected essays*. New Brunswick, NJ: Transaction Publishers.

McChesney, R. (2009). Public scholarship and the communications policy agenda. In Schejter, A. (Ed.), *And Communications for All* (pp. 41–46). Lanham, MD: Rowman and Littlefield.

McClenaghan, P. (2000). Social capital: Exploring the theoretical foundations of community development education. *British Educational Research Journal, 26*, 565–582. doi:10.1080/713651581

McDonald, D. W. (2002). *Social issues in self-provisioned metropolitan area networks*. Paper presented at the ACM Conference on Human Factors in Computing Systems. Minneapolis, MN.

McGarty, T. P., & Bhagavan, R. (2002). *Municipal broadband networks: A revised paradigm of ownership*. Retrieved from http://www.lus.org/uploads/Municipal-BroadbandNetworksStudy.pdf

McKnight, J. L., & Kretzmann, J. P. (1996). *Mapping community capacity*. Chicago, IL: The Neighborhood Innovations Network.

Meegan, R., & Mitchell, A. (2001). It's not community round here, it's neighbourhood: Neighbourhood change and cohesion in urban regeneration policies. *Urban Studies (Edinburgh, Scotland), 38*(12), 2167–2194. doi:10.1080/00420980120087117

Meinrath, S. (2005). Community wireless networking and open spectrum usage: A research agenda to support progressive policy reform of the public airwaves. *The Journal of Community Informatics, 1*(2), 204–209.

Meinrath, S. D., & Calabrese, M. (2008). White space devices & the myths of harmful interference. *New York University Journal of Legislation and Public Policy, 11*(3), 495–518.

Merriam v. Moody's Executors, 25 Iowa 163 (1868).

Merton Group. (2003). *Municipal broadband report: Feasibility study report for Hanover, NH*. Retrieved from http://www.telmarc.com/Feasibility/Hanover%20Feasibility%20Study.pdf

MetroFi. (2008). *Notice to customers*. Retrieved from http://www.metrofi.com/press/20060130b.html

Meyer, D. (2005). Social movements and public policy: Eggs, chicken, and theory. In Meyer, D., Jenness, V., & Ingram, H. (Eds.), *Routing the Opposition: Social Movements, Public Policy, and Democracy* (pp. 1–26). Minneapolis, MN: University of Minnesota Press.

Middleton, C., & Potter, A. B. (2008). *Is it good to share? A case study of Fon and Meraki approaches to broadband provision.* Paper presented at the 17th Biennial International Telecommunications Society Conference. New York, NY.

Middleton, C., Clement, A., Crow, B., & Longford, G. (2008). *ICT infrastructure as public infrastructure: Connecting communities to the knowledge-based economy & society.* Final Report of the Community Wireless Infrastructure Research Project. Retrieved January 26, 2012, from http://www.cwirp.ca/files/CWIRP_Final_report.pdf

Middleton, C., & Crow, B. (2008). Building wi-fi networks for communities: Three Canadian cases. *Canadian Journal of Communication*, *33*, 419–441.

Mikrotik Inc. (2003). *RouterBoard products.* Retrieved August 27, 2012, from http://routerboard.com/

Mikrotik Inc. (2012). *The dude.* Retrieved April 12, 2012, from http://www.mikrotik.com/thedude.php

Millonzi, K. (2011). *New municipal broadband limitations.* Retrieved from http://canons.sog.unc.edu/?p=4967

Ministry of Economic Development. (2008). *Digital strategy 2.0: Smarter through digital.* Retrieved from http://www.beehive.govt.nz/release/digital-strategy-20-%E2%80%93-smarter-through-digital

Ministry of Economic Development. Ministry of Health, Ministry of Research Science and Technology, Ministry of Education, Department of Labour, The NZ National Library, et al. (2004). *Digital strategy: A draft New Zealand digital strategy for consultation.* Retrieved from www.beehive.govt.nz/Documents/Files/Digital%20Strategy.pdf

Ministry of Finance. (2009). *Position paper on telecom sector in India.* Retrieved on January 14, 2012 from http://www.pppinindia.com/pdf/ppp_position_paper_telecom_122k9.pdf

Ministry of Social Development. (2010). *The social report: Te purongo tangata 2010 - Social connectedness.* Retrieved August 28, 2012, from http://www.socialreport.msd.govt.nz/social-connectedness/index.html

Minneapolis, W. (2011). *Website.* Retrieved from http://www.ci.minneapolis.mn.us/wirelessminneapolis/

Minshall, T., Mortara, L., Valli, R., & Probert, D. (2010). Making Asymmetric partnerships work. In *Research Technology Management* (pp. 53–63). Industrial Research Institute Inc.

Miracle Group. (2001). *Company milestones.* Retrieved May 30, 2012, from http://miracle.cz/czech/group_mezniky.htm

Mitchell, C. (2010). *Florida muni dunnellon building FTTH network.* Retrieved from http://www.muninetworks.org/content/florida-muni-dunnellon-building-ftth-network

Mitchell, C. (2012). *North Carolina county turns on first white spaces wireless network in nation.* Retrieved from http://www.muninetworks.org/content/north-carolina-county-turns-first-white-spaces-wireless-network-nation

Mitchell, C. (2012). *Big bucks: Why North Carolina outlawed community networks.* Retrieved from http://www.muninetworks.org/content/big-bucks-why-north-carolina-outlawed-community-networks

Mitchell, C. (2012c). *Broadband at the speed of light: How three communities built next-generation networks.* Retrieved from http://download.broadband.gov/plan/national-broadband-plan.pdf

Moon, I., Songatikamas, T., & Wall, R. (2009). *Antenna project.* Retrieved from http://sjsulug.engr.sjsu.edu/rkwok/projects/Omni_and_Biquad_antenna_2009.pdf

Morton, C. (2010). *Building the wifi parabolic antenna by dxzone.com.* Retrieved August 22, 2012, from http://www.dxzone.com/cgi-bin/dir/jump2.cgi?ID=17641

Mosco, V. (1996). *The political economy of communication.* London, UK: Sage.

Moura, B., & Fearey, S. (2006). *Joint venture: Silicon Valley network and SAMCAT announce vendor for wireless Silicon Valley.* Retrieved from http://www.jointventure.org/inthenews/pressreleases/090506wirelessvendor.html

Moy, P., Scheufele, D., & Holber, P. (1999). Television use and social capital: Testing Putnam's time displacement hypothesis. *Mass Communication & Society*, *2*(1/2), 27–46.

Mozer, Ł., Zapalski, M., Antoniuk, R., Drewicz, K., Machniak, A., Chiliński, T., et al. (2012). *LMS - LAN management system*. Retrieved April 12, 2012, from http://www.lms.org.pl/?lang=en

Mugume, K. E. (2006). *Telecentre assessment surveys: Nakaseke and Buwama multipurpose community telecentres. Technical Report*. Kampala, Uganda: Makerere University.

Municipal Wireless Networks. (2012). *Wikipedia*. Retrieved from http://en.wikipedia.org/wiki/Municipal_wireless_network

Myslík, V. (2008). *Crusader FSO link*. Retrieved July 8, 2012, from http://crusader.eu/

Nair, K. G. K., & Prasad, P. N. (2002). Development through information technology in developing countries: Experiences from an Indian state. *Electronic Journal of Information Systems in Developming Countries, 8*(2), 1–13.

Napoli, P. (1999). The marketplace of ideas metaphor in communications regulation. *The Journal of Communication, 49*(4), 151–169. doi:10.1111/j.1460-2466.1999.tb02822.x

Nasscom. (2012). *Women take the lead in rural BPOs*. Retrieved on January 14, 2012 from http://www.nasscom.in/women-take-lead-rural-bpos

National Cable and Telecommunications Association v. Brand X Internet Services, 545 U.S. 967 (2005).

National Conference on Communications. (2011). *Website*. Retrieved from http://www.cedat.mak.ac.ug/ncc

National Rural Health Mission. (2010). *Rural health statistics in India*. Retrieved on January 14, 2012 from http://mohfw.nic.in/BULLETIN%20ON.htm

National Telecommunications and Information Administration. (2010). *Digital nation: 21st century America's progress toward universal broadband internet access*. Washington, DC: NTIA.

National Telecommunications and Information Administration. (2011). *Exploring the digital nation: Computer and internet use at home*. Washington, DC: US Department of Commerce.

National Telecommunications and Information Administration. (2012). *About BroadbandUSA*. Retrieved from http://www2.ntia.doc.gov/about

Nepal, O. L. E. (2011). Our projects. *OLPC Project*. Retrieved July 29, 2011 from http://www.olenepal.org/olpc_project.html

NET Service Solution. s.r.o. (2012). *ISPadmin: Systém pro správu sítí*. Retrieved April 12, 2012, from http://www.ispadmin.eu/

Neumann, J. (2007). Community wi-fi in UK and Germany: A roundup. *Nettime Discussion Board*. Retrieved February 14, 2008, from http://www.nettime.org/Lists-Archives/nettime-l-0710/msg00034.html

New Zealand Government - Ministry of Communications and Information Technology. (2007). *Digital strategy: Creating our digital future*. Retrieved 2 November, 2007, from http://www.digitalstrategy.govt.nz/

Newman, K. (2008). *Connecting the clouds: The internet in New Zealand*. Auckland, New Zealand: Activity Press in association with InternetNZ.

NextWLAN. (2006). *Proposal in response to San Francisco tech connect RFP*. San Francisco, CA: City of San Francisco.

Nielsenwire. (2011). Swiss lead in speed: Comparing global internet connections. Retrieved November 09, 2011, from http://shar.es/q7ur4

Nixon v. Missouri Municipal League, 541 U.S. 125 (2004).

Nolan, C. (2005). *Taking sides in the municipal wireless showdown*. Retrieved from http://www.eweek.com/c/a/Government-IT/Taking-Sides-for-the-Municipal-Wireless-Showdown

Nortel. (2005). *Bell Canada and Nortel deliver advanced broadband services and applications to northern Ontario community*. Retrieved December 27, 2011, from http://www2.nortel.com/go/news_detail.jsp?cat_id=-8055&oid=100190608

Nortel. (2011). *Chapleau case study*. Retrieved December 26, 2011, from http://www2.nortel.com/go/news_detail.jsp?cat_id=-9252&oid=100204737&locale=en-US&NT_promo

Notícias, U. O. L. (2011). Brazilian children are the youngest children to access social networks. Retrieved November 09, 2011, from http://migre.me/66Oo8

Novák, P. (2009). 1993 - Lidé a firmy. *Orcave Company Blog*. Retrieved July 8, 2012, from http://blog.orcave. com/index.php?itemid=48&catid=1

NREGA. (2012). *Smart cards: Biometrics scanning for signatures and handheld devices*. Retrieved on January 14, 2012 from http://www.nrega.net/ict/ongoing-ict-projects

NTA. (2003). *Management information system*. Retrieved July 25, 2011, from http://www.nta.gov.np/articleimages/ file/NTA_MIS_1.pdf

NTA. (2010). *License fee*. Retrieved July 23, 2011, from http://www.nta.gov.np/en/content/index.php?task=articl es&option=view&id=46

NTA. (2010). *Ten year master plan (2011 - 2020 AD): For the development of telecommunication sector in Nepal*. Retrieved from http://www.nta.gov.np

NTA. (2011). *Management information system*. Retrieved July 27, 2011 from http://www.nta.gov.np/articleimages/ file/NTA_MIS_51.PDF

Null, E. (2013). Municipal broadband: History's guide. *I/S: A Journal of Law and Policy for the Information Society*.

O'Loughlin, D. S. (2006). Preemption or bust: Fear and loathing in the battle over broadband. *Cardozo Law Review*, *28*(1), 479–510.

OECD. (2008). *Broadband growth and policies in OECD countries*. Paris, France: OECD.

OECD. (2009). *Broadband and the economy*. Seoul, South Korea: OECD.

OECD. (2010). *Fixed broadband penetration and density*. Retrieved from http://www.oecd.org/sti/broadbandandtelecom/39574903.xls

OECD. (2011). *Average advertised download speeds*. Retrieved from http://www.oecd.org/dataoecd/10/54/39575095.xls

OECD. (2012). *OECD broadband portal*. Retrieved from http://www.oecd.org/document/54/0,3746, en_2649_34225_38690102_1_1_1_1,00.html

OLPC. (2011). Over 2 million children and teachers in 42 countries are learning with XO laptops today. *About the Project/Countries*. Retrieved July 24, 2011, from http:// one.laptop.org/about/countries

Olson, M. (1971). *The logic of collective action public goods and the theory of groups*. Cambridge, MA: Harvard University Press.

Omaha Wireless. (2012). *About us*. Retrieved from http:// omahawireless.unomaha.edu/index.html

Onyx, J., & Bullen, P. (2000). Sources of social capital. In Winter, I. (Ed.), *Social Capital and Public Policy in Australia* (pp. 105–135). Melbourne, Australia: Australian Institute of Family Studies.

Orange & Green. (2012). *CIBS - Clever ISP business service*. Retrieved April 12, 2012, from http://www.cibs.cz/

Ordower, G. (2005). Broadband quest cost $300,000. *Daily Herald*. Retrieved from http://lafayetteprofiber. com/imagesNRef/Docs/TriCitiesCosts.html

Organisation for Economic Co-Operation and Development. (2009). *African economic outlook 2009*. Geneva, Switzerland: OECD.

Organization for Economic Cooperation and Development. (2011). *OECD broadband statistics*. Retrieved from http://www.oecd.org/sti/ict/broadband

Ovum Consulting. (2009). *Broadband policy development in the Republic of Korea*. Retrieved from http://unpan1. un.org/intradoc/groups/public/documents/un-dpadm/ unpan041139.pdf

Pade-Khene, C., & Swery, D. (2011). Toward a comprehensive evaluation framework for ICT for development evaluation – An analysis of evaluation frameworks. In *Proceedings of the European Conference on Information Management & Evaluation*. IEEE.

Paldam, M., & Christoffersen, H. (2004). Privatization in Denmark, 1980-2002. *Social Science Research Network*. Retrieved April 13, 2009, from http://papers.ssrn.com/ sol3/papers.cfm?abstract_id=514242

Paragould Cablevision v. City of Paragould, 930 F.2d 1310 (8th Cir. 1991).

Parikh, T. S. (2009). Engineering rural development. *Communications of the ACM*, *52*(1), 54–63. doi:10.1145/1435417.1435433

Parker, E. (2000). Closing the digital divide in rural America. *Telecommunications Policy*, *24*(4), 281–290. doi:10.1016/S0308-5961(00)00018-5

PC Engines GmbH. (2007). *WRAP - Wireless router application platform*. Retrieved August 25, 2012, from http://www.pcengines.ch/wrap.htm

PC Engines GmbH. (2008). *ALIX system boards*. Retrieved August 25, 2012, from http://pcengines.ch/alix.htm

Peha, J. (2009). A spectrum policy agenda. In Schejter, A. (Ed.), *And Communications for All: A Policy Agenda for a New Administration* (pp. 137–152). Lanham, MD: Lexington Books.

Pehe, J. (1994). Civil society at issue in the CR. *RFE/RL Research Report, 3*(32).

Perry, B. (2012). *Household incomes in New Zealand: Trends in indicators of inequality and hardship 1982 to 2011*. Retrieved from http://www.msd.govt.nz/about-msd-and-our-work/publications-resources/monitoring/household-incomes/index.html

Peterka, J. (2001). *Historie naší liberalizace, díl III: Liberalizace Internetu v ČR*. Retrieved February 27, 2012, from http://www.earchiv.cz/b01/b1023001.php3

Peterka, J. (2002). *ČTÚ zakázal Telecomu jeho ADSL!* Retrieved March 7, 2012, from http://www.earchiv.cz/b02/b0702001.php3

Petitioner's Brief, City of Abilene v. FCC, 164 F.3d 49 (D.C. Cir. 1998)

Petr Anderle. (2005). *Máme na čem stavět: Útržky z historie občanské společnosti*. Retrieved March 8, 2012, from http://www.cs-magazin.com/index.php?a=a2005032099

Pettigrew, A. (1987). Context and action in the transformation of the firm. *Journal of Management Studies, 24*(6), 649–670. doi:10.1111/j.1467-6486.1987.tb00467.x

Pettigrew, A. (1992). The character and significance of strategy process research. *Strategic Management Journal, 13*, 5–16. doi:10.1002/smj.4250130903

Pigg, K. E., & Crank, L. D. (2004). Building community social capital: The potential and promise of information and communications technologies. *The Journal of Community Informatics, 1*(1).

Pithart, P., Suk, J., Přibáň, J., Čarnogurský, J., Marek, J., & Motej, O. (2006). *Transformation: The Czech experience*. Prague, The Czech Republic: People in Need.

Portal G1. (2011). Brazilians are the ones who most watch videos and TV on the internet. Retrieved November 09, 2011, from http://migre.me/66O8V

Portal, I. G. (2010). *Brazilian users are the ones who spend most time on the internet*. Retrieved November 09, 2011, from http://goo.gl/RJ8Ou

Poster, M. (1995). *CyberDemocracy: Internet and the public sphere*. Retrieved April 16, 2009, from http://www.hnet.uci.edu/mposter/writings/democ.html

Postill, J. (2008). Localizing the internet beyond communities and networks. *New Media & Society, 10*(3), 413–431. doi:10.1177/1461444808089416

Pot, M. (2005). *Home-brew Compact 6dBi collinear antenna*. Retrieved August 22, 2012, from http://martybugs.net/wireless/collinear.cgi

Pourebrahimi, B., Bertels, K., & Vassiliadis, S. (2005). A survey of peer-to-peer networks. In *Proceedings of the 16th Annual Workshop on Circuits, Systems and Signal Processing*. Veldhoven, The Netherlands: IEEE.

Powell, A. (2009). *Last mile or local innovation: Canadian perspectives on community wireless networking as civic participation*. Retrieved from http://www3.fis.utoronto.ca/iprp/cracin/publications/workingpapersseries.htm

Powell, A. (2011). *#Fail: What we learn from failed tech projects*. Retrieved January 24, 2012, from http://lse.academia.edu/AlisonPowell/Talks/49255/_FAIL_What_we_learn_from_failed_community_technology_projects

Powell, A., & Shade, L. R. (2006). Going wi-fi in Canada: Municipal and community initiatives. *Government Information Quarterly, 23*(3/4), 381–403. doi:10.1016/j.giq.2006.09.001

Pritchard, W. (2004). *Wireless networks: Opportunities and challenges for foothill college*. White Paper. Retrieved from http://fhdafiles.fhda.edu/downloads /etsfhda/WirelessWhite Paper.pdf

Procházka, V. (2012). *Vczela*. Retrieved from http://www.czela.net/pub/czela/info/

Putala, C. (2006, June 14). *Testimony before the committee of the judiciary, US Senate, hearing on reconsidering our communications laws: Ensuring competition and innovation*. Washington, DC: US Senated.

Putnam, R. D. (1995). Bowling alone: America's declining social capital. *Journal of Democracy*, *6*(1), 65–78. doi:10.1353/jod.1995.0002

Putnam, R. D. (1996). The strange disappearance of civic America. *The American Prospect*, *7*(24).

Putnam, R. D. (2000). *Bowling alone: The collapse and revival of American community*. New York, NY: Simon & Schuster.

Putnam, R. D. (2002). Bowling together. *The American Prospect*, *13*(3).

Putnam, R. D. (2007). E pluribus unum: Diversity and community in the twenty-first century. *Scandinavian Political Studies*, *30*(2), 137–174. doi:10.1111/j.1467-9477.2007.00176.x

Pyramid Research. (2005). Municipality wi-fi: Despite EarthLink, Google, viability remains unclear. *Pyramid Research Analyst Insight*. Retrieved from http://www.pyramidresearch.com/documents/AI-Wi-Fi.pdf

Quan-Haase, A., & Wellman, B. (2002, November 12). How does the internet affect social capital. *IT and Social Capital*.

Quinn, P. (2006). Community wireless and the digital divide. *Center for Neighborhood Technology*. Retrieved from http://www.cnt.org/repository/WCN-AllReports.pdf

Quinn, P. (2006). Community wireless networks: Cutting edge technology for Internet access. *Center for Neighborhood Technology*. Retrieved November 1, 2011 from http://www.cnt.org/repository/WCN-AllReports.pdf

Rafaeli, S., & LaRose, R. (1993). Electronic bulletin boards and 'public goods' explanation of collaborative mass media. *Communication Research*, *20*(2), 277–297. doi:10.1177/009365093020002005

Ramachandran, T. V. (2002). *Building on the initial success of cellular: The Indian experience*. ITU.

Ramirez, R. (2007). Appreciating the contribution of broadband ICT with rural and remote communities: Stepping stones toward an alternative paradigm. *The Information Society*, *23*, 85–94. doi:10.1080/01972240701224044

Ramos, M. C. O. (2000). Às margens da estrada do futuro: Comunicações, políticas e tecnologias. Sao Paolo, Brasília: UnB.

Rao, B., & Parikh, M. (2002). *Wireless broadband experience: The U.S. experience*. Paper presented at the First International Conference on Mobile Business. Athens, Greece.

Rao, B., & Parikh, M. (2003). Wireless broadband networks: The U.S. experience. *International Journal of Electronic Commerce*, *8*(1), 37–53.

RBI. (2006). *Circular RBI/2005-06/288, January 25, 2006*. Retrieved on January 14, 2012 from http://rbi.org.in/scripts/BS_CircularIndexDisplay.aspx?Id=2718

RBI. KYC Norms. (2002). *Circular: DBOD.AML. BC.18/14.01.001/2002-03*. Retrieved on January 14, 2012 from http://www.rbi.org.in/scripts/NotificationUser.aspx?Id=819&Mode=0

Readhead, A., & Trill, S. (2003). The role of ad hoc networks in mobility. *BT Technology Journal*, *21*(3), 74–80. doi:10.1023/A:1025159115207

Reardon, M. (2007). *Earthlink to lay off 900*. Retrieved from http://news.cnet.com/8301-10784_3-9767410-7.html

Rediff Business. (2010). *Low cost ATMs: Vortex's pioneering gift to India*. Retrieved on January 14, 2012 from http://business.rediff.com/slide-show/2010/sep/16/slide-show-1-innovation-vortex-gift-to-rural-india.htm

Rediff News. (2007, November 29). *DesiCrew: A girl's pioneering vision for rural BPOs*. Retrieved on January 14, 2012 from http://specials.rediff.com/money/2007/nov/29sld1.htm

RedTap. (2006). *Executive summary of proposal in response to San Francisco TechConnect RFP*. San Francisco, CA: RedTap.

Rehm, G. (2007). *How to build a tin can waveguide antenna*. Retrieved August 22, 2012, from http://www.turnpoint.net/wireless/cantennahowto.html

Reinwand, C. (2006, June 26). *Letter to Brian Moura, Chairman, SAMCAT*. San Mateo, CA: San Mateo County Telecommunications Authority.

Rheingold, H. (1993). *The virtual community*. Reading, MA: Addison-Wesley Publishing Co.

Rich, S. (2012). *New Hanover county, N.C., first in nation to deploy 'super wi-fi' network*. Retrieved from http://www.govtech.com/e-government/New-Hanover-County-NC-Super-Wi-Fi-Network.html?elq=cc575b8cbd0a44588d262afb34031bbb

Rideout, V., & Reddick, A. (2005). Sustaining community access to technology: Who should pay and why! *The Journal of Community Informatics, 1*(2), 45–62.

Robinson, J. P. (1976). Interpersonal influence in election campaigns: Two step-flow hypotheses. *Public Opinion Quarterly, 40*(3), 304–319. doi:10.1086/268307

Robison, B. (2012). *WaveLAN cards*. Retrieved from http://en.wikipedia.org/wiki/WaveLAN

Rodriguez del Corral, A. (2011). Financing high speed broadband—What do you think? *European Broadband Portal*. Retrieved July 16, 2012, from http://broadbandeurope.wordpress.com/2011/05/01/financing-high-speed-broadband-what-do-you-think/

Rowell, L. (2007). Can mesh networks bring low-cost, wireless broadband to the masses? *netWorker, 11*(4), 26–33. doi:10.1145/1327512.1327514

RTBI. et al. (2010). *Evaluating a real time biosurveillance program – A pilot project report of interim findings and discussion workshop*. Retrieved on January 14, 2012 from http://rs.rtbi-iitm.in/BioSurvey/doc/Workshop/RTBP-FindingsWorkshop-REPORT.pdf

RTBI. IIT Madras. (2012). Technology innovations. *Gramateller*. Retrieved on January 14, 2012 from http://www.rtbi.in/innovate.html

Rubio, F. (2009). Public computer centers program – Sustainable adoption program. *Broadband USA*. Retrieved November 1, 2011 from http://digitalelpaso.com/documents/Sustainability%20(Non%20Infrastructure)%20Grant%20App%20-%2020090819.pdf

Rural Communications Development Fund. (2012). *Website*. Retrieved from http://bit.ly/ucc_rcdf

Rural Health Statistics. (2010). *RHS bulletin*. New Delhi, India: Government of India.

Ruth, S., & Schware, R. (2008). *Pursuing truly successful e-government projects: Mission impossible?* Information Technology in Developing Countries, 18*(3)*. Retrieved July 25, 2011 from http://www.iimahd.ernet.in/egov/ifip/oct2008/stephen-ruth.htm

Ryan, R. M. (1982). Control and information in the intrapersonal sphere: An extension of cognitive evaluation theory. *Journal of Personality and Social Psychology, 43*, 450–461. doi:10.1037/0022-3514.43.3.450

Saguaro Seminar. (2007). *Civic engagement in America*. Retrieved January 3, 2009, from http://www.hks.harvard.edu/saguaro/index.htm

Salinas v. United States, 522 U.S. 52 (1997).

Sandvig, C., Young, D., & Meinrath, S. (2004). *Hidden interfaces to "ownerless" networks*. Paper presented to the 32nd Conference on Communication, Information, and Internet Policy. Washington, DC.

Sandvig, C. (2004). An initial assessment of cooperative action in wi-fi networking. *Telecommunications Policy, 28*, 579–602. doi:10.1016/j.telpol.2004.05.006

Schaffer, B. S. (2007). The nature of goal congruence in organizations. *Super Vision, 68*(8), 13–17.

Schement, J. (2009). Broadband, internet and universal service. In Schejter, A. (Ed.), *And Communications for All: A Policy Agenda for a New Administration* (pp. 3–27). Lanham, MD: Lexington Books.

Schmidt, T., & Townsend, A. (2003). Why wi-fi wants to be free. *Communications of the ACM, 46*(5), 47–52. doi:10.1145/769800.769825

Seakay, Cisco, & IBM. (2006). *Proposal in response to San Francisco TechConnect RFP*. San Francisco, CA: Government of San Francisco.

Sein, M. K., Ahmad, I., & Harindranath, G. (2008). Sustaining ICT for development projects: The case of grameenphone CIC. *Telektronikk, 104*(2), 16–24.

Sennett, R. (2006). *The culture of the new capitalism*. New Haven, CT: Yale University Press.

Settles, C. (2011). Should public-private partnerships be in your broadband future? *Government Technology*. Retrieved October 20, 2011 from http://www.govtech.com/wireless/Public-Private-Partnerships-Broadband-Future.html

Shaffer, G. (2007). Frame-up: An analysis of arguments for and against municipal wireless initiatives. *Public Works Management & Policy, 11*(3), 204–216. doi:10.1177/1087724X06297347

Shaffer, G. L. (2010). *Peering into the future: How wifi signal sharing is impacting digital inclusion efforts.* Philadelphia, PA: Temple University.

Shah, D. V., & Scheufele, D. A. (2006). Explicating opinion leadership: Nonpolitical dispositions, information consumption, and civic participation. *Political Communication, 23*(1), 1–22. doi:10.1080/10584600500476932

Shapiro, M. (2006). Municipal broadband: The economics, politics and implications. *Pike and Fisher.* Retrieved from http://www.pf.com/marketResearchPDInd.asp?repId=397

Shukla, R. (2010). The official poor in India summed up. *Indian Journal of Human Development, 4*(2). Retrieved on January 14, 2012 from http://xa.yimg.com/kq/groups/12632651/1180166002/name/The+Official+Poor+in+India+Summed+Up_Rajesh+Shukal.pdf.pdf

Shukla, R. K. (2011). *NCR analysis (limited circulation).* New Delhi, India: National Council of Applied Economic Research.

Sicker, D. C., Grunwald, D., Anderson, E., Doerr, C., Munsinger, B., & Sheth, A. (2006). Examining the wireless commons. In *Proceedings of the 34th Research Conference on Communication, Information.* IEEE.

Silva, J. (2010). What is net neutrality and why you should worry about it. Retrieved November 09, 2011, from http://www.trezentos.blog.br/?p=6465

Silverstone, R. (2004). Regulation, media literacy and media civics. *Media Culture & Society, 26*(3), 440–449. doi:10.1177/0163443704042557

Simpson, L. (2005). Community informatics and sustainability: Why social capital matters. *The Journal of Community Informatics, 1*(2), 102–119.

Simpson, L., Wood, L., & Daws, L. (2003). Community capacity building: Starting with people not projects. *Community Development Journal, 38*(4), 277–286. doi:10.1093/cdj/38.4.277

Sinan, A., Escobari, M., & Nishina, R. (2001). *Assessing network applications for economic development.* Cambridge, MA: Harvard University.

Siochrú, S. O., & Girard, B. (2005). community-based networks and innovative technologies: New models to serve and empower the poor. *The United Nations Development Program.* Retrieved from http://propoor-ict.net

Siochru, S. O., & Girard, B. (2005). *Community-based networks and innovative technologies: New models to serve and empower the poor.* Geneva, Switzerland: UNDP.

Sirbu, M. (2006). Evolving wireless access technologies for municipal broadband. *Government Information Quarterly, 23*, 480–502. doi:10.1016/j.giq.2006.09.003

Skidmore, S. (2008, July 19). MetroFi ending wi-fi service in Calif., Ore., Ill. *San Jose Mercury News.* Retrieved from http://www.mercurynews.com/ci_9637287?IADID

Šlinz, P. (2012). *Issues of highly stressed wireless networks based on 802.11b/g standards.* (MSc. Thesis). Masarykova univerzita v Brně. Retrieved from http://is.muni.cz/th/208329/fi_m/

Smithfarm, L., Lemmio, J., & Voženílek, P. (2011). Eurotel company: The history. *Wikipedia.* Retrieved March 5, 2012, from http://en.wikipedia.org/wiki/Eurotel

Soman, S. (2010, November 8). Talking their way to success. *The Times of India Chennai,* p. 3.

South Carolina. (2002). *South Carolina code Ann. § 58-9-2620.* Retrieved from http://www.scstatehouse.gov/code/t58c009.php

Souvirón, M., & María, J. (1999). *The process of liberalization and the new telecommunications regulations.* Granada: Editorial Comares.

Sova, M. (2009). *Plzeňské PilsFree trnem v oku komerčním providerům: Komu slouží dezinformace?* Retrieved July 8, 2012, from http://www.internetprovsechny.cz/plzenske-pilsfree-trnem-v-oku-komercnim-providerum-komu-slouzi-dezinformace/

Sova, M. (2010). Soumrak volného 10 GHz pásma aneb kdo má zájem na jeho zarušení? *Internet pro všechny.* Retrieved July 8, 2012, from http://www.internetprovsechny.cz/soumrak-volneho-10-ghz-pasma-aneb-kdo-ma-zajem-na-jeho-zaruseni/

Spector, D., Lubin, G., & Giang, V. (2012). *The 15 most disliked companies in America*. Retrieved from http://www.businessinsider.com/the-most-hated-companies-in-america-2012-6?op=1

Spoonley, P., Peace, R., Butcher, A., & O'Neill, D. (2005). Social cohesion: A policy and indicator framework for assessing immigrant and host outcomes. *Social Policy Journal of New Zealand, 24*, 85–110.

Sporek, J. (2005). *Server, router and WiFi AP providing access to university network for students and guests*. (BSc. Thesis). FAI UTB ve Zlíně.

Statistics New Zealand. (2006). *QuickStats about Clendon South*. Retrieved August 9, 2011, from http://www.stats.govt.nz/Census/2006CensusHomePage/QuickStats/AboutAPlace/SnapShot.aspx?id=3524822&type=au&ampParentID=1000002

Statistics New Zealand. (2006). *QuickStats about Papakura East*. Retrieved August 9, 2011, from http://www.stats.govt.nz/Census/2006CensusHomePage/QuickStats/AboutAPlace/SnapShot.aspx?id=3525610&type=au&ampParentID=1000002

Statistics New Zealand. (2009). *Measuring New Zealand's progress using a sustainable development approach: 2008*. Retrieved from http://www.stats.govt.nz/searchresults.aspx?q=social%20cohesion&mp=20&sp=0&sort=r

Statistics New Zealand. (2012). *Census 2006 households with school-aged children*. Retrieved from http://www.stats.govt.nz

Strover, S., Chapman, G., & Waters, J. (2003, September). Beyond community networking and CTCs: Access, development and public policy. *TPRC*, p. 24.

Sundfeld, C. A. (2006). A regulação das telecomunicações: Papel atual e tendências futures. Revista Eletrônica de Direito Administrativo Econômico, *8*. Retrieved from http://www.direitodoestado.com/revista/REDAE-8-NOVEMBRO-2006-CARLOS%20ARI%20SUNDFELD.pdf

Sunnyvale. (2008). *Website*. Retrieved from http://sunnyvale.ca.gov/Departments/Library/Community+Profile.htm

Super Micro Computer. (2009). *Supermicro solutions based on Intel® Atom™ processors*. Retrieved August 27, 2012, from http://www.supermicro.com/products/nfo/atom.cfm

Superinteressante. (2010). Broadband incites economic growth. Retrieved November 09, 2011, from http://goo.gl/e7ADR

Surana, S., Patra, R., Nedevshi, S., Ramos, M., Subramanian, L., Ben-David, Y., & Brewer, E. (2008). Beyond pilots: Keeping rural wireless networks alive. In *Proceedings of NSDI 2008: 5th USENIX Symposium on Networked Systems Design and Implementation*. Retrieved January 26, 2012, from http://www.usenix.org/event/nsdi08/tech/full_papers/surana/surana.pdf

Sviták, J. (2007). *Modeling, simulation and throughput analysis for the 802.11 protocols family*. (MSc. Thesis). FAI UTB in Zlin.

Sweetland, S. R. (1996). Human capital theory: Foundations of a field of inquiry. *Review of Educational Research, 66*(3), 341–359.

Swirbul, C. (2006). *State broadband battles*. Retrieved from http://www.baller.com/pdfs/APPA_Broadband_Battles.pdf

TAGITM. (2008). *2008 award recipients*. Retrieved November 10, 2011, from http://www.tagitm.org/?page=2008Awards

Takahashi, T. (Ed.). (2000). *Green book of the information society in Brazil*. Sao Paolo, Brasília: MCT. Retrieved from http://www.mct.gov.br/index.php/content/view/18878.html

Tapia, A. (2009). Municipal broadband. In Schejter, A. (Ed.), *And Communications for All*. Lanham, MD: Lexington Books.

Tech Law Journal. (2005). *8th circuit rules in north Kansas City municipal broadband case*. Retrieved from http://www.techlawjournal.com/topstories/2005/20051229.asp

Telecom Regulatory Authority of India. (2011). *Report*. Retrieved on January 14, 2012 from http://www.trai.gov.in/Default.asp

Thapa, D., & Sæbø, Ø. (2011). Demystifying the possibilities of ICT4D in the mountain regions of Nepal. In *Proceedings of the 44th Hawaii International Conference on System Sciences (HICSS 44)*, (pp. 1-10). Kuai, HI: IEEE Computer Society.

Thapa, D., & Sein, M. K. (2010). *ICT, Social Capital And Development: The case of a mountain region in Nepal.* Paper presented at third annual SIG GlobDev workshop. Saint Louis, MO.

Thapa, D. (2011). The role of ICT actors and networks in development: The case study of a wireless project in Nepal. *The Electronic Journal of Information Systems in Developing Countries, 49*(1), 1–16.

Thapa, D., Sein, M. K., & Sæbø, Ø. (2012). Building collective capabilities through ICT in a mountain region of Nepal: Where social capital leads to collective action. *Information Technology for Development, 18*(1), 5–22. doi:10.1080/02681102.2011.643205

Tilly, C. (1978). *From mobilization to revolution.* Reading, MA: Addison-Wesley.

Touraine, A. (1985). An introduction to the study of social movements. *Social Research, 52*(4), 749–787.

Toyama, S. (2007). *Local area SNS and community building in Japan.* Retrieved from http://www.ccnr.net/prato2007/archive/toyama%20135.pdf

TRAI. (2004). *Press release 47/2004.* Retrieved on January 14, 2012 from http://www.trai.gov.in/WriteReadData/trai/upload/PressReleases/160/Press%20Release-%2030%20July-04-Final.pdf

Travis, H. (2006). Wi-fi everywhere: Universal broadband access as antitrust and telecommunications policy. *The American University Law Review, 55*(6), 1697–1800.

Tuck, E. (2009). Suspending damage: A letter to communities. *Harvard Educational Review, 79*(3), 409–427.

Turek, L. (2009). *Řízení toku v přístupových bodech bezdrátové sítě IEEE 802.11.* (Diplomová Práce). Praha: MFF UK. Retrieved from http://8an.cz/papers/Rizeni-TokuWiFi.pdf

Turek, L. (2012). *VisualOSPF.* Retrieved May 31, 2012, from http://intra.praha12.net/ospf/

Turner, S. D. (2005). *Broadband reality check.* Washington, DC: Free Press.

Tynan, D. (2012). White spaces: The next generation of wireless broadband has landed. *PCWorld.* Retrieved from http://www.pcworld.com/article/248847/white_spaces_the_next_generation_of_wireless_broadband_has_landed.html

Ubilium. (2012). *Project Chapleau.* Retrieved June 13, 2012 from http://www.ubilium.com/chapleau.html

Ubiquiti Networks, Inc. (2012). *AirControl beta.* Retrieved April 12, 2012, from http://www.ubnt.com/aircontrol

UbuntuNet Alliance. (2009). *Overview of fibre infrastructure opportunities in the UbuntuNet region.* Kampala, Uganda: UbuntuNet Alliance.

UnArt. O. S. (2009). *UnArt revue.* Retrieved from http://unart.cz/casopis.htm

UNESCO. (2010). Nepal en route for introducing ICT in education. Communication and Information Sector's News Service. *Retrieved July 21, 2011, from* http://portal.unesco.org/ci/en/ev.php-URL_ID=30622&URL_DO=DO_TOPIC&URL_SECTION=201.html

Universal Service. (2012). *Wikipedia free encyclopedia.* Retrieved March 12, 2012 from http://en.wikipedia.org/wiki/Universal_service

Unwin, T. (2005). *Partnerships in development practice: Evidence from multi-stakeholder ICT4D partnership practice in Africa.* Paris, France: UNESCO.

Unwin, T. (2009). *ICT4D: Information and communication technologies for development.* Cambridge, UK: Cambridge University Press.

Uppal, M., Nair, S. K. N., & Rao, C. S. R. (2006). India's telecom reform: A chronological account. *NCAER.* Retrieved on January 14, 2012 from http://www.iipa.org.in/common/pdf/PAPER%2015%20-%20TELECOM%20REFORM.pdf

US Census Bureau Quick Facts. (2010). *Website.* Retrieved from http://quickfacts.census.gov/qfd/states/06/06075.html

Utah Code Ann. §§ 10-18-202, -203, -204, -302 (West, Westlaw current through 2012).

Utz, A., & Dahlman, C. (2007). *Promoting inclusive innovation: Unleashing India's innovation: Toward sustainable and inclusive growth.* Retrieved on January 14, 2012 from http://siteresources.worldbank.org/SOUTHASIAEXT/Resources/223546-1181699473021/3876782-1191373775504/indiainnovationchapter4.pdf

Vágner, A. (2011). *Real performance of devices operating on 802.11n.* (MSc. Thesis). Brno University of Technology. Retrieved from https://www.vutbr.cz/en/studies/final-thesis?zp_id=40262

vanDijk, J. (1999). *The network society: Social aspects of new media* (Spoorenberg, L., Trans.). London, UK: Sage.

Varshney, U. (2007). Pervasive healthcare and wireless health monitoring. *Mobile Networks and Applications, 12*(2-3), 113–127. doi:10.1007/s11036-007-0017-1

Venkatesh, V., Morris, M. G., Davis, G. B., & Davis, F. D. (2003). User acceptance of information technology: Toward a unified view. *Management Information Systems Quarterly, 27*(3), 425–478.

Vergunst, P. (2006). *Community cohesion: Constructing boundaries between or within communities-of-place?* Paper presented at the Rural Citizen: Governance, Culture and Wellbeing in the 21st Century Conference. New York, NY.

Vermont Office of Secretary of State. (2009). *Municipal law basics.* Retrieved from http://www.sec.state.vt.us/municipal/pubs/municipal_law_basics.pdf

View, M. (2008). *Google provides free wi-fi.* Retrieved from http://www.mountainview.gov/services/learn_about_our_city/free_Wi-Fi.asp

Vilímovský, M. (2005). *Community computer networks.* (BSc. Thesis). Vysoká škola ekonomická v Praze, Fakulta informatiky a statistiky, Katedra informačních technologií, Praha.

Virginia Regulatory Issues Pertaining to Municipal Broadband. (2004). *Website.* Retrieved from http://top.bev.net/tamp/7-Common_Appendices/Main_Project_Papers/Virginia_Regulatory_Issues_Pertaining_to_Municipal_Broadband.pdf

Vishwanath, A. (2006). The effect of the number of opinion seekers and leaders on technology attitudes and choices. *Human Communication Research, 32*(3), 322–350. doi:10.1111/j.1468-2958.2006.00278.x

Vision 2 Mobile. (2008). *Portland seizes MetroFi muni wi-fi gear.* Retrieved from http://www.vision2mobile.com/news/2008/10/portland-seizes-metrofi-muni-wi-fi-gear.aspx

Vondráček, M. (2004). *CZFree forum - First fiber optics project in Czela.net.* Retrieved May 30, 2012, from http://czfree.net/forum/showthread.php?s=&threadid=10139&perpage=16&highlight=&pagenumber=1

Vos, E. (2005). *Colorado anti-muni bill passed by house local govt committee.* Retrieved from http://www.muniwireless.com/2005/04/07/colorado-anti-muni-bill-passed-by-house-local-govt-committee

Vos, E. (2005). *Reports on municipal wireless and broadband projects.* Retrieved from http://www.muniwireless.com

Vos, E. (2008). *Oklahoma City rolls out world's largest muni wi-fi mesh network.* Retrieved from http://www.muniwireless.com/2008/06/03/oklahoma-city-deploys-largest-muni-Wi-Fimesh-network/

Vos, E. (2010). Philadelphia city council committee approves purchase of citywide wi-fi network. *MuniWireless.* Retrieved March 27, 2012 from http://www.muniwireless.com/2010/06/10/philly-approves-purchase-of-citywide-wifi-network/

Vos, E. (2010). *Wilmington, NC uses white for smart city, eco-friendly wireless applications.* Retrieved from http://www.muniwireless.com/2010/02/24/wilmington-uses-white-spaces-for-smart-city-ecofriendly-wireless-applications

Vos, E. (2011). *Newton, NC deploys free downtown wi-fi service on the cheap.* Retrieved from http://www.muniwireless.com/2011/09/27/newton-nc-deploys-free-downtown-wifi-service-on-the-cheap

Vos, E. (2011). *Cleveland, Ohio neighborhood deploys large outdoor free wi-fi network.* Retrieved from http://www.muniwireless.com/2011/11/10/cleveland-ohio-neighborhood-deploys-large-outdoor-free-wi-fi-network

Vroom, V. C. (1964). *Work and motivation.* New York, NY: Wiley.

Walker, B. (2009). Wireless traffic management network for downtown renewal. *Industrial Ethernet Book, 52,* 31–32.

Warschauer, M. (2003). *Technology and social inclusion: Rethinking the digital divide.* Cambridge, MA: MIT Press.

Waterman, H. (1981). Reasons and reason: Collective political activity in comparative and historical perspective. *World Politics, 32*, 554–589. doi:10.2307/2010135

Wellman, B. (2001). *The persistence and transformation of community: From neighbourhood groups to social networks.* Paper presented to the Law Commission of Canada. Ottawa, Canada.

Wellman, B. (Ed.). (1999). *Networks in the global village: Life in contemporary communities.* Boulder, CO: Westview Press / Perseus Books Group.

Wellman, B., & Berkowitz, S. (Eds.). (1988). *Social structures: A network approach.* Cambridge, UK: Cambridge University Press.

White Space, F. C. C. (2012). *Website.* Retrieved from http://www.fcc.gov/topic/white-space

Wikipedia. (2004). Capacitor plague. *Wikipedia, the free encyclopedia.* Retrieved from http://en.wikipedia.org/w/index.php?title=Capacitor_plague&oldid=508636355

Wikipedia. (2012). Linksys WRT54G series. *Wikipedia, the free encyclopedia.* Retrieved from http://en.wikipedia.org/w/index.php?title=Linksys_WRT54G_series&oldid=508160017

Wilco, R. (2004). *Austin wireless city project.* Retrieved from http://www.lessnetworks.com/static/AustinWirelessCityProject.ppt

Williams, J. (2009). *Connecting people: Investigating a relationship between internet access and social cohesion in local community settings.* (Unpublished Doctoral Thesis). Massey University. Palmerston North, New Zealand.

Williams, D. (2006). On and off the net: Scales for social capital in an online era. *Journal of Computer-Mediated Communication, 11*(2). doi:10.1111/j.1083-6101.2006.00029.x

Wilson, M., & Kraatz, J. (2008). *Wireless internet company serving Cupertino, Sunnyvale shutting down.* Retrieved from http://www.siliconvalley.com

Wireless Africa Alliance. (2009). *Website.* Retrieved from http://www.wireless-africa.org

Wireless Gumph. (2002). *Easy homemade 2.4 Ghz omni antenna - Gumph.* Retrieved August 22, 2012, from http://wireless.gumph.org/articles/homemadeomni.html

Wireless Minneapolis. (2006). *Municipal broadband business case.* Retrieved October 9, 2011 from http://www.ci.minneapolis.mn.us/wirelessminneapolis/Mpls-Wireless_BusinessCase_V3.pdf

Wireless Networking in the Developing World. (2006). *A practical guide to planning and building low-cost telecommunications infrastructure.* Kampala, Uganda: Limehouse Book Sprint Team.

Wireless Silicon Valley. (2006). *Requests for proposals for a regional broadband wireless network for Silicon Valley.* Retrieved from http://www.jointventure.org/programs-initiatives/smartvalley/projects/wirelesssv/documents

Wireless Silicon Valley. (2008). *Wireless Silicon Valley adds Covad Communications to the team and plans test in San Carlos.* Retrieved from http://www.jointventure.org/programs-initiatives/wirelesssiliconvalley/updates.html

Women of Uganda Network. (2006). *Uganda country-based research, policy support and advocacy partnerships for pro-poor ICT.* Kampala, Uganda: Women of Uganda Network.

Wong, M. (2007). *Wireless broadband from backhaul to community service: Cooperative provision and related models of local signal access.* Paper presented at the 35th Research Conference on Communication, Information and Internet Policy. Arlington, VA.

Wong, M., & Clement, A. (2007). *Sharing wireless internet in urban neighbourhoods.* Paper presented at the Third Communities and Technologies Conference. East Lansing, MI.

Wong, M. A. (2007). Community wireless: Policy and regulation perspectives. *Journal of Community Informatics, 3*(4). Retrieved from http://owl.english.purdue.edu/owl/resource/560/10

Wonki Min. (2010). *Broadband policy in Korea.* Retrieved from http://siteresources.worldbank.org/BELARUSEXTN/Resources/1.77koreabb.pdf

World Health Organization. (2007). *Not enough here, too many there: Health workforce in India.* Geneva, Switzerland: World Health Organization.

WSIS. (2003). *Declaration of principles and plan of actions.* Retrieved on September 19, 2012, from http://www.itu.int/wsis/docs/promotional/brochure-dop-poa.pdf

WSIS. (2005). *Tunis agenda for the information society*. Retrieved on September 2012, from http://www.itu.int/wsis/docs2/tunis/off/6rev1.html

Youtie, J. (2000). Field of dream revisited: Economic development and telecommunications in LaGrange, Georgia. *Economic Development Quarterly, 14*(2), 146–153. doi:10.1177/089124240001400202

Žáček, P. (2007). *Measurement and optimalization of routing performance*. FAI TBU in Zlin.

Zandl, P. (2001). Internet proti monopolu - Tři roky poté. *Lupa.cz*. Retrieved August 29, 2012, from http://www.lupa.cz/clanky/internet-proti-monopolu-tri-roky-pote/

Zheng, Y., & Walsham, G. (2008). Inequality of what social exclusion in the e - society as capability deprivation. *Information Technology & People, 21*(3), 222–243. doi:10.1108/09593840810896000

Zwimpfer, L. (2010). Building digital communities: A history of the 2020 trust. In Toland, J. (Ed.), *Return to Tomorrow: Fifty Years of Computing in New Zealand*. Wellington, New Zealand: New Zealand Computer Society.

About the Contributors

Abdelnasser M. Abdelaal is Assistant Professor of Information Technology at Ibri College of Applied Sciences, Sultanate of Oman. He earned his PhD in Information Technology from University of Nebraska at Omaha. His research focuses on quality of service, call admission control, business models, service pricing, and socioeconomics of community wireless networks. Abdelaal is one of the founders of the Omaha Wireless Network. He organized mini-tracks and panels on community wireless networks in major international conferences and summits. He has published more than a dozen papers on community wireless networks in international conferences and journals. In addition, he managed and participated in a number of Information Technology for Development (IT4D) and community empowerment projects. He also provides consultation regarding broadband policy and IT4D.

* * *

Sylvie Albert is the Dean of the Faculty of Business and Economics at the University of Winnipeg and was an Associate Dean and Professor of Strategy at Laurentian University. Dr. Albert arrived at Laurentian in 2004 with 20 years of experience in the public and private sector as the Director of two economic development corporations and as a consultant providing planning services on community development projects. During her 10 years in consulting practice under her firm, Planned Approach Inc., Dr. Albert built feasibility studies and public-private partnerships toward new intelligent/smart communities. Dr. Albert used her research on smart/intelligent communities to educate communities and regions on this topic and help those that were interested in building innovative collaborative programs on e-learning, e-health, e-collaborations, and more. She published two books on the topic of intelligent cities and has been the lead jurist for the International Intelligent Community Award (ICF is a New York-based think tank on intelligent cities).

Francisco Paulo Jamil Almeida Marques is a Lecturer at the Federal University of Ceará, Brazil. Marques has a PhD in Communication Studies (Federal University of Bahia, Brazil). In 2006, he was a Visiting Scholar at Saint Louis University, USA. Over the last years, his research interests have focused on political journalism and on the political uses of new media. His latest publications include "Government and e-Participation Programs: A Study of the Challenges faced by Institutional Projects" (*First Monday Journal*, 2010), and a co-edited book titled *Internet and Political Participation in Brazil*. The author wishes to thank the Brazilian National Council for Scientific and Technological Development (CNPq) for partially funding his research.

Michal Bližňák was born in Slavičín, the CR, received the M.Sc. degree in Automated Control of Technological Processes from the Brno University of Technology, Faculty of Technology, in Zlin, in 2001, and his Ph.D. degree in Technical Cybernetics from Tomas Bata University, in Zlin, in 2008. Since 2003, he is affiliated with the Faculty of Applied Informatics, Tomas Bata University, in Zlin. His main activities are cross-platform programming, parallel algorithms and programming, and embedded systems development.

Julius Butime is a Lecturer with the Department of Electrical and Computer Engineering at Makerere University, where he is the Head of Department and a Researcher with the Community Wireless Resource Centre (CWRC). Dr. Butime holds a B.Sc. in Electrical Engineering from Makerere University, Uganda, and a Ph.D. in Electrical Engineering from the University of Navarra, Spain. His research interests are in computer vision, radio frequency engineering, and wireless networks.

Giovanni Camponovo graduated in Management Sciences in 2002 and holds a Ph.D. in Business Information Systems from the University of Lausanne. He is currently Teacher and Researcher at the Department of Business and Social Sciences at the University of Applied Sciences of Southern Switzerland (SUPSI). He also teaches at the Universities of Geneva, Lausanne and Lugano, and the Swiss Academy of Accounting. His research interests include information systems, innovation management, management accounting, business models, and electronic and mobile business.

Lorenzo Cantoni graduated in Philosophy and holds a PhD in Education and Linguistics. He is full Professor at the Università della Svizzera Italiana (University of Lugano, Switzerland), Faculty of Communication Sciences. He is Dean of the Faculty and Director of Institute for Communication Technologies (ITC). He is Scientific Director of the Laboratories eLab, webatelier.net, and NewMinE Lab. His research interests are where communication, education, and new media overlap, ranging from computer-mediated communication to usability, from eLearning to eTourism, and from ICT4D to eGovernment.

Tomáš Dulík was born in Slavičín, the CR. He received his Master's degree from Computer Science in 1998 at Brno University of Technology, Faculty of Electrical Engineering, Department of Computer Science, and Ph.D. degree in 2012 from Tomas Bata University in Zlin, Faculty of Applied Informatics. His research is focused on the baseband signal processing in wireless network devices.

Heather E. Hudson is a Professor and Director of the Institute of Social and Economic Research and Professor of Public Policy at the University of Alaska – Anchorage. Previously, she was founding Director of the Communications Technology Management and Policy Program at the University of San Francisco. Her research focuses on applications and effects of information and communication technologies, regulation and policy issues, and strategies to extend affordable access to new technologies and services. She has planned and evaluated communication projects in northern Canada, Alaska, and more than 50 developing countries. She is the author of many conference papers and articles and several books, and has testified as an expert witness on communications policy issues. She has consulted for the private sector, government and international agencies, and consumer and indigenous organizations. She has held a Fulbright Policy Research Chair at Carleton University and a Fulbright Distinguished Lectureship for the Asia/Pacific. She has also been a Sloan Foundation Industry Fellow at Columbia University's Institute for Tele-Information, an Honorary Research Fellow at the University of Hong Kong, Senior Fellow at CIRCIT in Australia and at the East-West Center in Hawaii, and an IEEE Distinguished Lecturer.

Roman Jašek received the M.Sc. degree in Computer Science from the Palacký University in Olomouc, Faculty of Science, and his Ph.D. degree in Technology from Charles University in Prague in 2000. His work experience is focused on enterprise security. Since 2000, he has been affiliated with Faculties of Management and Economics and Applied Informatics in Tomas Bata University in Zlin. His main activities are related to computer security science and artificial intelligence.

Ashok Jhunjhunwala is a Professor in IIT Madras, where he leads the Telecommunications and Computer Networks Group (TeNeT), which works with industry in the development of technologies relevant to India. It has incubated over 35 companies in the last twenty years. He chairs Rural Technology and Business Incubator (RTBI) at IIT Madras and Mobile Payment Forum of India (MPFI). Dr. Ashok Jhunjhunwala has been awarded Padma Shri in the year 2002 apart from a multitude of other awards. He has published many papers, notably in IEEE, *Journal of IETE, Journal of Optics, Journal of Optoelectronics*, UN Department of Economic and Social Affairs ICT Task Force Series 4, *Journal of Community Informatics, Euromicro Journal of Microprocessing, Computer Networks and ISDN, Journal of System Architecture, Journal of Rural Development, Indian Journal of Radio and Space Physics*, to name a few. Dr. Jhunjhunwala is a Director in the Board of TTML, Polaris, 3i Infotech, Sasken, Tejas, Tata Communications, and Exicom. He is member of Prime Minister's Scientific Advisory Committee.

Yasuhiko Kawasumi is the Rapporteur for ITU-D SG2 Q10-3/2 on "Telecommunications/ICTs for Rural and Remote Areas" since 2002 (ITU is the leading UN agency for information and communication technology issues: http://www.itu.int). He is now Special Advisor to the ITU Association of Japan http://www.ituaj.jp/. He participated in projects funded by Asia-Pacific Telecommunity (APT) http://www.apt.int/ in Nepal, Marshall Islands, Micronesia, and Bhutan. He started his business career at Japanese monopoly overseas telecommunication operator Kokusai Denshin Denwa Co. Ltd. in 1961 after graduating from KEIO University. His experienced the construction of Transpacific Submarine Cable No. 1, its maintenance and operation, satellite systems, network management. He attended ITU's various conferences and meetings. He participated in Maitland Commission meetings as advisor to Dr. Koji Kobayashi (NEC Chairman) in Arusha (Tanzania) and Bali (Indonesia) in 1984. He was appointed as rapporteur of the Focus Group on "New Technologies for Rural Communications" of ITU-D Study Group 2 in 1999 and submitted the final report on "New Technologies for Rural Applications" in 2001. He has been the rapporteur for the Question of ITU SG2 "Telecommunications/ICT for Rural and Remote Areas" for the three study cycles, i.e. 2002-2006, 2006-2010, and 2010-2014.

André Lemos is an Associate Professor of the Faculty of Communication at the Federal University of Bahia, Brazil, and Coordinator of the Cybercity Research Group (GPC) with a PhD in Sociology, Université Paris V, Sorbonne (1995). He is the former President of the Brazilian Association of Communications, PhD Programs, the former Chair of the Department of Communication (Facom/UFBa), and the former Director of PhD Program in Communication and Culture (Facom/UFBa). André Lemos has 10 (ten) published books, more than 30 chapters in other published books, and more than 40 articles in peer-reviewed scientific journals worldwide. He was also a Visiting Scholar at the University of Alberta and McGill University, Canada, 2007-2008. He is now working on mobile communication and locative media studies.

N. Neeraja is a Researcher at IIT M's Rural Technology and Business Incubator, and her research interests are in the area of financial inclusion and mobile banking for financial inclusion.

Eric Null is a consultant and advocate for community and municipal wireless networks. He received his B.A. in Political Science from the University of Vermont. He received his J.D., with concentration in Intellectual Property and Communications Law, from Cardozo Law School in 2012. His research interests focus on policy, regulation, and legal issues facing community and municipal wireless networks, the telecommunications industry, and the Internet in general. He has published articles about the legal aspects of municipal broadband, network neutrality, and ICANN's NewgTLD Program. He has worked closely with prominent Intellectual Property professors Susan Crawford and Brett Frischmann, providing research for their respective books.

Dorothy Okello is a Lecturer with the Department of Electrical and Computer Engineering at Makerere University, where she is also Director of the Community Wireless Resource Centre (CWRC). She has over 15 years of diverse experience in teaching, researching, and in conducting projects and contributing to policies in the ICT sector at national and international levels. Dr. Okello is also an activist in the area of getting more women, small-scale enterprises, and rural communities engaged in the information society for development—via policy advocacy and via program implementation and monitoring and evaluation. She is Founder of Women of Uganda Network (WOUGNET), established in 2000, with a mission to promote and support the use of Information and Communication Technologies (ICTs) by women and women organizations in Uganda. Dr. Okello holds a B.Sc. in Engineering (Electrical – First Class Hons.) from Makerere University, Uganda, an M.Sc. in Electrical Engineering from the University of Kansas, United States (Fulbright Scholar), and a Ph.D. in Electrical Engineering from McGill University, Montreal, Canada (Canadian Commonwealth Scholarship Recipient). Her research interests are in wireless networks and radio resource management.

Anna Picco-Schwendener graduated in 2002 with a major in Communication Technologies. After her degree, she worked for several years in the communication and IT field of an important steel trading company (Duferco SA). Currently, she is a Researcher at the eLab and NewMinE Lab, where she is involved in different projects. Since May 2010, she is enrolled as a PhD student at the Communication Sciences Department of the Università della Svizzera Italiana under the supervision of Prof. Lorenzo Cantoni. Her research interests include wireless communities, motivations, Wi-Fi usages/needs in different contexts, municipal wireless networks, mobile city and uni apps, mobile learning and digital copyright for eLearning.

Evelyn Posey is a Professor of Rhetoric and Writing Studies at the University of Texas at El Paso (UTEP). With Kate Mangelsdorf, she has co-authored three composition textbooks, including *Choices: A Writing Guide with Readings* (5th ed.), for Bedford/St. Martin's, 2012, and *The World of Writing: A Guide*, for Pearson Longman, 2011. Posey has also published in *Computers and Composition*, the *Journal of Developmental Education*, *Teaching English in the Two-Year College*, and the *Writing Center Journal*. Posey has served UTEP as Chair of the Department of English, Associate Dean of the College of Liberal Arts, and Associate Vice President for Instructional Technology. In addition to her work in composition, Posey received a 3.4 million National Science Foundation grant for the advancement of women in science and engineering academic disciplines and now serves as a consultant to universities wishing to improve the recruitment and retention of women faculty.

Mahabir Pun has a Master degree in Educational Administration, University of Nebraska at Kearney, Nebraska, USA. He is the Team Leader of Nepal Wireless Networking Project since 2002. He has received a number of awards for his work on community development and the latest is the Innovation on Access Provision Award from Information Society Innovation Fund, Australia, in 2012.

Janani Rangarajan is a researcher at IIT M's Rural Technology and Business Incubator. Her research interests are in the area of rural development, education, and energy.

Gwen Shaffer is an Assistant Professor in the Department of Journalism and Mass Communication at California State University, Long Beach. Her research examines how to best incorporate telecommunications policy and economic realities into the Internet architecture. Specifically, the research considers telecommunications company mergers, expanding the application of open Internet principles, and wireless device attachment. Previously, Gwen was a postdoctoral researcher in the Computer Science Department at University of California, Irvine. She earned a PhD in Mass Media and Communication from Temple University in Philadelphia in 2009. Her dissertation examined the potential for peer-to-peer networking to help bridge the digital divide.

Barbara Walker has over 30 years of experience in the computer industry in a variety of areas including management, software application development, systems analysis, systems project management, business process reengineering, strategic planning, and networking. Ms. Walker is a graduate of the University of Texas at El Paso with a BBA in Accounting. She served as a Co-Chair and Co-Founder of Digital El Paso. Currently, Ms. Walker is a Key Account Manager for Cisco Systems, managing business with West Texas public sector clients.

Jocelyn Williams is Head of the Department of Communication Studies at Unitec Institute of Technology in Auckland, New Zealand, where she manages academic programs, teaches, and supervises communication degree students. She also researches the relationship between Internet use and community outcomes. In particular, since 2002, she has worked with "Computers in Homes" (CIH), which aims to build stronger digital capability and social capital within communities in New Zealand by providing economically disadvantaged families with a computer and a free Internet connection. In 2011, Dr. Williams accepted an invitation to chair the Auckland CIH Steering Group, which is responsible for governing delivery on the objectives of the 2020 Communications Trust through its work with Computers in Homes. Dr. Williams' longitudinal case studies investigate the relationship between social cohesion and Internet access. Her other academic roles have included being President of the Australian and New Zealand Communication Association (ANZCA) in 2008-2009 and a continuing role on the Executive of ANZCA in subsequent years.

Index